Scorer, R. S.
(Richard Segar),
1919-

The satellite as
microscope.

$69.95

DATE			

SATELLITE AS MICROSCOPE

To the greens, the reds, and the blues who wish to preserve the vegetation, the soil, and the clean sky respectively, by first understanding what the air does with our outpourings and by learning to interpret the wonderful pictures which science and technology have given us.

ELLIS HORWOOD SERIES IN ENVIRONMENTAL SCIENCE

Series Editor: R. S. SCORER, Emeritus Professor and Senior Research Fellow in Mathematics and Environmental Technology, Imperial College of Science and Technology, University of London

A series concerned with nature's mechanisms — how earth and the species which inhabit it fit together into a dynamic whole, and the means by which evolution has taught them to survive.

We are *not* primarily concerned to exploit the environment to human advantage, although that may happen as a result of understanding it.

We are interested in the basic nature of the physical world, the special forms that it takes on earth, the style of life species which exploit special aspects of nature as well as the details of the environment itself.

ATMOSPHERIC DIFFUSION, 3rd Edition
F. PASQUILL and F. B. SMITH, Meteorological Office, Bracknell, Berks
AIRBORNE PESTS AND DISEASES
D.E. PEDGLEY, Centre for Overseas Pest Research, London
LEAD IN MAN AND THE ENVIRONMENT
J.M. RATCLIFFE, Visiting Scientist, National Institute for Occupational Safety and Health, Cincinnati, Ohio
THE PHYSICAL ENVIRONMENT
B. K. RIDLEY, Department of Physics, University of Essex
CLOUD INVESTIGATION BY SATELLITE
R. S. SCORER, Imperial College of Science and Technology, University of London
SATELLITE AS MICROSCOPE
R. S. SCORER, Imperial College of Science and Technology, University of London
GRAVITY CURRENTS: In the Environment and the Laboratory
JOHN E. SIMPSON, Department of Applied Mathematics and Theoretical Physics, University of Cambridge

On related subjects

SATELLITE MICROWAVE REMOTE SENSING
T.D. ALLAN (Ed.), Institute of Oceanographic Sciences, Wormley
ENVIRONMENTAL AERODYNAMICS
R. S. SCORER, Imperial College of Science and Technology, University of London

SATELLITE AS MICROSCOPE

R. S. SCORER M.A., Ph.D., A.R.C.S.(Hon.)
Emeritus Professor and Senior Research Fellow
in Mathematics and Environmental Technology
Imperial College of Science and Technology, University of London

ELLIS HORWOOD
NEW YORK LONDON TORONTO SYDNEY TOKYO SINGAPORE

First published in 1990 by
ELLIS HORWOOD LIMITED
Market Cross House, Cooper Street,
Chichester, West Sussex, PO19 1EB, England

A division of
Simon & Schuster International Group

Typeset in Times by Ellis Horwood Limited
Printed and bound in Great Britain
by Hartnolls, Bodmin

British Library Cataloguing in Publication Data

Scorer, R.S. (Richard Segar) *1919–*
Satellite as microscope
1. Clouds. Observation. Use of artificial satellites
Title
551.57′6

ISBN 0–13–791344–3

Library of Congress Cataloging-in-Publication Data

Scorer, R. S. (Richard Segar), 1919–
Satellite as microscope / R. S. Scorer.
p. cm. — (Ellis Horwood series in environment science)
ISBN 0–13–791344–3
1. Air–Pollution–Remote sensing. 2. Artificial satellites in air pollution control. 3. Satellite meteorology.
I. Title. II. Series.
TD890.S28 1989
628.5′3–dc20 89–24430
CIP

ACKNOWLEDGEMENTS

Special thanks are due to the Natural Environment Research Council, who have made the pictures available by supporting the Dundee University Meteorological Satellite Laboratory, whose own staff have made this kind of study possible by their care and high standard of pictures. This work would not have been possible without the support of the Central Electricity Generating Board through their Research Laboratory at Leatherhead, who have shown a consistent interest in improving human understanding of the effect of human activity on the atmosphere.

Old and New Clouds, 1140,24.10.88, Agfachrome

The spectrum of droplet sizes in a cloud is in a continual state of evolution. If the vertical motion is slow, an almost steady, old cloud state is reached, provided that the cloud is not evaporated. Rapid upward motion and its greater degree of supersaturation causes a higher proportion of smaller droplets. In the picture the line of rapidly growing cumulus is new cloud in northwest Germany, possibly over industrial complexes in the Ruhr area, but it is surrounded by old cloud which is a dull grey. Thus the small droplets can be seen to be responsible for skyward scatter of sunshine (see section 4.2).

It has often been suggested that the darker clouds obtain their colour from dirty pollution. On the whole, air pollution is composed of very small particles which act as efficient, and often hygroscopic, condensation nuclei; thus pollution from ships produces ship trails (see Chapter 6) which are much whiter than the surrounding clouds, and this is typical of pollution — it increases the cloud albedo and only makes clouds temporarily dirty-looking when copious black smoke particles coagulate, and are later deposited by their size in a smog.

Table of contents

Preface

The word 'pollution' was originally intended to appear in the title of this book, but it was gradually realized that what is now called cloud physics was equally involved because ordinary clouds, which had been thought of as well-understood, did not always appear in their stereotype guise in satellite pictures. For instance, aircraft contrails and orographic cirrus sometimes appear white, and sometimes black, and sometimes some black and some white in the same picture; and Ch 4 sometimes shows desert dust and sometimes sees through it, so that the dust is of very variable composition even when the picture in visible light seems the same.

Without doubt the most interesting channel is Ch 3. There has been a tendency to avoid it because it presents very unfamiliar pictures, with enormous variations in cloud brightness depending mainly on particle size. It is quite wrong to mix it with other channels and to present the results or the channel alone in false colour without knowledge of its properties. I have tried to present the material so that a newcomer to the use of the pictures can acquire the knowledge quickly and have a reference book to refer to when working with this valuable source of information. To help an understanding of the way in which the Ch 3 component of sunshine is scattered back into space all the pictures in this channel have been presented in photopositive, like Ch 2, which it often, but not always, simulates. This means that the rule is 'bright is white', except for Ch 4, in which 'hottest is blackest'. Ch 5 has not been used at all because its differences from Ch 4 are small and insufficient as a basis for deducing anything about particle size, for which the channels must be much narrower than the difference between them. It is thus much to be regretted that we lost, through old age in 1986, the Coastal Zone Color Scanner, which has been very useful in pollution studies.

The change to the present title was made because particle size (in condensation clouds or pollution) plays a major role in determining the scattering properties of clouds. In an important sense deductions about the particulate composition from what we see on the very much greater scale is analogous to the study of cloud particles through meteorological optics and sky colour, except that we are not concerned with bright and sometimes coloured spots, arcs, and lines, because the moving point of observation in the making of satellite pictures prevents such phenomena from appearing; but we are concerned with the relative brightness of different parts of the same cloud or cloud formation, and we are very much concerned with seeing through clouds to what lies below.

Several challenges remain, such as the mechanics of spike clouds, and the details of the mixture of thermal emission and scatter of sunshine in Ch 3 when the sun is very low. Much of the effort by qualified scientists has been expended on schemes using the latest technology to achieve purposes thought about before the technology was available. The present purpose has been to understand as much as possible of the incoming information before subjecting it to computerized transformation for particular uses. It has required the simple principles of scattering by clouds of particles and an appreciation of clouds as live entities which are continually being transformed and their particle size spectrum changed by the mechanisms which make them grow and eventually evaporate. The spirit is that of Minnaert: to understand our environment in more detail and to enjoy more deeply its complex and beautiful evolutions.

1
Introduction

Cloud physics and dynamics have been illuminated by the understanding of the behaviour of the particles of which the clouds are composed. Their fallspeed and optical properties are very important in determining the appearance of cloud. Fallstreaks, haloes and their associated phenomena, rainbows, cloud bows, polarization, and colour all require knowledge of the properties of the individual particles before they can be fitted into a total picture. And conversely, by careful observation of all that can be seen, much can be learned about the particles at the microscopic level, even if only crude eye observations are used.

In particular, the colours of bows and arcs contain information from which particle sizes may be deduced, and the range of sizes may be obtained from the details of the supernumerary arcs of the bows. The dark edges of cumulus towers seen tangentially are in stark contrast with the bright silver lining seen in different illumination. The transparency of clouds enabling the sun's disk to be seen sharply defined; the brightness of mock suns in some contrails; the shadows rapidly changing among clouds of various heights. Such beams and shades tell us about quite subtle details of the particles and their distribution in space at the microscopic level even though the eye observations are simple and large in scale.

It is no surprise, therefore, that satellite pictures, with a pixel size of about a kilometre or more, contain information about the particles of which the clouds are composed. Coloured arcs do not appear in satellite pictures because of the line-by-line continuous strip structure of the pictures, but we have the same scene surveyed by several different wavelengths and from different directions, and from our knowledge of the physics of clouds of particles we can deduce what they are like individually.

Fortunately there exists in the Dundee University archive a record of about a decade of a selection of the passes by the Coastal Zone Color Scanner (CZCS) which was designed to make observations of the colour of the sea. Because the sea is darker than clouds and land the radiometers were very sensitive. Frequently, even usually, they were saturated by the sunshine reflected from land or clouds and then no detail is obtained in those areas in the visible wavelengths. But they happen to have made a very good record of haze, which is much less bright than clouds. This is of great value even though the haze often frustrates the original purpose — to record sea colour — and it can serve as a guide in the design of future radiometers. These observations have the additional advantage that glint from the sea surface was eliminated by tilting the camera about 20° towards the pole. This has the disadvantage that it distorts the picture in relation to conventional map projections. The Dundee archive is also the source of all the other pictures used in preparing this book. It is a high quality record and the laboratory has produced excellent prints.

Many different-looking pictures can be obtained from the same satellite message by controlling the emphasis given to different intensities; these cover a far greater range than can be simultaneously presented on photographic paper. Thus at sunset the contrast between the strong illumination of some high clouds and those further east which are only illuminated by skylight cannot all be shown in full detail together. The shadows which reveal the contours of a cloud top may use up all the shades the eye can see in one glance so that even pale clouds look black. Thus the area of a cloud may be apparently increased or decreased according to the emphasis given to the shades at its edges.

In detailed studies of temperature variations, whether in the sea surface or on the tops of cumulonimbus or cyclones, it may be desirable to have as accurate measurements as possible. This is not difficult in a serious research investigation of a particular case or type of weather, but it is not attempted here because the emphasis is upon what can be done rather than consume all the time in working up the details. The aim is to give quicker and wider understanding of what is contained in the pictures. A good example is provided by the penetration of cumulonimbus into the stratosphere, which is very important in the dynamics of the cloud as a whole; we shall be concerned only with the very small scale mechanisms which transform the appearance of the cloud tops (Chapter 5).

Another example of the limitation of the approach of this book is provided by the modelling of the transport of air pollution. We are concerned with the opportunities the satellite observations provide to test a mathematical model of the transport by actually observing the displacement with time. We are not concerned with the dubious process of 'validating' a particular model. Anyone who has been involved in working out the details for a particular occasion will understand that unless the case is of some economic importance there is a point beyond which the effort is not well-spent once the physics and mechanics of the occasion have been understood.

Some cloud forms have been discovered by satellite. For example, the spike clouds (which, I am advised by an international authority, should be referred to as 'cirrus spissatus spicus', although it is not clear how this helps to understand them) called 'frontal instability' in [1], have still not been given a satisfactory explanation, and their appearance in different wavelengths is helpful in defining the problem needing solution, or it may define a task for the SPOT or some other satellite.

The wavelengths available for the present investigation were as follows:

Satellite	Channel code	Wavelength band (μm)
NOAA 6–10	Ch 1	0.55–0.68
	Ch 2	0.725–1.10
	Ch 3	3.55–3.93
	Ch 4	10.5–11.5
	Ch 5	11.5–12.5
CZCS	CZ 1	0.433–0.453
	CZ 2	0.510–0.530
	CZ3	0.540–0.560
	CZ 4	0.660–0.680
	CZ 5	0.700–0.800

Ch 1 and CZ 1–4 are included in the range visible to the human eye (0.4–0.7 μm approximately). At ultraviolet (UV) wavelengths less than 0.3 μm sunshine is completely absorbed in the stratosphere, mainly by ozone, and the shorter of the visible wavelengths are absorbed by aerosol and clouds to some extent and also scattered so that the sky appears blue. The blue of the skylight is almost uniform and therefore may be subtracted from the picture so that we see only the scatter by aerosol. The blue sky does not seriously interfere with the observation of haze.

Ch 2 and CZ 5 are in the near infrared (IR), but are treated like visible light. Indeed Ch 2 is widely referred to as 'VIS'. Together with Ch 3 and CZ 5 it is included in 'sunshine', but not in 'sunlight'. The reflective properties of land surfaces are often unfamiliar in the near IR where the ground is mostly brighter, relative to clouds, than in the visible channels, Ch 1 and CZ 1–4. For all sunshine the terminology of optics is freely used in discussion.

Ch 3 is a special and most interesting case. Its intensity is very small, being about 0.6% of the intensity of visible green (see **5.8.1**), so that much amplification is required to provide pictures like those of other channels. This has meant that many Ch 3 pictures have been subject to a stripey form of interference which spoils some pictures, although the detail is often good in spite of this. At night the radiometer receives enough of the earth's emission to make good pictures in which the intensity is related to temperature, but this is almost all swamped by the scattered sunshine by day, although the emission can sometimes be detected in uniform dark areas, particularly the sea. The absorption of this waveband by water and ice is very large, so that the beam is reduced to below detection level by passage through less than 0.2 mm of water. Consequently the reflectivity of many clouds is reduced to zero for practical purposes, and they appear black in pictures. By contrast, the reflection from a flat water surface is much more intense than in other bands, and the glint from the sea surface saturates the radiometer, appears uniformly white over large areas, and extends further from the sun's mirror image point in a more or less calm sea than in other channels. This wavelength is large compared with the size of haze particles, and so haze is penetrated by it almost as effectively as by Chs 4 and 5. In this work we have used pictures in photopositive for Ch 3 because all the studies are by day, and 'bright is white' is the rule as in CZ 1–5 and Chs 1 and 2.

Chs 4 and 5 are far outside the range in which scattered sunshine is measurable, and the pictures only show the earth's emission, and shades are therefore related to temperature. The convention of showing these pictures in photonegative, in which 'bright is black', has become universally accepted because it shows the high clouds of important weather systems as bright, which makes them more immediately intelligible beside the photopositive pictures of Chs 1 and 2. These deep IR channels penetrate almost all haze, and are useful in the context of cloud physics by distinguishing cloud heights by their temperature differences, a task which is sometimes difficult in other channels. For our purposes there is no difference between Ch 4 and Ch 5 and they do not help much with problems of particle size. They are absorbed as intensely as Ch 3 by water and ice, which prevents any image from below penetrating any but a very thin cloud layer, whereas in Chs 1 and 2 we can often see details of a lower layer through an upper layer (see Chapter 3).

Under each satellite picture we place the following information with the punctuation shown:

Chapter.Section.Number and Time(GMT),Day.Month.Year,Channel

Thus '**3.3.3** 0836,1.7.88,3' means 'Chapter 3, Section 3, Picture No 3; taken at 0836 GMT on 1st July 1988 in Channel 3'.

AVHRR channel numbers have no prefix, those of CZCS have the prefix CZ, while in a very few pictures of the era before AVHRR we refer to the two VHRR channels as 'VIS' and 'IR'. These last two were very broad bands with the ranges 0.6–0.7 μm and 10.5–12.5 μm respectively.

Other satellite channels exist in quite large number. The Landsat pictures are often very spectacular, but are not available daily for a given point, and are too near to aerial photography to have anything much more to say about cloud and haze. The hemispherical pictures made by the geostationary meteorological satellites have a great advantage in providing much more frequent pictures, but there exists no archive which makes them easily available for inspection to the lone worker at acceptable expense; nor are they much use beyond about 55°N, and at present they have no channel as interesting as Ch 3. These circumstances may change in the future, but for the present the Dundee archive provides the material in a most suitable and accessible form.

Some reviewers may be inclined to say it is a pity that this or that bit of investigation was not carried a bit further. But it must be remembered that none of the satellites was designed for the purposes described in this book. Only simple aspects of the phenomena, almost any of which might be studied to the level of a doctoral thesis, are described here — perhaps just to whet the appetite. One might define them as those phenomena which are obvious in satellite pictures to the observant naked eye, which can be the basis of knowledge about things the naked eye cannot see.

Reference to other books is made numbers in square brackets.

2
The scattering of sunshine

There are several mechanisms by which sunshine may reach, or be prevented from reaching, the eye of an observer or the radiometer of a satellite, and which may modify the beam.

2.1 SPECULAR REFLECTION

By this term we mean the reflection from a smooth, i.e. shiny, surface which, in the present context, usually means a water surface. When the surface is flat and calm the reflection is very bright at (and close to, when the calm is not perfect) the point of mirror reflection, but the surface is very dark elsewhere. When it is disturbed the reflection is spread within a cone which becomes wider the rougher the surface becomes; at the same time the brightest area becomes less bright and the illumination is spread over a larger area. A very calm patch within a rough area becomes a black patch even if it is quite close to the point of mirror reflection. A stronger wind in a bright area makes it duller, while it may brighten a dull area.

Although the satellite makes an exposure of each pixel for less than 0.002 s, the intensity is the average over the whole pixel, which is much larger than the waves on the sea which scatter the reflection. Consequently we see no evidence of the rapid fluctuations which we freeze in an ordinary snapshot of 1/500 s. Nevertheless the wind direction can often be seen in glint areas because the spatial variations of wind strength are often longer than the pixel diameter (1.1 km), and are more analogous to cloud streets than to the waves we see with the naked eye. Air navigators recognize the 'wind lanes' as representing the surface wind direction, but they are usually made visible by foam lines on the sea surface with spacing much smaller than a pixel.

2.2 DIFFUSE REFLECTION

This refers to the illumination we receive from a matt or roughened surface. Such surfaces on land usually have selective absorption of colours, and most materials have their own reflected colours, which are the least absorbed and therefore the most strongly reflected. Diffuse reflection is in all directions away from the surface. Thus a plume of iron oxide dust emitted from a steel works has the same colour, red, as a lump of the same material, the other visible colours being absorbed. Many objects appear relatively light in some wavelengths and dark in others: and relative to the brightness of clouds most forms of ground (rock, soil, vegetation, etc.) appear brighter in the near IR channels (Ch 2 and CZ 5) than in the blue (CZ 1), green (CZ 2), or orange-red (Ch 1 and CZ 3 and 4), although this is not an invariable rule, particularly with forests which are usually dark in the near IR.

2.3 SCATTER BY REFRACTION

Transparent materials such as cloud droplets or ice crystals change the direction of incident rays passing through them, the angle being very variable because it is dependent on the angle of incidence at the points of entry and emergence of the ray. Since rays may also be internally reflected they may emerge in almost any direction. We must not be misled by the prominence of the rainbow, which is at the boundary of the region into which rays may be scattered after only one internal reflection. The secondary (and subsequent) rainbows are progressively very much more feeble, and most of the light which might form a coloured bow impinges on other droplets and is smudged to form an arc glowing with shades of white, if it can be seen at all, on a cloud top.

2.3.1
Inside the rainbow the raindrops are scattering light by refraction and one internal reflection, so that the sky looks bright. We see the rainbow at the edge of the illuminated area because the position of the edge is determined by the refractive index of water, which varies slightly according to colour. At sunset (when the end of the rainbow is almost vertical) the atmosphere has scattered out the blue end of the spectrum so that the yellow/red colours dominate. The secondary bow is feeble because there is almost no rain in the required position, and we do not see the glow outside it (like the glow inside the primary), which would be due to scatter by two internal reflections and would not be as bright as that by a single reflection.

2.4 PRIMARY AND SECONDARY SCATTER

Primary scatter is light received from the first particle on which a ray was incident. The scatter from a tenuous haze is, for practical purposes, all primary scatter because the intensity of direct sunshine on a haze particle is much more than the illumination by scatter from all the other particles. Most of the sunlight usually passes through the haze without impinging on any of the haze particles; but in a very dense haze and in almost all clouds no rays pass right through without impinging on a particle, and in

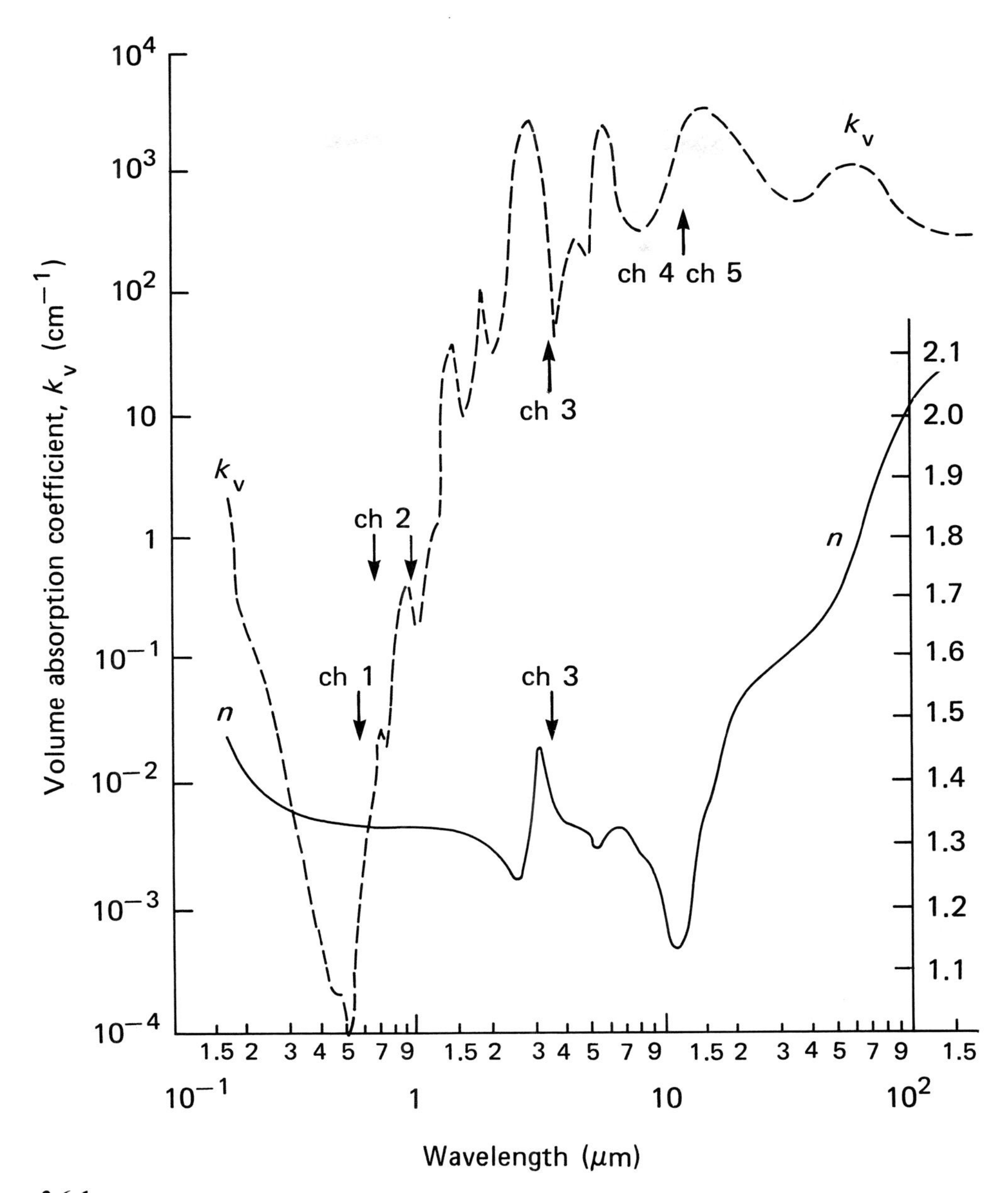

2.6.1
The absorption coefficient k and refractive index n of water (based on Goody, 1964 [2]).

Radiation of the wavelengths of Chs 3, 4 and 5 is strongly absorbed and reduced to negligible intensity by passage through 200 μm of water. That of Chs 1 and 2 is only slightly absorbed by passage through a dense cloud but appears to be reduced because it is scattered in all directions. The positions of the channels are indicated by arrows, which show the limits in the case of the wide Ch 2.

The 'blip' in the refractive index at the Ch 3 wavelength causes a very strong specular reflection at a water surface.

Although ice has similar absorptive properties there is no evidence that Ch 3 is as strongly reflected at an ice/air interface.

that case secondary and higher-order multiple scatter is important. Most of the radiation coming from a bright white cloud in sunshine is secondary or multiple scatter by refraction and reflection, the rays having been reflected by or passed through several droplets.

2.5 SCATTER BY DIFFRACTION

In geometrical optics every object has a shadow area, which is very sharply defined in the case of a point source of light. The sun produces a penumbra because it is a source of finite size, and in this context should be regarded as a collection of point sources. The wave fronts on the boundary of the conical shadow region spread into the shadow to an extent determined by the distance from the object, the wavelength of the radiation, and the size and shape of the object. The intensity of the radiation which is thus diverted from a straight-line path is very small when the wavelength is large compared with the object causing the shadow, so that long wavelengths can pass unmodified through a haze of very small particles, and we say that the wavelength can be used to 'see through' the haze. But none of the satellite wavelengths can pass through ordinary clouds because the droplet size is not small compared with the wavelength. Haze particles are typically around 0.1 μm or less while cloud particles are from about 1 μm up to around 100 μm, and larger ones are prevented from falling to the ground only if there are strong upcurrents.

2.6 ABSORPTION BY WATER AND ICE

Visible light is only very feebly absorbed by water or ice, whereas very short UV radiation is very strongly absorbed. The IR wavelengths of CZ 5 and Ch 2 are weakly absorbed, while absorption of the longer wavelengths of Ch 3 and Chs 4 and 5 is very strong. A layer of cloud absorbs all the radiation coming from below it in Chs 3,4 and 5, but in Chs 1 and 2 some may penetrate through a thin layer to give an image composed of bright and dark areas when seen from above, but much stronger variations when seen from other directions.

Because of absorption all we see in a satellite picture are the properties of the particles close to the upper surface. In Chs 1 and 2 the base of a cloud glows with the light we normally see scattered through from above. But in Ch 3 no sunshine penetrates because of the strong absorption, and consequently cloud shadows are very dark and sharply defined.

Thus the amount of visible light penetrating clouds is considerable as we can see in daytime, but the absorption of the longer wavelengths of the earth's emission is complete in all thick and most thin layers

2.7 ABSORPTION BY ATMOSPHERIC GASES

The solar spectrum fits very neatly into the wavelength band to which the atmosphere is largely transparent. At UV wavelengths less than about 0.3 μm most oxygen molecules undergo dissociation and recombination, and combination into ozone (O_3) takes place to form the so-called ozone layer. Ozone itself completes the absorption of the UV shorter than 0.3 μm, but the range between about 0.3 and 0.4 μm is only partly absorbed in the stratosphere, and a large proportion of what reaches the troposphere is absorbed by cloud and haze.

The visible and near IR ranges are only partly absorbed, as indicated in Fig. 2.7.1. At the longer wavelengths of the solar spectrum many bands are absorbed, mainly by H_2O and CO_2. Other gases with more than two atoms per molecule absorb narrow bands at these longer wavelengths, and the observed spectrum reaching the earth's surface can be used to determine the total load of these gases, individually, above the observing station by comparison with the intensity of neighbouring unabsorbed

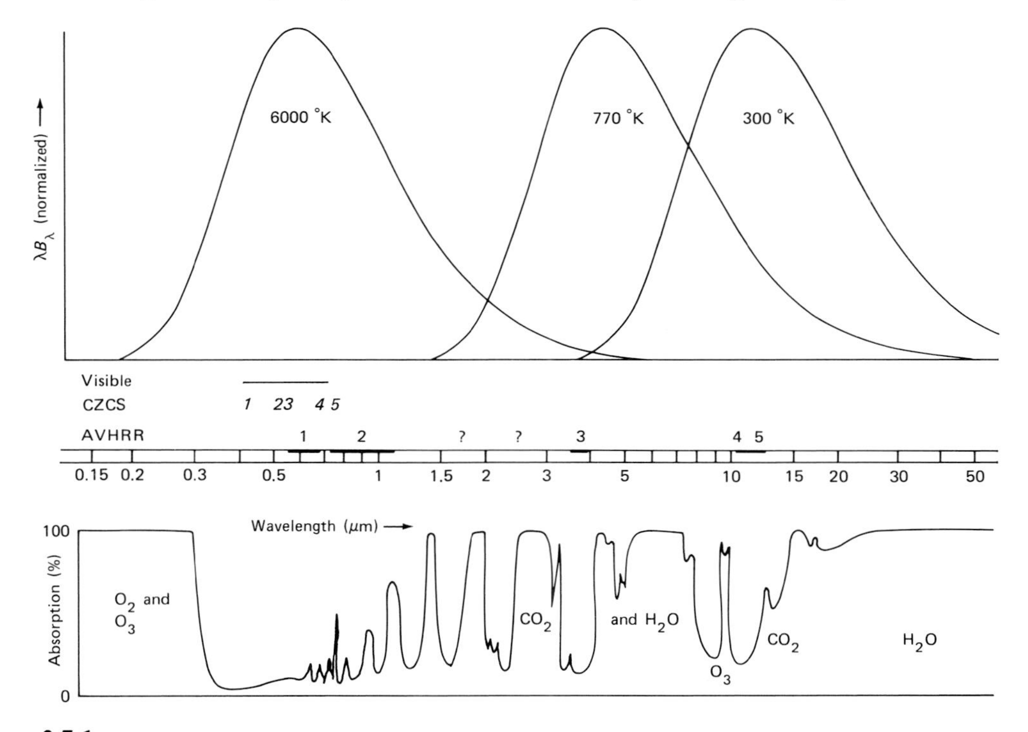

2.7.1
Emission spectra and atmospheric windows (based on Goody, 1964).

At the top are the theoretical black-body emission spectra for temperatures shown, which represent the sun, dull red heat, and typical earth values.

Against the logarithmic wavelength scale in the middle the satellite channels and the visible range are marked.

The bottom section indicates the absorption due to one passage vertically through a typical atmosphere, which must be doubled for reception of scattered sunshine by a satellite.

Absorption of UV is complete for wavelengths less than 0.3 μm, due to ozone, and CO_2 and H_2O absorb all wavelengths greater than about 15 μm. In between, the absorption is very variable, the bands least absorbed being termed 'windows'. The absorption peaks are due to H_2O, CO_2, N_2O, CH_4, O_3 and a variety of lesser components, such as SO_2 which have various residence times depending on their chemical reactivity.

The windows not yet exploited by meteorological satellites are at 1.3, 1.6, 2.3 and 8.5 μm. Ch 3 is situated in the small overlap range of the solar and terrestrial emissions, and is at the maximum of the emission spectrum for a body at dull red heat (see Ch 5).

The areas under the intensity curves are the same because the incoming and outgoing radiation are the same globally and in the long term, but that is not very meaningful in practice because of the great variability of insolation with latitude and time of day and of scatter with albedo. The colder clouds and polar regions may be at temperatures closer to 230 K, and at this temperature there is negligible overlap with the solar spectrum.

wavelengths. The most important of these gases are SO_2 and CH_4, both of which have a large turnover. Being very reactive, they are rapidly removed. The other most important are the chlorofluorocarbons (CFCs) which do not undergo chemical decomposition until they have diffused up into the stratosphere and therefore have a very much longer lifetime in the troposphere; in spite of their relatively small concentrations they do have a significant absorptive effect on the IR radiative balances.

The satellites are designed to observe the earth through the windows, i.e. the wavebands which are only slightly absorbed by the atmospheric gases. The water vapour windows are the most important for Chs 2 and 3. The longer wavelength bands in which H_2O absorption is large are important in determining the temperature of the lower layers of the troposphere, but CO_2 is much more important higher up because its proportion is approximately the same throughout 99.9% of the atmosphere.

Clouds, except for thin layers which have holes in them, are opaque to the wavelengths of the earth's emission, and absorb them completely.

2.8 THE COLOUR OF HAZE

When haze particles are less than about 0.1 μm in size the scatter is much stronger towards the violet end of the solar spectrum, and so the haze has a blue appearance. Still smaller particles may scatter more violet, but the intensity is less because that component of sunshine is weaker than the blue and the cross-section area of the particles is less.

A very dense haze of coloured particles exhibits their natural colour just as it would if the particles were much bigger.

Smoke is composed of solid particles or droplets condensed out of hot vapours which are usually an unburnt component of the fuel. If the particles are around 0.1 μm in size smoke appears bluish in sunshine, but if we look at the sun through a smoke plume we see a magenta colour which is the appearance of sunshine with the blue removed. Haze in a valley with trees which provide a dark background has a blue tinge, due to the light it scatters, but from the bottom of the valley the sky appears much whiter than from a mountain top, while the sun itself takes on an orange hue because the blue is scattered sideways and backwards.

2.9 THE DIRECTION OF SCATTER

The intensity of scatter varies according to the relative magnitude of radiation wavelength and scattering particle size. **2.9.1** is similar to the figure given by Minneart [3] and is based on Mie theory for dielectric spheres such as water droplets. The scattered radiation is a mixture of reflected, refracted and diffracted rays. More accurate calculations have been performed for conducting, absorbing or opaque particles using modern computers, but the atmosphere contains such a variety of particle shapes and sizes that no set of calculations can include more than a minority of particles present, and even then would only treat the case of primary scatter. As we shall see, much of the discussion concerns the particle sizes present on the outside of clouds, where evaporation, condensation and lively motion is often occurring. We can make some fairly precise statements, and discover the mechanisms at work, but it

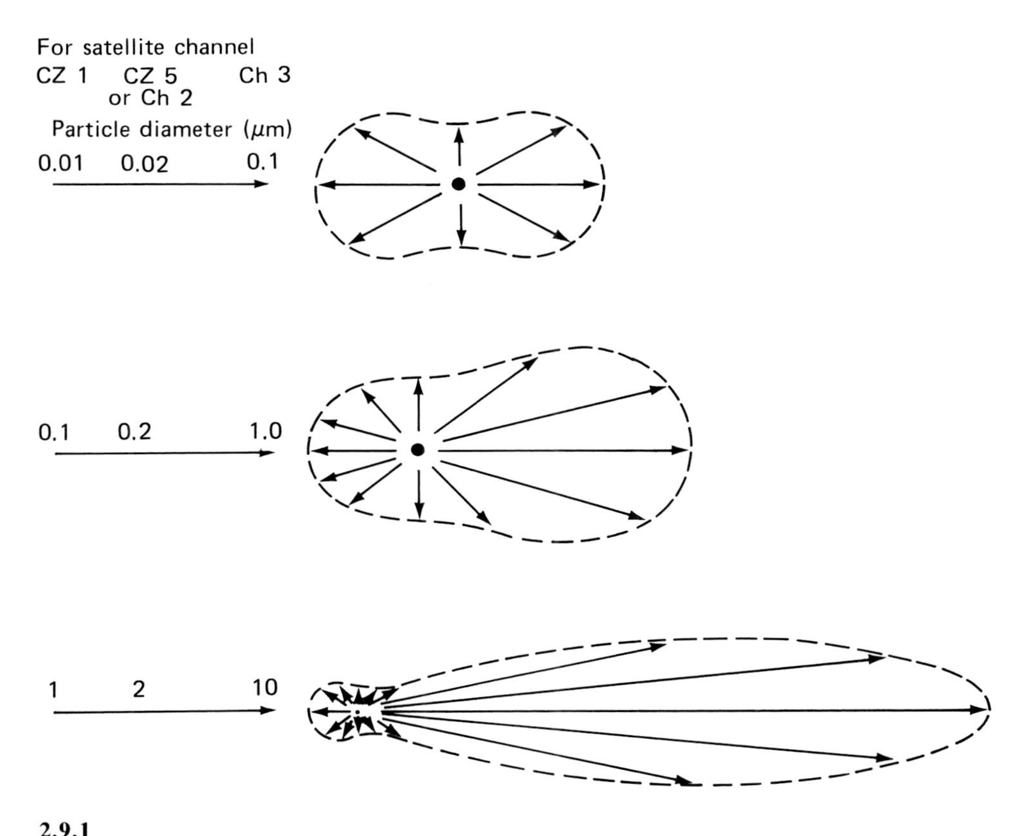

2.9.1
Scattered radiation according to Mie theory for dielectric spheres showing dependence on particle size for different satellite channels.

For incident radiation from the left the numbers are the particle sizes for the different channels which give the scatter distribution shown on the right.

In reality particle sizes are not grouped close to the values shown, the intensity may be drastically reduced by absorption, and the particles are not necessarily spherical and may be present in a wide range of sizes simultaneously. These complexities do not obscure many interesting features of observed intensities, particularly in the visible wavelengths which detect the smallest aerosol particles and Ch 3 in which absorption distinguishes particle size very effectively (see Chapter 5).

is difficult to go further into generalities without satellite observations designed specifically for this purpose.

In the case where backward and forward scatter are of comparable magnitude, the sideways scatter is strongly polarized. This is easily seen by eye in the visible range using a polarizing screen (or, in the case of the blue sky, by observing Haidinger's brushes, which requires no optical aid); but polarization is of little consequence in clouds because it is smudged by multiple scatter in most cases. Rainbows and cloud bows which involve only one or two reflections are strongly polarized, being generated by single-particle scatter.

In Ch 3 illumination, where absorption by clouds is strong, we find that among the larger particles of ice or water, rays would be almost completely absorbed by traversing very few particles (at least two, and probably not more than about five). Most of the scatter must therefore come from the smallest cloud particles close to the cloud surface. Haze particles are usually so small that Ch 3 radiation passes through with negligible scatter and it cannot be used to detect haze. Clouds in Ch 3 sunshine are brightest when the scatter is mostly forwards or backwards from particles smaller than the wavelength (3.7 μm). Further discussion of this channel is given in Chapter 5.

2.10 SHADOWS

When seen from satellites a cloud may appear well-separated from its shadow. This depends on the height of the cloud above the shadow and the angles of the sun and of the satellite as seen at the cloud. In Chs 1, 2 and 3 cloud shadows are most easily seen if there is a cloud layer below which otherwise appears white. The cloud casting the shadow may be almost indistinguishable from the cloud below if it is small and of similar texture and brightness; it is then only recognised by its shadow. This is often the case with small patches of cirrus or contrails.

In Chs 4 and 5 there are no shadows cast on any surfaces because the rays are all absorbed and what we see is emissions only. Small patches of cirrus or contrails which are much higher than the next layer below appear much whiter (cooler) and often significantly displaced from the position of their shadow if it is visible in any of Chs 1, 2 or 3.

In Ch 3 a fragment of cirrus composed of particles of size about 10 μm upwards appears black because it absorbs all the incident sunshine in that waveband. Its shadow therefore appears as a dark duplicate of the cloud. Since there is no scatter by aerosol haze, and almost no clouds scatter brightly, there is no skylight. This makes shadows relatively much darker than in Chs 1 and 2.

3.1.1, 3.1.2, 3.1.3, 3.1.4
The ice edge off the south coast of Baffin Island is partially covered by a layer of very thin stratus which is partly transparent in Chs 1 and 2, but since the ice is black in Ch 3 it does not show through. Similarly the temperature contrast at the edge is very small, so that it could scarcely show in Ch 4 even if the stratus were transparent.

The cellular nature of the stratus is seen best in the southeast corner of the Ch 2 picture, but the ship trail shows best in Ch 3 because it is a dropsize phenomenon (see Chapter 6 and **5.7.7–9** of the previous day).

The cirrus is duplicated by its shadow in Ch 3, but is not as clearly visible as its shadows in Chs 1 and 2 , and has no shadows in Ch 4.

See also **13.1.6** where a high cloud is transparent in Ch4 at night.

3
Vision through clouds

3.1 PARTIAL OBSCURATION BY SPARSE CLOUDS

If some rays coming from an object below a layer of cloud pass through it without impinging on any cloud particles, a fairly clear image of the object may be seen through the layer. This means that it has straight ray paths through the cloud,

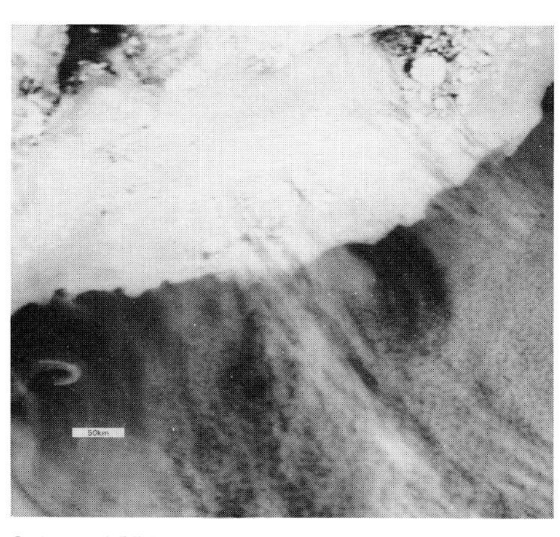

3.1.1 1557,12.6.82,1

3.1.2 1557,12.6.82,2

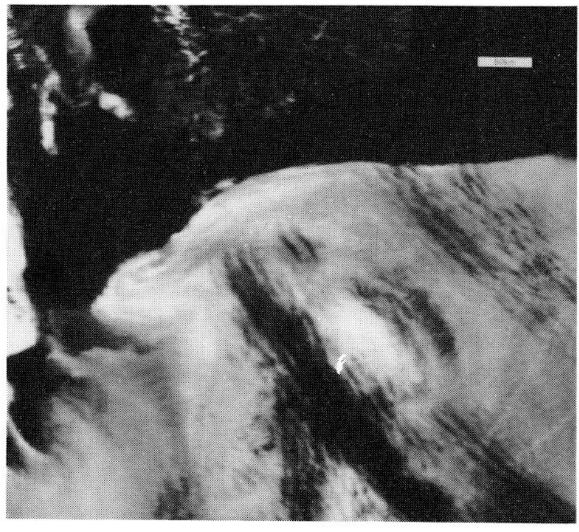

3.1.3 1557,12.6.82,3

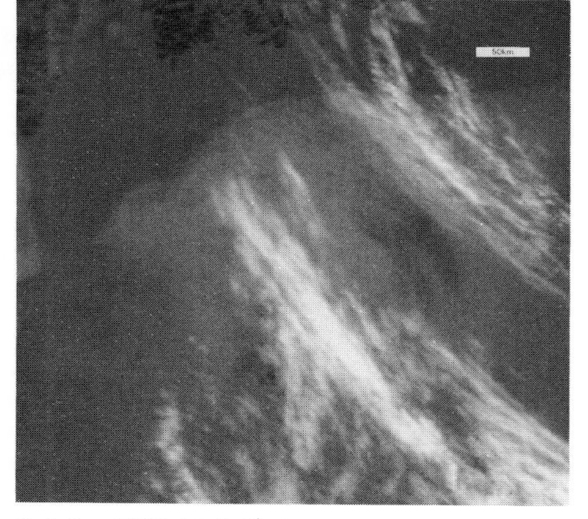

3.1.4 1557,12.6.82,4

although the holes may be very small and mostly much smaller than a pixel. Thus the texture of a low layer of cloud, may be seen through very sparse cirrus in all wavelengths.

In Ch 3 the cirrus appears as dark streaks partially obscuring low cloud, which may be white or a duller grey. In Ch 4 (in photonegative) the cirrus is cold white and the low cloud a warmer grey. In both Chs 3 and 4 the direct illumination from the low cloud to the satellite passes between the cirrus particles but is absorbed by any on which it impinges.

In Ch 2 the cirrus is sometimes scarcely visible and may even appear as a dark area because its backward scatter is often very feeble if the multiple scatter is weak. The same high cloud may appear bright white when seen from the ground because the forward scatter from a few ice crystals is nevertheless strong. Sometimes the cirrus shadows on lower cloud layers can be seen by satellite even when the cloud is almost invisible.

3.2 OPTICAL TRANSMISSION OF IMAGES THROUGH CLOUD LAYERS

If there exist no direct ray paths between the cloud particles then no images of objects below can be 'seen' from above in Chs 3, 4 or 5, because the rays are all absorbed. Even if the first droplet encountered does not absorb a ray completely, it needs very few more encounters for complete absorption.

In Chs 1 and 2 variations of brightness of the world below a cloud layer may penetrate through so as to be visible above. The effect depends on the height of the layer above the objects and the pixel size. If the cloud height is less than a pixel size, contrast may be lost but not detail. With increasing height of the obscuring layer, detail is gradually smudged until it is lost completely. This is well-illustrated by showing pictures through the same layer of frosted material at various heights (**3.2.1–3**).

Obviously thicker layers diffuse the image of the scene below, as seen from above, more than thin layers at the same height because every bundle of direct rays is spread out in more directions. Thus thick fog completely obscures detail on which it is resting.

In the case of a layer through which detail below can be seen in Ch 2 or 1, we can determine whether this is because it has holes in it smaller than the pixel size or because it is low and thin, by seeing whether or not the same detail can be seen in Ch 3 and 4.

3.2.1, 3.2.2, 3.2.3 (see **5.4.10–15** for discussion of the clouds)
The three pictures are all of the same original taken with a tracing paper overlay (i) on the surface, (ii) with the overlay at the equivalent of 17 km, and (iii) at 34 km, according to the 100-km scale on the picture, above the picture to represent a thin cloud layer. The top part remains uncovered in all pictures. It is seen that the height of the obscuring layer determines the scale of the detail transmitted.

3.2.1 1445,1.7.88,4

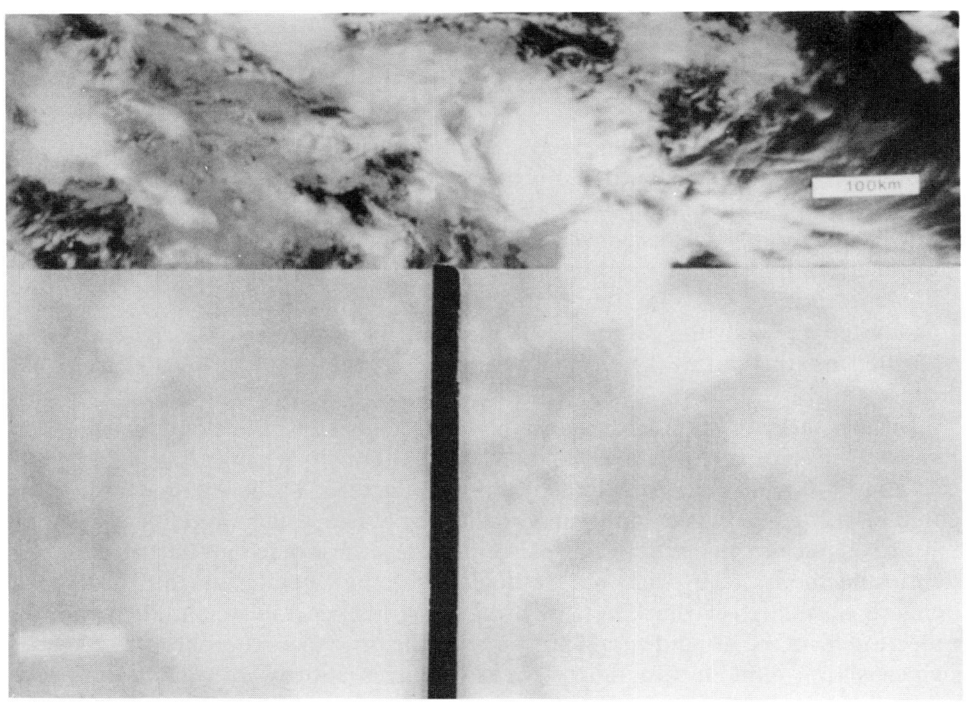

3.2.2 1445,1.7.88,4

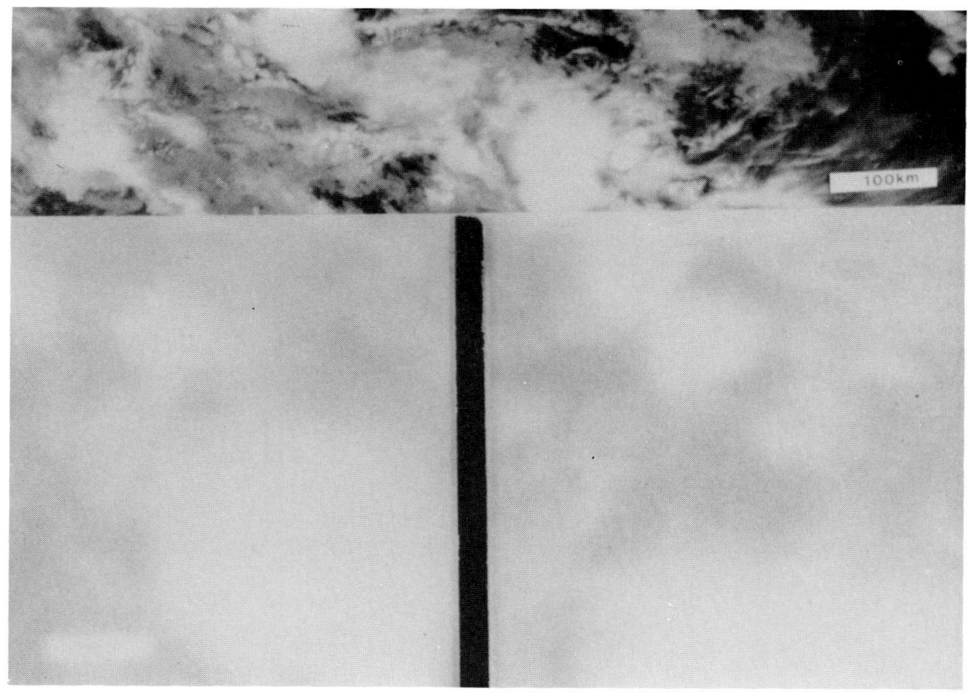

3.2.3 1445,1.7.88,4

3.3 APPARENT TRANSMISSION OF IMAGES THROUGH LOW CLOUD LAYERS

If the reflectivity of the top of a cloud layer is changed by circumstances taking place below, and those circumstances apply in only part of the scene, that part may appear to be differently illuminated. For example, a layer of stratus lying across a coast is subjected to more vigorous convection over land by day if the ground is warmed by sunshine to a higher temperature than the sea. The sunshine consists of the whole solar spectrum as seen at the bottom of the atmosphere, a large fraction is transmitted through the stratus to the ground, where again a large fraction is absorbed and this warms the ground. We next consider what is seen in each channel from above.

In Ch 1 the satellite sees the sunshine scattered back by the cloud which is very much of the same intensity over land and sea. Although a change in appearance of the cloud top in so far as it is influenced by an increase in the thermal convection could be recognized by an observer flying not far above it, this would be a structure pattern on a scale smaller than a pixel in a cloud layer less than one-third of a pixel in depth and therefore it could not be detected in the satellite image. The ground beneath is usually very dark in Ch 1 if there is good vegetation, and therefore the reflection from the ground would not enhance the brightness compared with the sea (which shows glint due to the bright cloud layer above). It would, therefore, probably be impossible to locate the coastline in the Ch 1 image.

In Ch 2 the main difference from Ch 1 is that the ground is much brighter relative to clouds. This would therefore contribute significantly to the illumination of the cloud base and a fraction of this would emerge from the top. The coastline may therefore be located by the increase in cloud brightness on crossing inland.

In Ch 3 the sunshine is absorbed so that the land is in the very dark shadow of the cloud. All we see from above is the radiation scattered upwards by the smaller droplets in the uppermost metre or two of the cloud. Over the sea the convection is weak and slow so that the drop size distribution at the cloud top is a mature one in which the smallest droplets have evaporated by the action of surface tension. The cloud looks a dull grey. Over the land the higher surface temperature increases the strength of the convection, but the cloud is confined below a strong inversion generated overnight (by cooling at the cloud top) and does not rise significantly higher. Having reached the top more quickly, cloud elements contain a greater proportion of the smaller droplets and scatter a greater intensity of Ch 3 upwards. The cloud top over land therefore appears brighter, as is the case in Ch 2 to a lesser degree, but the cause is not the same.

In Ch 4 and 5 pictures we do not see sunshine but a measure of the cloud top temperature; and since this is the same at the cloud top over both sea and land there is no clue as to the position of the coast.

Although Ch 2 is the only channel in which we receive a different message radiated from the land and sea beneath the cloud, the brightest indication of the position of the coast may be in Ch 3. Later in the day the heating over land may produce a convective structure with elements larger than the pixels, and then the coast can be seen in Ch 1, and perhaps Ch 4 also.

In general there are many possibilities, depending on the nature of any image which may be transmitted through cloud. We present a case which poses several problems indicating the difficulty of making a simple statement as to what we shall see (**3.3.5/6/7/8**). There is a narrow band of orographic cirrus in the lee of the mountains of southern Norway, the wind being almost exactly along the line from Bergen to Skagen, i.e. from the northwest.

The absence of good shadows in Ch 3 indicates that the low sun, at about 1400 solar time over Jutland, was contributing, by reflection from the ground, very little of the radiation coming from below the clouds in that channel. There were no constrasts over Denmark as sharp as those over Norway, and the appearance of haze was produced. The lack of illumination of the cloud off the southern tip of Norway in Ch 3, except in the sun-facing edge, emphasizes the absence of skylight in this channel (see section 5.5) compared with Chs 1 and 2, particularly in Ch 2 where the skylight contributes to the ground illumination. This results in the absence of detail in Chs 3 and 4 below cloud where there are no significant temperature contrasts.

3.3.1, 3.3.2, 3.3.3, 3.3.4

The northeast coast of England is covered by North Sea stratus with some high ground protruding through inland. In Ch 1 the land is dark and there is no indication in the cloud where the coastline is. In Ch 2 the land is brighter and because the cloud is very low this additional light source makes the cloud brighter over the land.

The same appears to be the case in Ch 3 but the radiation is absorbed by the cloud so we see only what is scattered by the droplets at the upper surface of the cloud in the sunshine. There are stronger upcurrents over the land, which has been warmed by the sunshine penetrating the cloud. Consequently smaller droplets are carried to the cloud top than over the sea, and this causes more backscatter in Ch 3.

In Ch 4 we see that the land is warmer than the sea and the shallow cloud above. We can see it is shallow by the way it has penetrated up some valleys. In central England the cloud is rapidly evaporating.

It is noticeable that Chs 1 and 2 do not characterize the top of the cloud over the sea in the same way as Ch 3, the shading being different.

3.3.5, 3.3.6, 3.3.7, 3.3.8

In Ch 1 there is a dark shadow over the towns of Skien and Larvik, to the northeast of the upwind end of the orographic cirrus cloud, and over the land of Norway the cloud is brighter than over the sea. The coastline of north Jutland (Jylland) can just be seen through the cloud by the brightness of the beach, but the cloud is not brighter over the land of Denmark.

In Ch 2 the part of the cirrus trail above Jutland is significantly brighter than over the sea, and the part over Norway is much brighter. It is also brighter over Zealand (Sjaelland) and nearby Skane, the Swedish south coastline from Oland to Helmstad being discernible as the boundary of the brighter land. The islands of Laeso and Anholt in the Kattegat are also visible in Ch 2, and many cloud shadows over land are well contrasted.

In Ch 3 there is no excess of brightness of the cirrus over any land, and the cirrus is a good example of bright cirrus with small particles (see 5.2 and 8.1) in a sky where there is a mixture of bright and dark high clouds. There is also an appearanace, both in Ch 3 and Ch 4, of a hazy obscuration between the cloud and the land below, which increases southwards from the Skagerrak. There are pinpoint hot spots (see 5.8) at Aalborg (in Jutland), near Vanersborg (at the southwest end of Vanern in Sweden), and near Holbaek (Zealand), but all shadows are very weak, indicating very little scatter of sunshine from the ground in Ch 3.

In Ch 4 the very hazy appearance around Denmark is even stronger than in Ch 3. There are of course no shadows in Ch 4 but over Norway the cloud is less dark than over the Skagerrak, indicating transmission of the surface emissions through the cloud. But as in Ch 1, the same appearance does not occur over north Jutland.

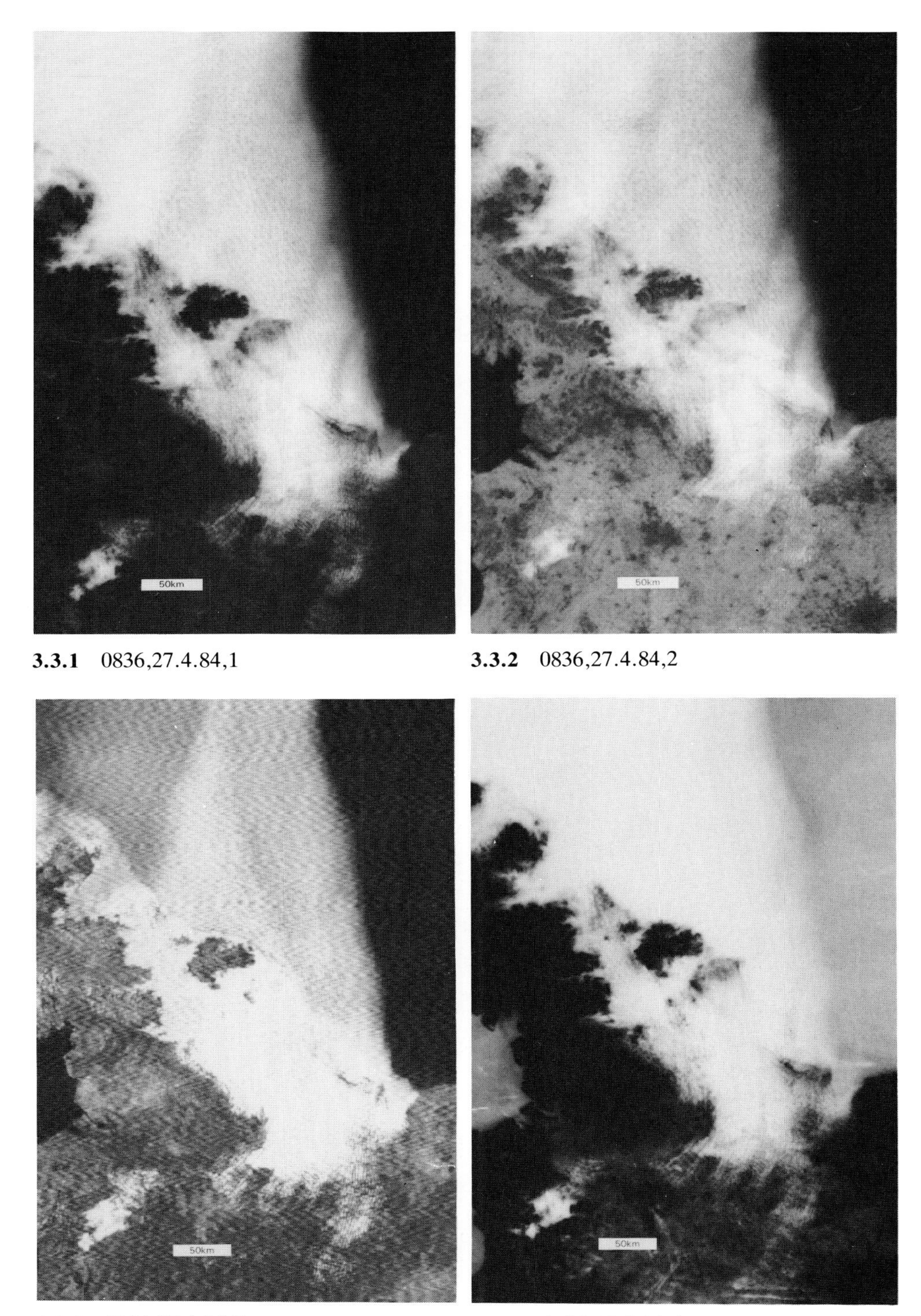

3.3.1 0836,27.4.84,1

3.3.2 0836,27.4.84,2

3.3.3 0836,27.4.84,3

3.3.4 0836,27.4.84,4

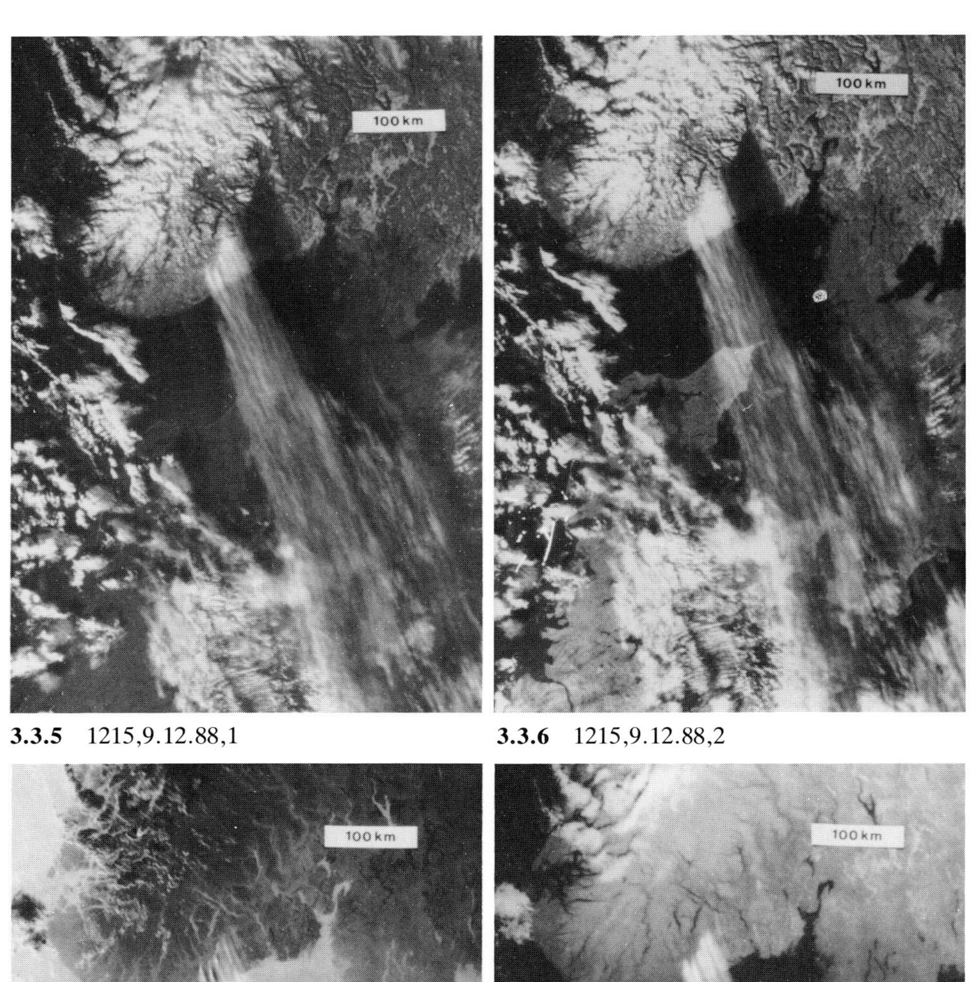

3.3.5 1215,9.12.88,1

3.3.6 1215,9.12.88,2

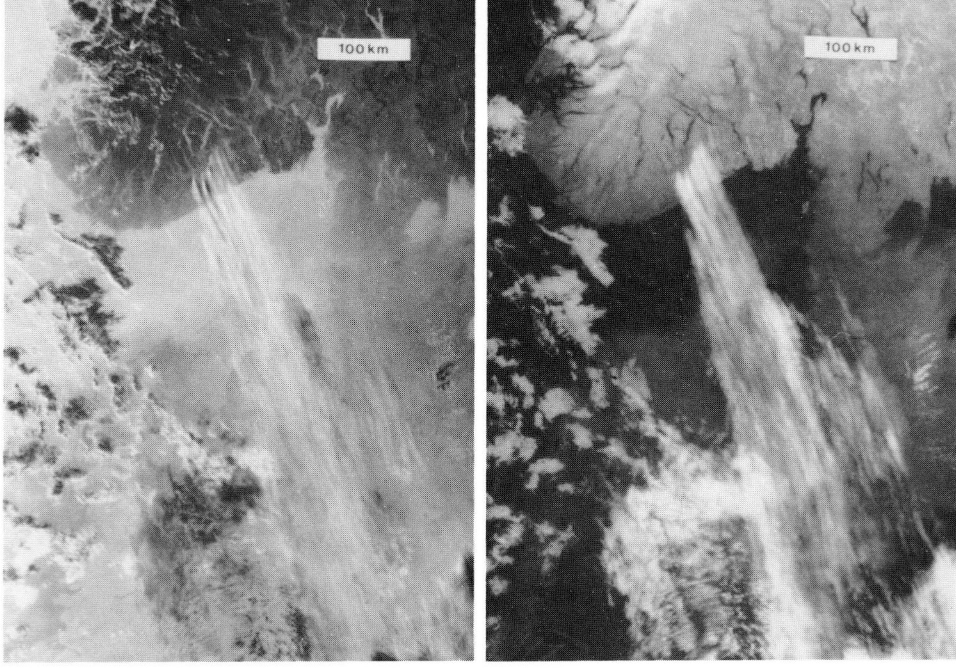

3.3.7 1215,9.12.88,3

3.3.8 1215,9.12.88,4

4

Basic cloud physics

In this chapter we draw attention to the aspects of cloud physics which are important in optimum interpretation of meteorological satellite pictures.

4.1 CONDENSATION NUCLEI

Experiments by Aitken showed that in the air there are usually a very large number of nuclei on to which water vapour condenses to form water droplets at very small water supersaturation pressures. The most numerous have been called 'Aitken' nuclei, and C. T. R. Wilson showed that with patience these can be removed, by gravitational deposition of the droplets formed on them, until a point is reached where large supersaturations can be achieved without condensation. In nature it does not appear that great supersaturations can be attained because there are always plenty of nuclei available, and a meteorologically significant absence of nuclei never occurs on a wide scale. However, most of the time there are present large numbers of other, larger nuclei, which greatly affect the sizes of droplets found in clouds.

Most of these larger nuclei come from the land in a great variety of forms — from combustion, from blown dust, from vegetation — and from the sea — when it is disturbed by wind, algae or some other life forms, which cause the release of gaseous bubbles under water. There are also large areas which are covered with snow or ice, where no nuclei are produced if the wind is light, and large areas of the Pacific Ocean where the surface waters contain no nutrients and which are therefore 'deserts' as far as the production of nuclei by life forms is concerned. In these unproductive areas there are usually a few of the larger nuclei present, and although droplets initially form on Aitken nuclei they soon evaporate because of surface tension and their water is recondensed on the few larger nuclei. This produces clouds of fewer but larger than average droplets, with some interesting consequences described in Chapter 6.

In Antarctica there are high mountains and strong winds which from time to time raise many minute ice crystals from the surface. These grow in the air, which is often supersaturated for ice, and produce many frequent displays of optical phenomena in the summer, which are much less frequently seen elsewhere in the world. It is unlikely that any interesting phenomena arise there through lack of non-Aitken nuclei. But over the Arctic Ocean, which is flat and has long periods of light surface winds, a dearth of larger nuclei may be created. As we can see from the behaviour of fragments of cirrus, the fallout of ice crystals may be slow, but in the course of a few days the gravitational deposition of the particles can be very effective in cleaning the air.

4.2 THE EFFECT OF SURFACE TENSION: OLD AND NEW CLOUDS

Surface tension on the convex skin of a droplet increases the internal pressure and therefore also the vapour pressure at the surface. When cloud is formed in ascending air, particularly if the updraft is rapid, at first almost all the nuclei present create droplets. As the droplet radius increases, the supersaturation around the growing droplets decreases and vapour molecules diffuse from around the smaller droplets, where the vapour pressure is higher, and condense on the larger droplets, which exert a lower vapour pressure, particularly if they are hygroscopic.

The diffusion just described takes time. In any strong updraft new cloud is being rapidly created, and so there is a much greater proportion of smaller droplets than in weak updrafts or in stagnant cloud.

Close to the condensation level the amount of liquid water condensed is very small and so there are no large droplets anyway in an updraft; most of the droplets may be smaller than 1 μm diameter. This is 'new' cloud. When cloudy air mixes with unsaturated air on the outside the smallest droplets are the first to be evaporated, and the same happens in descending air. Evaporating cloud therefore has a dearth of the smallest droplets.

Because diffusion is a molecular mechanism, the continual evolution of the drop-size spectrum is an irreversible process. During this evolution clouds may possess a rapidly changing size spectrum. A cloud parcel far above its condensation level contains much more liquid water than parcels nearer the base. Droplets grow much more slowly by condensation than by collision. A droplet begins to make a significant number of collisions when it exceeds about 20 μm diameter. In a slow updraft, say less than 1 m^{-1}, it would take typically about 15 or 20 minutes from when the cloud first formed for a significant number of collisions to occur and speed up the production of larger droplets. Between 20 and 50 μm the size grows more quickly, but the fallspeed begins to produce fallout only when the particles exceed about 100 μm. The drizzle drops we often see and feel falling out of fog or very low cloud have reached a diameter of 100–200 μm. If they fall into clear air below cloud base they cool that air and reduce the strength of the convection in the cloud. In deeper clouds this is the beginning of the growth of 'warm' rain and if they freeze the growth is accelerated (see 4.3). This begins the redistribution of liquid or frozen water so that greater weights of it collect in the upcurrents through which these larger particles are falling and we get into the realm of rain and hail production. This is outside the present discussion.

For us the interesting area is the cloud top which the satellite sees. There, a parcel of cloud may be recently arrived new cloud from below with plenty of particles much smaller then 5 μm. If it is at the top of shallow cloud, only a short distance above its condensation level, as in low stratus or fog, it will contain no particles larger than about 10 μm. If it is far above its base it will contain many much larger particles, but only if it is still rising rapidly will it also contain any very small ones.

The case of a flat rather stagnant or only slow moving top of a cloud layer is of considerable interest. When there is clear sky above there is a continuous loss of heat into space in those wavelengths to which the atmospheric gases are transparent. Insofar as direct sunshine is absorbed by cloudy air it is absorbed over a depth of several hundred metres, but the amount of that heating is small compared with radiation emitted from the ground or cloud below, which is all absorbed. In any case

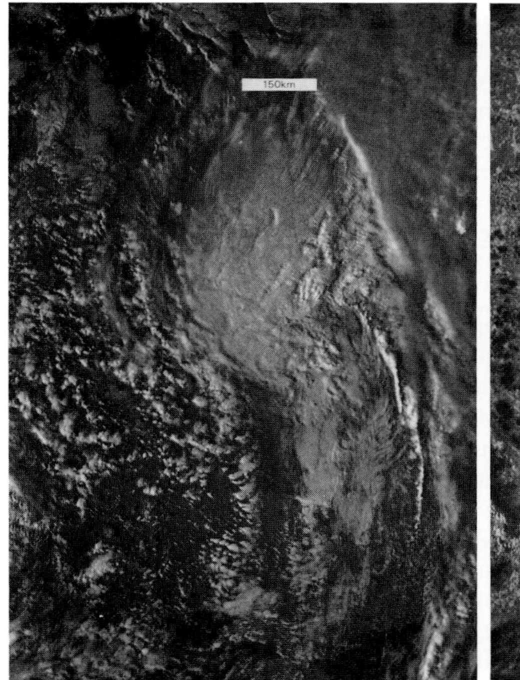

4.2.1 1331,27.9.84,2

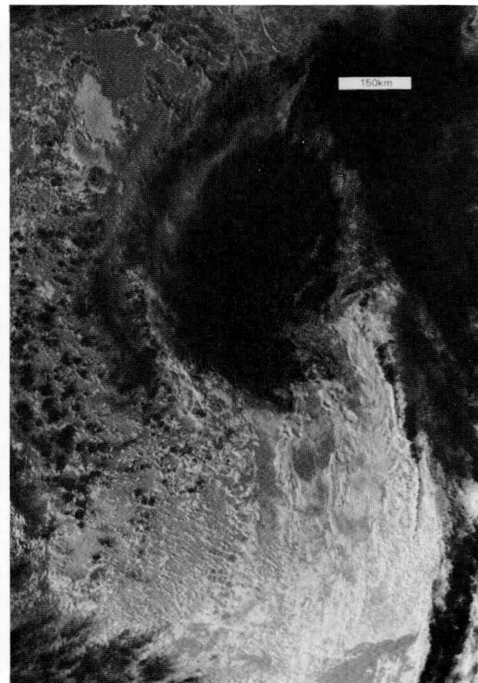

4.2.2 1331,27.9.84, 3

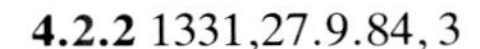

4.2.3 1331,27.9.84, 4

4.2.1, 4.2.2, 4.2.3
The unstable air mass behind a cold front in western Russia shows the effect of subsidence immediately behind the front with the creation of cumulus streets. Further northwest, where the cold air is deeper, the shadows lengthen as the clouds grow (Ch 2), the tops become colder (Ch 4), and the tops of the large clouds become black (Ch 3) because of the increased droplet size there. Some tops show a tendency to become fuzzy as ice anvils begin to be formed, and this is seen best in Ch 3.

the heat loss at the top causes a slow cooling of the layer below the top by downward convection. All layer clouds, therefore, have a cellular structure imposed by this buoyant convection, but it is not visible in cases where the cell or detail size is smaller than a pixel. In spite of the convective motion there is still time for the cloud to become 'old' cloud having very few of the very small droplets (see **3.3.3**).

If the convection is more vigorous because the land or sea surface below is warmer than the air just above it, but the cloud is nevertheless shallow because it is confined by a strong inversion above, we have new cloud with few large droplets because it is not very far above the cloud base.

At the other extreme, at the top of a cumulonimbus which has reached the tropopause, we again have a flat cloud top. When the upsurges of convection have ceased under part or all of the anvil, the average size of the particles is much larger because of the large amount of water that has been condensed as the cloud rose up to that height. The particles, furthermore are almost invariably all frozen and have grown large during the period in which they were present together with water droplets by the well known Bergeron–Findeisen mechanism. Most will therefore be larger than 20 μm, and many up to 100 μm in the earlier stages of the anvil. But as the top flattens out there will be a gravitational separation of the larger particles, leaving a predominance of the smallest at the cloud top. The falling larger particles are well known to fall out as mamma from the overhanging parts of the anvil (See **5.4.10–15**).

Old clouds look darker on the outside than new ones in direct sunshine because there is less multiple scatter on the outside. Evaporating fragments are typical old cloud. Each of the droplets of diameter 100 μm contains the same volume of water as a million droplets of 1 μm diameter, but the tiny droplets have 100 times the cross-section area, and this reduces the visibility within the cloud by a factor of 100. We can often see the sun's disk through an old bit of cloud while a new bit obscures it.

4.3 THE FORMATION OF ICE CLOUDS

Freezing of cloud particles does not usually begin until the temperature of the air and the particles is well below 0°C. Some clouds remain for several hours unfrozen while supercooled to around −15°C. What we are concerned with here is whether the ice particles are large or small when freezing first takes place, because that determines how much scatter there is in Ch 3.

In most cases, when freezing begins the ice crystals grow by direct sublimation of vapour as long as there remain some unfrozen water droplets present. Soon after freezing has begun all the droplets will have evaporated and the cloud will be composed of a smaller number of larger ice particles. Often they fall out as fallstreaks which can easily be seen by the eye, and the crystals may even grow in the clear air until they enter air that is not saturated for ice.

Freezing is spontaneous and immediate at temperatures below −40°C so that, if cloud forms, the droplets freeze immediately, there is no interregnum when droplets and crystals are present together, and the crystals can be much smaller than if they had grown in a cloud at warmer temperatures. If nothing else changed, the crystals would be larger than the original droplets which froze spontaneously, because the crystals exert a lower vapour pressure and therefore take more vapour from the air.

The case of orographic cirrus streamers (see **3.3.5/6/7/8** and **8.1.5**) illustrates all the possibilities. The cloud is first formed in the crest of a wave over a mountain when the air is lifted above its condensation level. Depending on the temperature, it may or may not freeze. If no freezing occurs, the cloud evaporates on returning to its original level on the lee side of the mountain. It might not actually return that far, and it then might form a streamer of water droplet cloud; this sometimes happens over isolated oceanic islands (see 6.6).

If the cloud freezes partially it may return to below the condensation level, where the water droplets will evaporate. The ice particles may form a streamer of ice cloud from the water banner cloud over the mountain if the air does not descend below the ice evaporation level. This is below the condensation level, but ice particles do not form when the air ascends through the ice evaporation level because there are no natural ice sublimation nuclei. Depending on how large the ice crystals have grown, they may or may not display evidence of having a significant fallspeed, and form mamma or fallstreaks.

If the cloud freezes completely, as would be the case if the cloud droplets were first formed at a temperature below −40°C, but the air does not ascend much further, the crystals will not grow much larger than the droplets out of which they were formed. Typically they could contain as much water substance as if they had remained unfrozen, the air had been lifted a further 300 m and they had grown so as to reduce the vapour pressure from that of water- to that of ice- saturation. A trail of 'small' particles is then produced, extending downwind from the wave.

Finally, if after condensation of some cloud the air returns to the condensation level, the particle size is determined by how far above the ice evaporation level the particles ultimately remain. The air might return below the evaporation level after a long distance, or the air might remain above it and the particles fall below it and evaporate there, in air which was not above its own ice evaporation level.

In all these different cases orographic cirrus would have a different appearance according to the viewing channel used and the direction of viewing in relation to the direction of sunshine.

The case of aircraft condensation trails presents very similar possibilities which are discussed in Chapter 6.

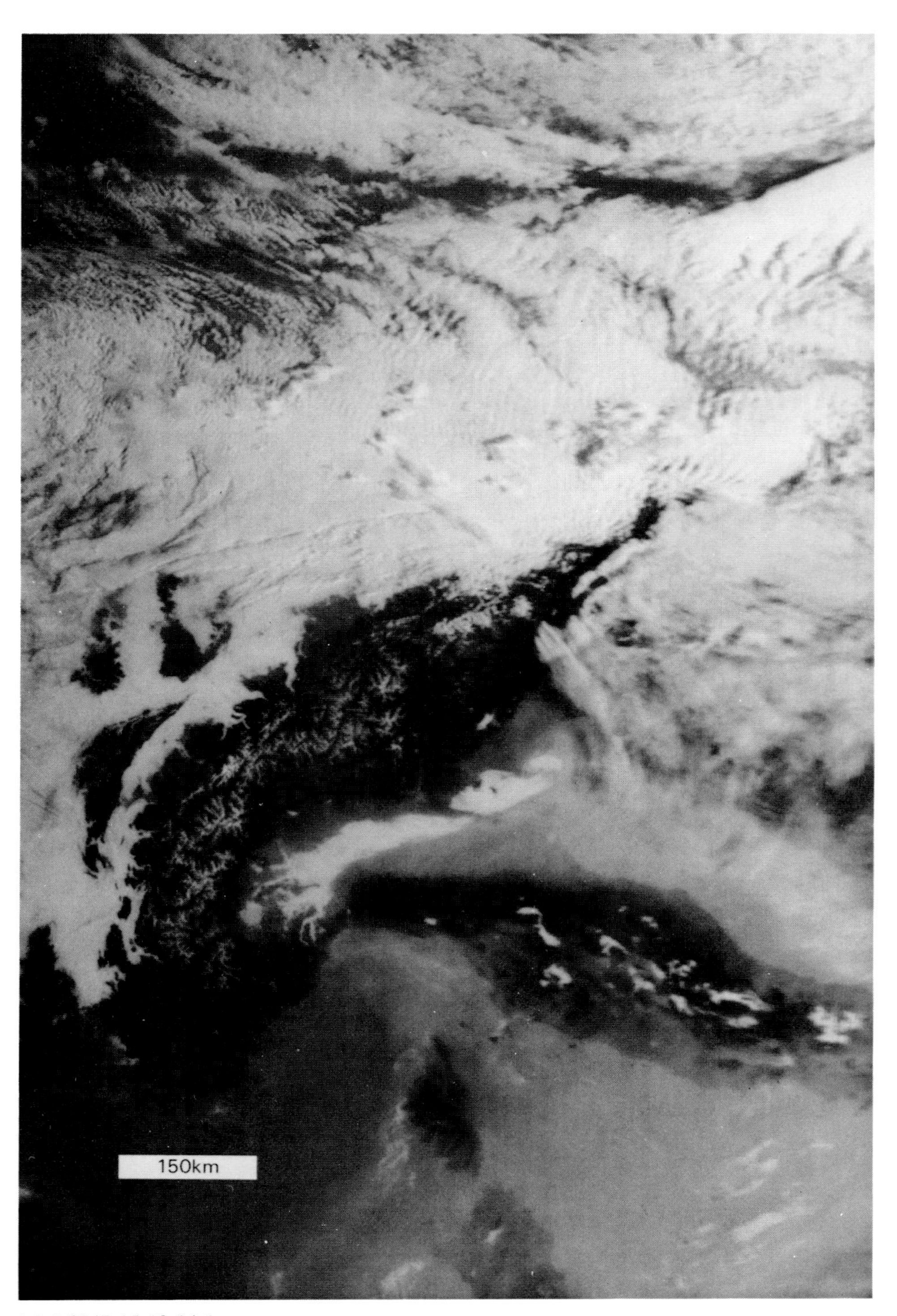

4.3.1 0847,15.10.84,1

4.3.2 0847,15.10.84,3

4.3.3 0847,15.10.84,4

4.3.1, 4.3.2, 4.3.3
A high-pressure area centered over Belgium with extension southwards into the Western Mediterranean had produced cold stagnant conditions with low stratus and fog in the west, but a wind from the north increasing with height over Eastern Europe.

Fog patches in Italy and low cloud in the Adriatic Sea with a few bits in northern Yugoslavia show up best in Ch 3 because of the good forward scatter from the early morning sunshine.

Glint is brightest in Ch 3 where it is variable along the Italian west coast because of the unevenness of the land breezes. There is a ship plume southeast of Elba, and another very small one southwest of Rome. The 'plumes' in the north wind at Savona, Genoa, and at La Spezia (**4.3.2**) are glint effects seen only in Ch 3.

The absence of skylight in Ch 3 produces much sharper shadows of the higher cloud on the stratus in Austria than in Ch 1.

The blockage of the cold air on the north side of the Alps gives rise to good orographic cirrus on the south side over Yugoslavia. There appears to be a slightly lower layer where the wind has a small easterly component not present in the higher layer in which the cirrus occurs further east. This eastmost cirrus appears to block the view of lower cloud, but that could be because the lower cloud is dark in Ch 4. Ch 3 shows the cirrus to be of particle size such that, while there is some forward scatter which we see as white, there is also a considerable absorption which makes the cirrus look black.

The western part of the orographic cirrus has a denser wave cloud at its upwind end where there are plenty of small water droplets. Ch 4 assigns a temperature as cold as the cirrus to the lee wave clouds on the south side of the Alps. The lee waves to the north are of shorter wavelength and are part of the low cloud system with a strong inversion at the top (see also **8.1.11/12/13**).

Along the east coast of Italy are many small land breeze jets (seen only in Ch 3) with small patches of glint at the coast, but extending only a very small distance seawards under the prevailing light northeasterly.

5
Channel 3

This interesting channel of the NOAA satellites was designed to be presented in photonegative pictures, like Ch 4, because at night it presents a very similar measure of temperature, but with sufficient differences to generate extra information. Such uses are not likely to have great precision because they are made at the short wavelength end of the earth's emission spectrum and the measurement is already reduced in quality by its low intensity (see **2.7.1** or **5.8.1**).

But we are concerned with daytime use, and it is therefore appropriate to present it as a photopositive picture, in which the brightest reflections of sunshine are the whitest, and it is then most easily compared with Chs 1 and 2, which are also reflecting sunshine. All the Ch 3 pictures in this book are in photopositive, and the most obvious differences from Chs 1 and 2 are that cirrus is almost all black and glint is extremely bright.

5.1 SEA GLINT

In **2.6.1** it will be seen that there is a 'blip' in the curve of the refractive index of water which overlaps the Ch 3 band of wavelengths. This has the effect of producing a very strong reflected beam when Ch 3 radiation impinges on a water surface. No evidence indicates a similar effect with ice. The reflection is so bright at low angles of incidence, such as occur towards the eastern horizon for the morning satellite, that the radiometer, which is required to be very sensitive to produce any picture at all, is saturated, and no detail is available in the glint area unless the sea is much rougher than usual. Conversely, when the sea is rough in an area of very bright glint, it is darkened (see **6.3.1**, off the Norwegian coast).

What follows in the rest of this section must be read in combination with section 5.5, where it is shown that there is no significant true skylight in Ch 3. The sun appears as a very bright source of Ch 3 radiation in an otherwise very dark sky, much as it appears on the moon. But the very small cloud particles at the base of clouds, of the order of 1 μm or less, have much less absorption of Ch 3 than the larger particles higher up in the cloud, and a significant forward scatter, which is very effectively reflected at the sea surface, occurs when the sunshine is almost horizontal.

Thus, when there is partial low cloud cover which allows the entry of the sun's rays almost horizontally, several reflections at the sea surface and scatter at the cloud base transmit the rays as if in a wave guide between the cloud and the sea. This is seen in the broken cloud area west of the terminator in **6.2.4**.

Cloud shadows are often very sharp and strong in areas of glint in Ch 3.

5.1.1 0830,20.8.84,2

5.1.1, 5.1.2

The area of very bright glint is more extensive in Ch 3 than in Ch 2. The brightest reflection seen in Ch 2 is from northern Jutland to the Frisian Islands to just east of Paris. But in Ch 3 the saturation intensity extends further west to the east coast of England, and further east into the Baltic Sea. Even quite near to the line of mirror reflection of the sun there are some calm areas

(*continued next page*)

5.1.2 0830,20.8.84,3

5.1.1, 5.1.2 (*continued*)
which are black, notably Oslo fjord, off the coast of Kent, and in the Channel opposite Boulogne.

A few hot spots (agricultural bonfires) are seen in England, and some in France; these are not seen in Ch 1. In Ch 3 the stratus is of very variable reflectance. The major lakes of Switzerland are very bright in Ch 3, but the oil slick southwest of Norway is black in both channels because of the placidity of the water there and to the south in the central North Sea.

5.1.3 0838,21.6.83,3

5.1.3
This is part of the remarkable picture **6.3.1** and it shows the 'gust' of shallow cold air which intruded into the North Sea from the Skagerrak behind some very clean low stratus in which oil platform plumes and ship trails were formed. The water, roughened by the gust, scattered the reflected sunshine over a wider angle with the result that the glint was darkened and marked with lines along the wind.

Notice the very much darker calm areas around the islands of Denmark, which contrast with the bright glint elsewhere.

5.2 ABSORPTION BY LARGE CLOUD PARTICLES

It is evident in **2.6.1** that the absorption by liquid water and ice has a value of about 100 per cm, which means that the intensity of a beam is reduced by a factor of $\exp(-1\times100)$ after passing through a length 1 cm of water. Thus passage through 100 μm of water reduces the intensity to about one-third. When droplets of the order of 10 μm are present the scatter is almost entirely forwards (see **2.7.1**), so that the rays entering a cloud penetrate deeply into it before any are turned round to emerge from it. The result is that no detectable fraction of the rays is scattered out from the cloud except those that impinge tangentially, and these emerge at an angle of 10–20°, or at most 30° from the incident direction of the sunshine.

Droplets of the order of 1 μm produce a wider scatter with a much greater proportion backwards, while those of order 0.1 μm would produce a still greater proportion backwards, but with a very small intensity because of the small ratio of particle size to wavelength. There is thus a rather small range of sizes which produce backward scatter, and sideways scatter is probably a mixture of primary sideways and general multiple scatter. The problem is a very complex one and it is difficult to be certain of the sizes of the particles when there is a high probability of a mixture of sizes over quite a wide range from 0.1 to 10 μm.

Many clouds have an abundance of droplets in the 10–50 μm range in which absorption is of great importance (see 5.4).

Variations in the size of ice particles cause large variations in the reflectance of aircraft trails and orographic cirrus (see Chapter 8).

5.3 DIRECTIONAL PROPERTIES OF CH 3 SCATTER

We can see the very marked consequences of directional properties of scatter and absorption of Ch 3 sunshine in the pictures below.

The implications of the pictures are as follows:

- Glint is very bright when the sun is low on the horizon and it saturates the radiometer in Ch 3.
- Clouds and their shadows are relatively very dark over glint areas.
- The relatively brightest clouds are low stratus viewed towards the sun.
- The next brightest clouds are towards the sun when it is higher in the sky.
- The dullest clouds are where the scatter is sideways.
- Clouds reflecting sunshine backwards are brighter than those where it is reflected sideways, but the backward intensity depends very much on the drop sizes present.
- Cirrus of frontal or anvil types does not usually reflect sunshine and appears black, casting very dark, sharply outlined shadows. The darkness of the shadows implies the absence of skylight (aerosol scatter) and low horizontal reflection from cumulus.

In addition to directional variations, reflectance depends significantly on particle size distribution in cloud tops (see also section 3.3).

It is to be noted that many clouds show directional variations in reflectance, both in the visible, as seen from aircraft, and in Chs 1 and 2. The reflectance varies with particle structure also (see Chapter 6).

It must be appreciated that the statements that certain effects are due to particle sizes and shapes are not proved by actually measuring the sizes by other means. The correctness of the explanations is inferred from their consistency. I do not suppose that anyone has measured the drop size distribution when they mentally accept the explanation of the supernumerary rainbows. In the case of satellite pictures we exploit the whole range of knowledge of optical phenomena and deduce the most probable behaviour of Ch 3 radiation in clouds of water and ice. It is preferable to go as far as possible this way, in the hope that a sceptic will mount some laboratory experiments with Ch 3 radiation to clear up any doubts. In the meantime we have a very good supply of satellite pictures among which to find cases which challenge our theories, and the best service is therefore to present good pictorial evidence.

5.3.1 1321.26.4.84,3

5.3.2 1500,26.4.83,3

5.3.3 1643,26.4.83,3

5.3.4 1643,26.4.83,2

5.3.1, 5.3.2, 5.3.3, 5.3.4, 5.5.5
Glint is a dominant feature when it occurs in Ch 3 pictures, and there is also a very strong directional variation in the intensity of sunshine scattered from low stratus clouds. This is because primary scatter from small droplets is the dominant mechanism. At 1324, when the satellite passed over Stockholm, Sweden was only darkly illuminated, while the North Sea stratus, seen looking towards the sun, and the low cloud over southern Russia, seen by primary backscatter, are typical of the brightest clouds. The pictures on the following two orbits (at 1500 and 1643) simply move the bright and dark areas 25° of longitude westwards, the solar time being roughly the same for each pass. The bright glint thus moves from the Adriatic to the Atlantic.

By comparison the illumination in Chs 2 and 4 is much more uniform. The northern part of the Adriatic is very calm and therefore has almost no glint in 5.3.1.

5.3.5 1643,26.4.83,4

5.4 REFLECTANCE CHANGES IN CUMULUS EVOLUTION

Small cumulus, which are situated only just above the condensation level and which are new and have not formed any large drops (larger than 20 μm), are bright in Ch 3 illumination. Over the sea they become black rather quickly, and when they grow big enough to produce anvils the anvils are equally dark.

Over land, cumulus appear to grow somewhat larger before becoming black, but the transition is nevertheless quite quick. No individual cloud has been observed in the act of changing from white to black because, at best, we do not have pictures at less than 100-min intervals (the orbit time). A cine (video) sequence of cumulus growth seen by Ch 3 would be an interesting enterprise. In a single picture the age of a cumulus cloud can be gauged for this purpose by its size.

The taller a cumulus grows, the greater is the proportion of large droplets close to the top, because of the increased quantity of condensed water. Over land the updraught strength is greater because of the rapid rise in temperature of the ground in sunshine compared with the steady sea temperature: typically, in the smaller cumulus, the thermals arriving at the top retain a greater proportion of the very small droplets which are evaporated by the action of surface tension, because there has been less time for them to be evaporated. The cloud is therefore more like new cloud, whereas it has begun to look like old cloud if it emerges at the top of a weak updraft in which a smaller degree of supersaturation has been maintained, as over the sea.

At the same time, the taller a cumulus grows, the wetter the cloud becomes. Inside a thermal the maximum updraft is about twice as strong as the rate of rise of the top of the thermal. At the top the updraft emerges into clear air, with which mixing takes place on a small scale, and this causes some evaporation, which quickly removes the smallest droplets. The fallspeed of a droplet of diameter 100 μm is about $\frac{1}{3}$ m s^{-1}, which is much less than the updraft inside a typical cumulus; and so droplets in the range 20–100 μm are brought to the upper surface of the cloud, where those in the range which most effectively scatter back Ch 3 radiation are quickly evaporated. It has been a common experience of glider pilots to find the windscreen liberally sprinkled with water at the top of a small cumulus. Equally, pilots flying just beneath small cumulus over the sea often find their windscreens made wet by droplets in the 100–300 μm range falling from the cloud bases with fallspeeds in the range 50–150 cm s^{-1}. Most of these drops evaporate before reaching the sea surface. (This last observation was frequently reported by pilots making searches of the sea surface over the ocean to the west of Gibraltar at a height just below cloud base during 1943–1944, and probably before and after). The stronger updrafts under cumulus over land sampled by glider pilots do not usually contain such fallout, although in the evening, when the updraft strength decreases, rain can often be seen falling out and, with luck, this can be confirmed by the appearance of a rainbow shortly before sunset (see [4] or [5]).

When the updraft is halted at an inversion and an anvil is formed, a formerly black water droplet cloud may begin to show a small reflectance as the larger droplets settle back into the cloud. At this stage significant scatter only occurs if either there were many smaller droplets already present in the cloud before the gravitational settling became effective or if there is sufficient mixing with the drier air above to decrease the size of the droplets at the top. Ice particles in a frozen anvil are almost without exception larger than the droplets from which they were formed and, since the absorption coefficient for ice is similar to that for water, the freezing of an anvil invariably makes it appear black to begin with. The following two mechanisms then operate to reduce the size of the particles at the top:

— Gravitational separation becomes important when the vertical velocity is reduced to zero. The larger particles can often be seen falling as mamma from the base of an anvil.

— Mixing with drier air from above the inversion which has caused the anvil formation. This is particularly important at the tropopause because the stable layer may have much less stability than an inversion within the troposphere such as causes the spreading out of moderate or small-sized cumulus. At the same time the stratospheric air may be very much drier than is found below the tropopause. The consequence is that old frozen anvils often begin to show an increased reflectance of Ch 3 while new parts remain intensely black; or alternatively, if evaporation is dominant, the newest tower, which is most often the highest, has intense evaporation and increase in scatter.

Many summer cumulonimbus seen over land display properties which cause an increase in the scatter of Ch 3 sunshine from their anvils, but no good examples have been found over the sea in temperate latitudes. This may be because the surface is never warm enough to make the tops penetrate the tropopause, and the phenomenon may occur in areas where tropical cyclones are generated.

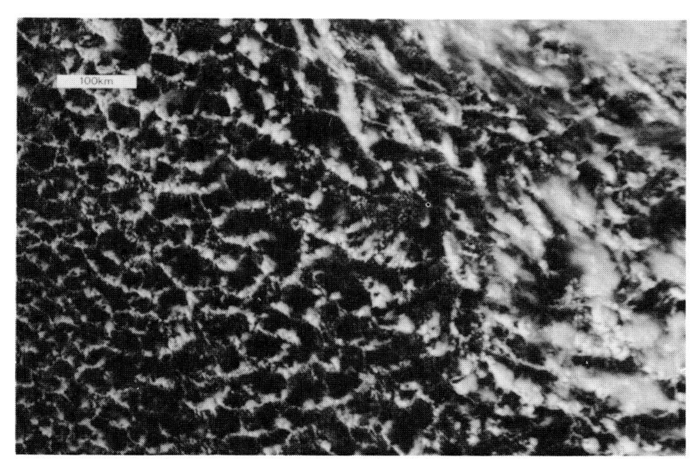

5.4.1 1621, 3.4.87,2

5.4.2 1621, 3.4.87,3

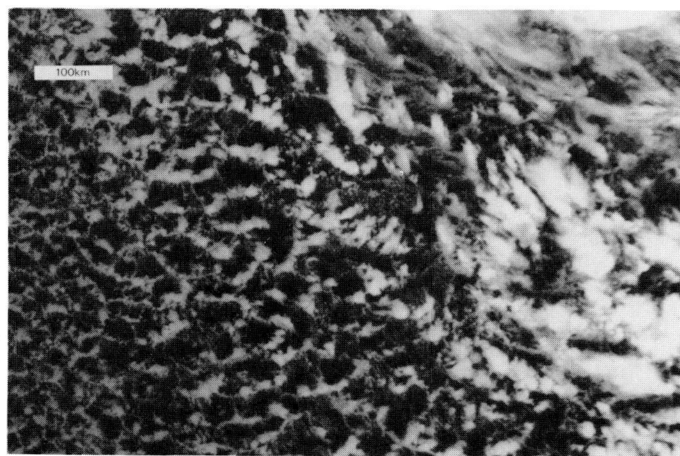

5.4.3 1621, 3.4.87,4

5.4.1, 5.4.2, 5.4.3
In the centre of the pass the glint is almost negligible so that the sea appears darker than the small beady cumulus, but the larger cumulus appear black in Ch 3. The beady cumulus are predominantly illuminated on the left (western) side which faces the sun.

In presenting this evidence it is borne in mind that the explanation is not necessarily correct merely because it seems to be possibly correct. The explanation is given because it fits into a consistent scheme. The observations were in many cases noted because they seemed to contradict previous explanations which were current at the time; for example it had been supposed earlier that ice was more absorbent of Ch 3 radiation than water, whereas there appears to be very little difference in the light of subsequent observations. The absence of reflection of Ch 3 by a cloud cannot be taken as evidence that some freezing has taken place.

With some amusement one may recall that it was supposed, over a century ago, that the spreading out of a cumulonimbus anvil was caused by the mutual repulsion of the ice particles, which were presumed, on account of the occurrence of lightening, to be highly charged with electricity of the same sign. We now know that they follow the air motion, with a small departure due to their fallspeed, and that if the electrostatic forces were strong enough to cause horizontal spreading they would cause visible vertical spread.

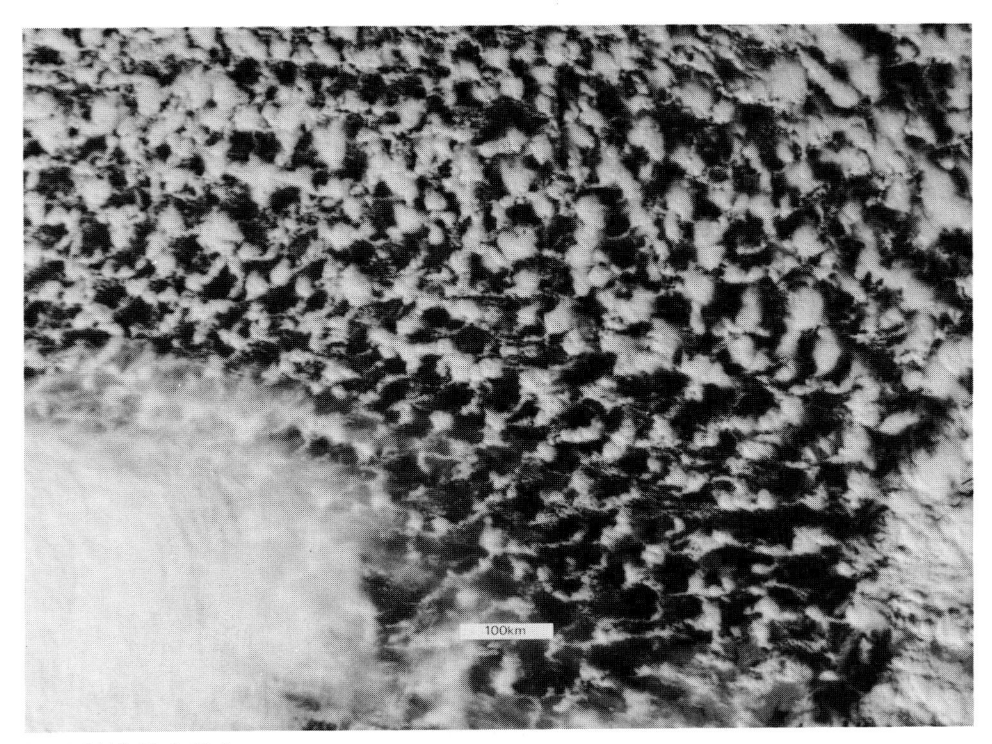

5.4.4 1410,11.3.82,2

5.4.4, 5.4.5, 5.4.6
In the western side of the pass, looking towards the sun, the glint is bright, but the small cumulus is basking in strong glint from the southwest. The cumulus is black in Ch 3.

5.4.5 1410,11.3.82,3

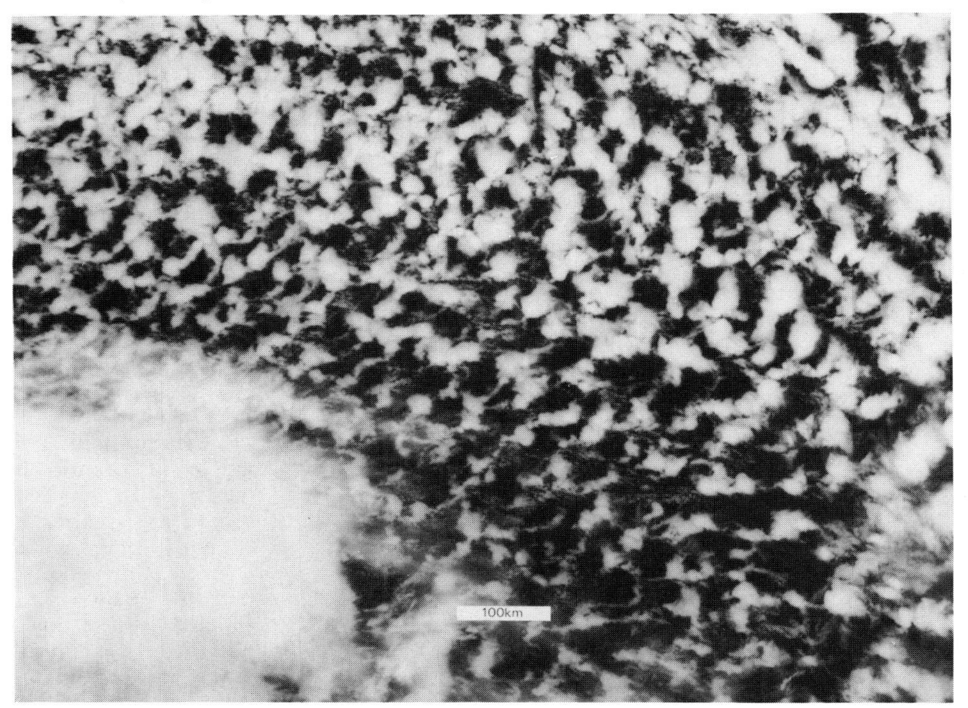

5.4.6 1410,11.3.82,4

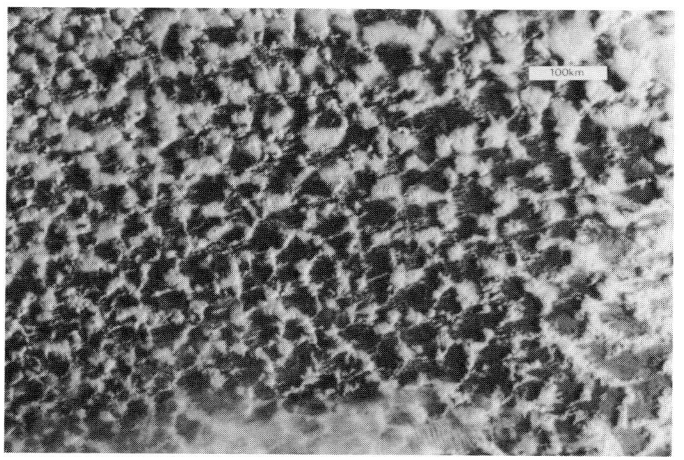

5.4.7 0934,11.3.82,2

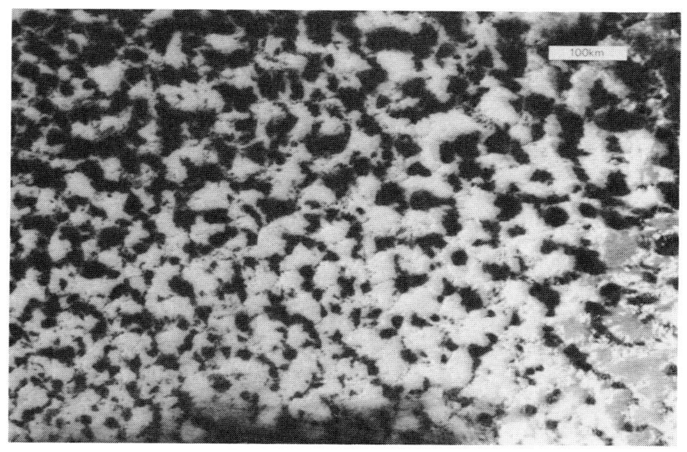

5.4.8 0934,11.3.82,3

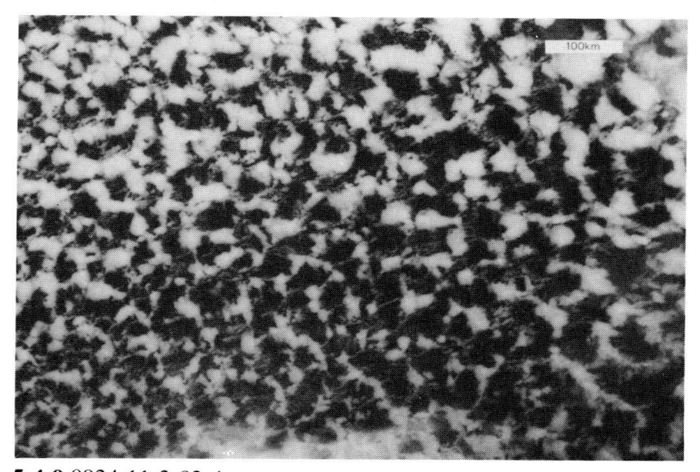

5.4.9 0934,11.3.82,4

5.4.7, 5.4.8, 5.4.9
In the morning the sun produces glint in the east, and in this picture, with a much lower sun than in the afternoon, the shadows of the dark clouds can be seen both on the glint and on the land (part of Ireland) in the east, in both Chs 2 and 3.

Note that the illumination in the shadows over the sea is about as intense as the unshaded land in Ch 3, but in Ch 2 the land is brighter.

In each of these three examples of Atlantic polar air cumulus with light showers, the pictures in Chs 2 and 4 are recognizably similar, although Ch 2 reveals the cloud top topography by shadows. But in Ch 3 cloud appears as the darkest, and the brightest, parts of the scene.

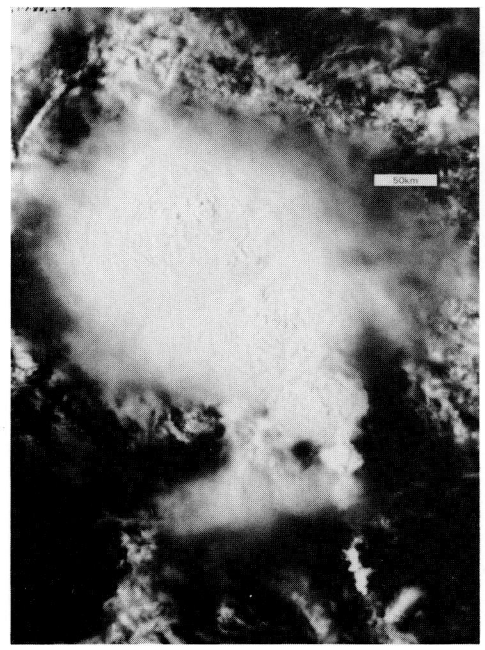

5.4.10 1445,1.7.88,2

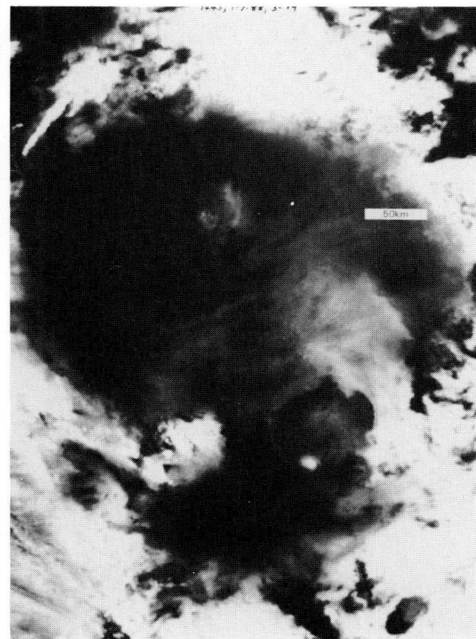

5.4.11 1445,1.7.88,3

5.4.10, 5.4.11, 5.4.12
The amount of detail discovered by Ch 3 in the tops of cumulonimbus is greater than in the other channels, and is of a different kind.

The tallest towers are the coldest in Ch 4, and top topography is revealed by the shadows in Ch 2. In Ch 3 we are observing the gradual dominance of the smaller ice crystals at the top: initially the tops are black, indicating complete absorption of sunshine, but the two mechanisms of sedimentation of the larger crystals and evaporation by mixing with dry stratospheric air reduce the size of the crystals present. The evaporation is the more effective of the two mechanisms because it does not depend on the pre-existence of the smaller crystals: it creates them. Sometimes a wind shear at the tropopause causes the smaller crystals to be carried off in a particular direction from the anvil.

This example occurred over East Germany and the situation in which they arose is described in the next case which occurred at the same time and about 430 km to the north-east of this scene.

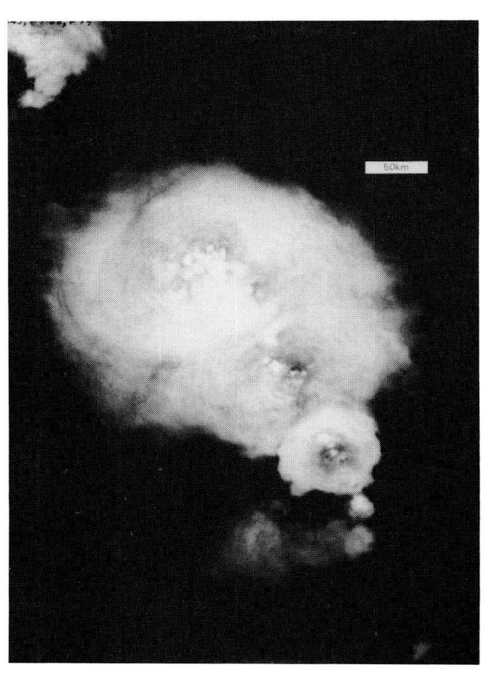

5.4.12 1445,1.7.88,4

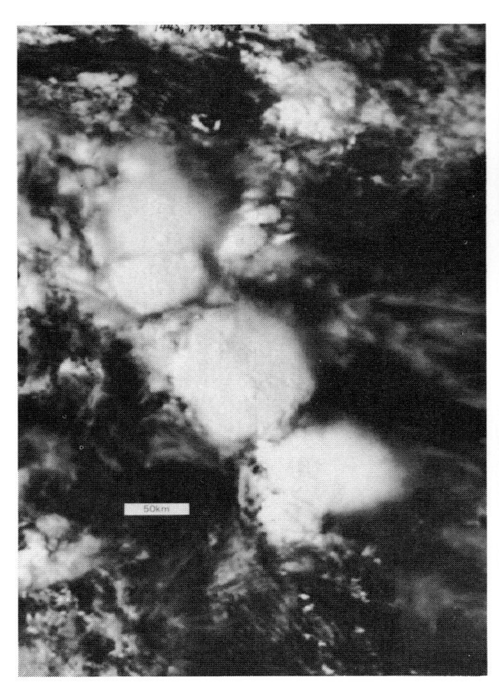

5.4.13 1445,1.7.88,2

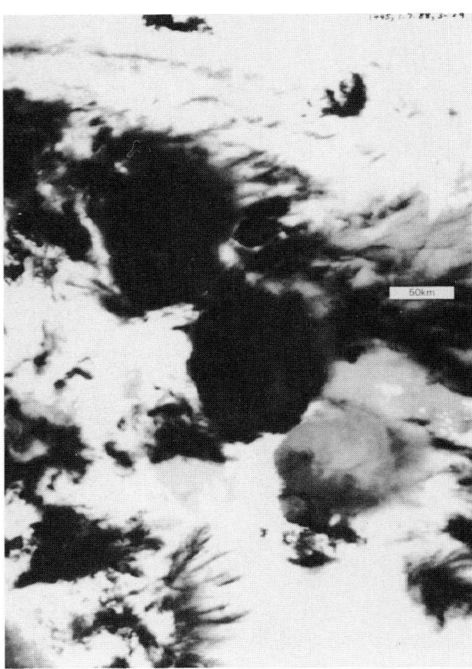

5.4.14 1445,1.7.88,3

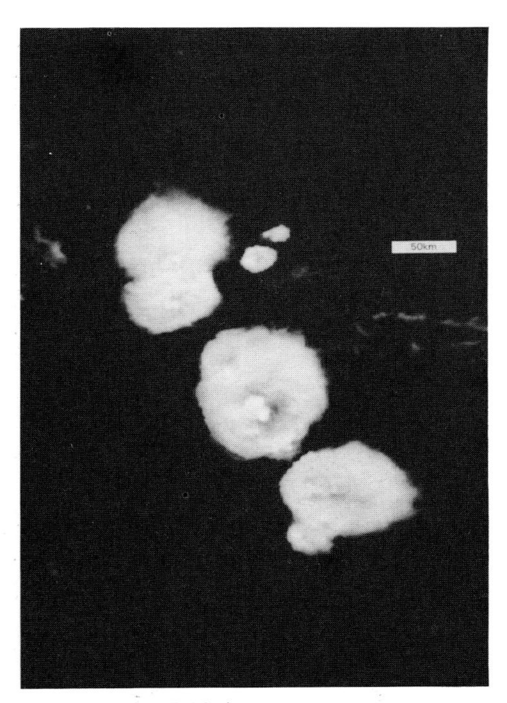

5.4.15 1445,1.7.88,4

5.4.13, 5.4.14, 5.4.15
These cumulonimbus lay on a trough line to the north and east of Szczecin in northwest Poland. The southernmost of the three was drifting northwards across the coast and appeared to have warmer air from the land fed into its base, which rose higher and became colder at the top. The top is therefore of much more varied shade, probably because having penetrated high into the drier air of the stratosphere the ice crystals have partially evaporated.

5.5 THE ABSENCE OF SKYLIGHT: THE TERMINATOR

The first clue to the absence of skylight is the very sharp boundary seen on the very dark shadows of cirrus on layers of low cloud. The fact that haze cannot be detected by Ch 3 indicates that it does not scatter it. Ch 3 illumination makes the scene appear more moonlike where most of the illumination is direct sunshine, and shadow areas are very dark.

The terminator (the boundary of the area in direct sunshine) is very difficult to identify in Ch 3 pictures because the thermal emission from the surface and clouds is comparable with the weak oblique sunshine in that area. It is easy to print the part of a picture in Ch 2 which is far into the area where the sun has not yet risen or has long since set, and some shadows can be seen because the illumination is mainly from the bright sky above the sun. But in Ch 3 no such shadows can be seen, and as we go further towards the dark area only the sun-facing edges of progressively higher clouds show any illumination. When the satellite is looking towards the low sun the glint is often very bright, while the sea below the satellite in the dark area glows more brightly than the clouds because it is warmer. This glow can be distinguished from glint by the absence of shadows of any low clouds such as cumulus.

At this point it is important to include mention of the wave-guide effect which may transmit Ch 3 illumination for long distances beyond the terminator into the dark region. This is discussed in the legend of pictures **6.2.3/4**, and only applies where there is broken low cloud over the ocean surface and negligible cloud layers at higher levels.

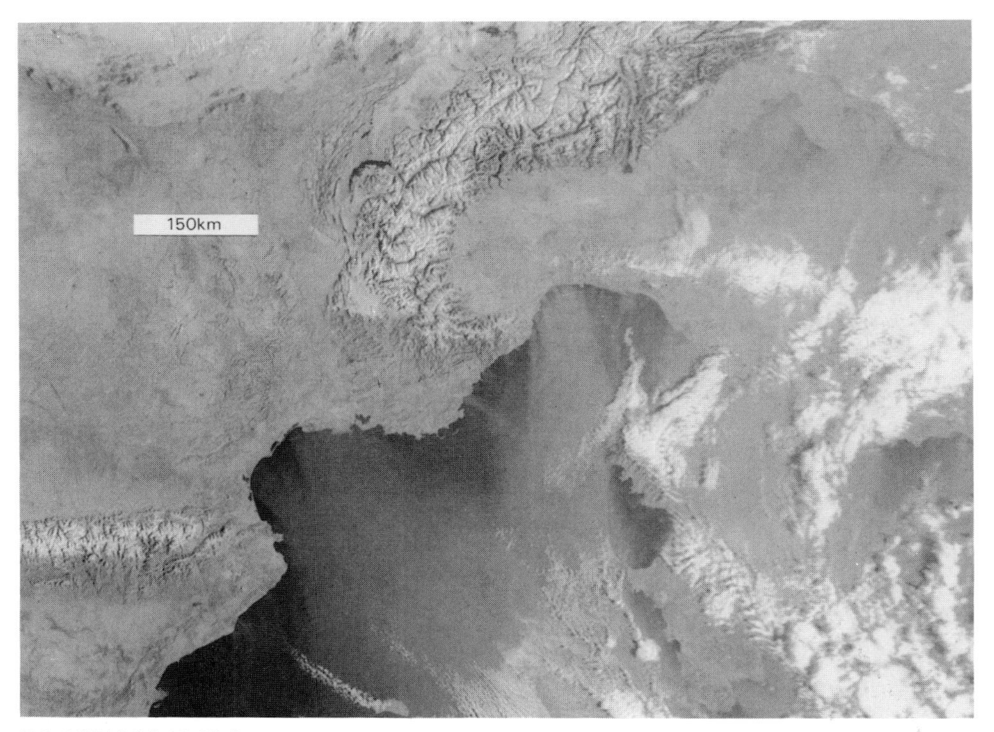

5.5.1 0810,20.10.80,2

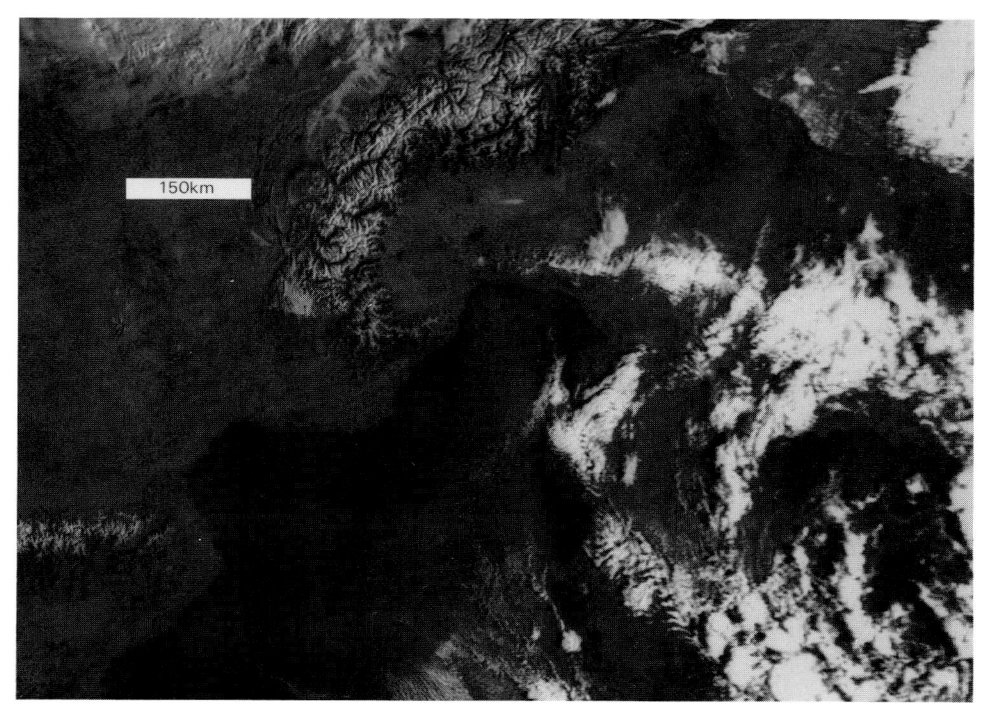

5.5.2 0810,20.10.80,2

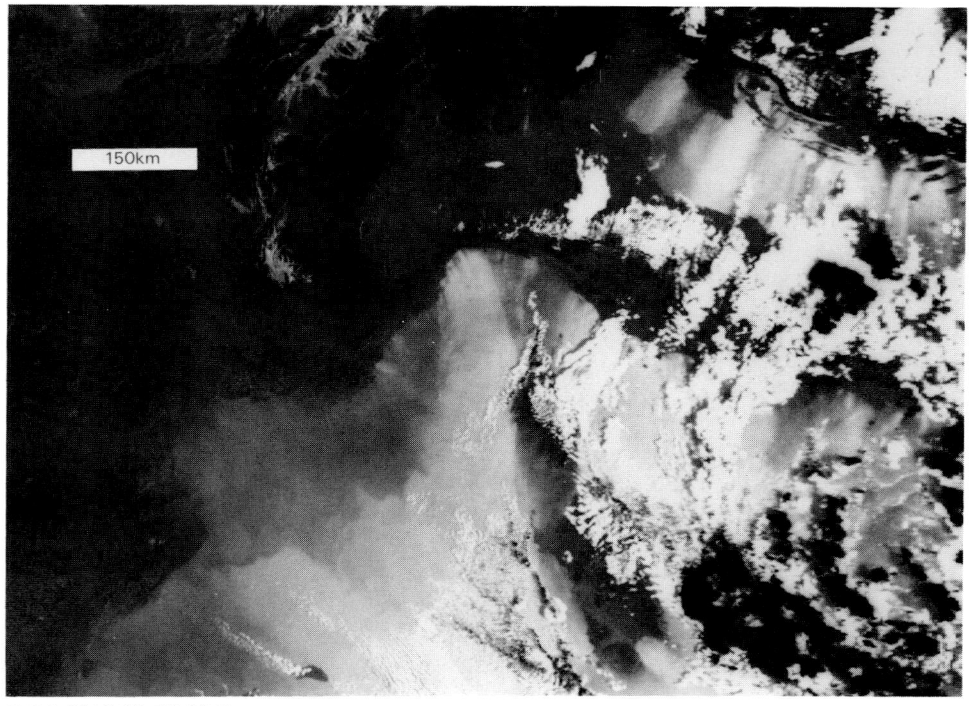

5.5.3 0810,20.10.80,3

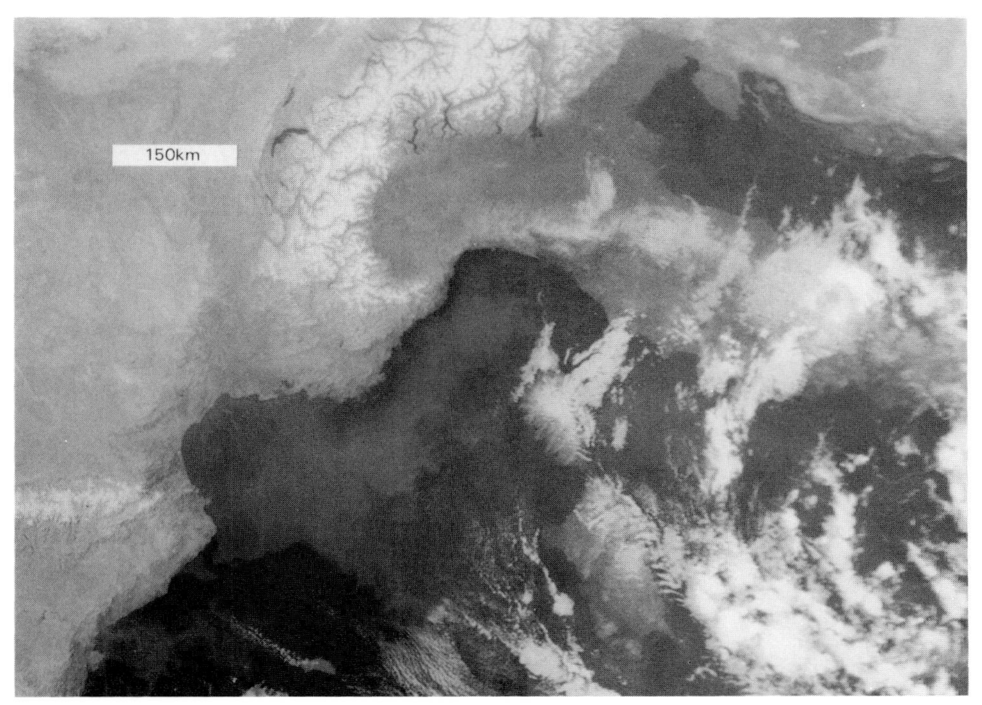

5.5.4 0810,20.10.80,4

5.5.1, 5.5.2, 5.5.3, 5.5.4
The mistral can often be seen by its effect on the glint off the south coast of France. In the case shown the terminator is in the middle of the picture. In the first picture of Ch 2, on the left, the wind is seen in the glint both in the Gulf of Lions and in the Gulf of Genoa. The second picture of Ch 2 is printed darker and with greater contrast, so that the glint on the west is absent because the illumination was darker, being due to skylight. That in the east was due to direct sunshine. The second picture resembles the Ch 3 version, in which the detail perceptible in the Gulf of Lions is entirely due to thermal emission and is the negative of Ch 4 in that area. Glint by direct sunlight in the Gulf of Genoa is due to direct sunshine and dominates the weaker thermal emission.

5.5.5 0831,19.10.79.2

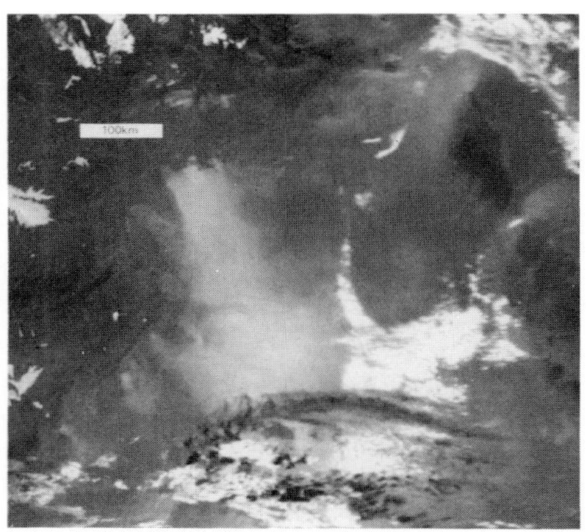

5.5.6 0831,19.10.79.3

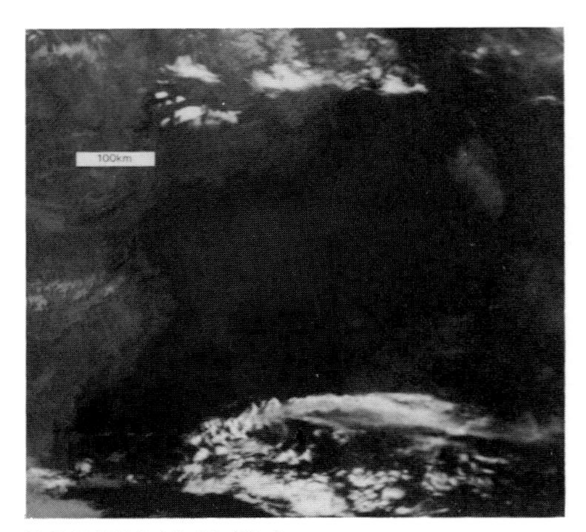

5.5.7 0831,19.10.79.4

5.5.5, 5.5.6, 5.5.7
The glint is clearly present in the Gulf of Lions at about the same time the previous year when the satellite pass occurred 21 min later and was therefore further west, viewed the area more obliquely and picked up the glint more intensely. The time recorded is that at which the satellite was closest to Dundee.

On occasions like this one, when the low stratus is producing intense forward scatter in Ch 3 and there is glint on the sea, some forward skylight may be significant on the cloud base.

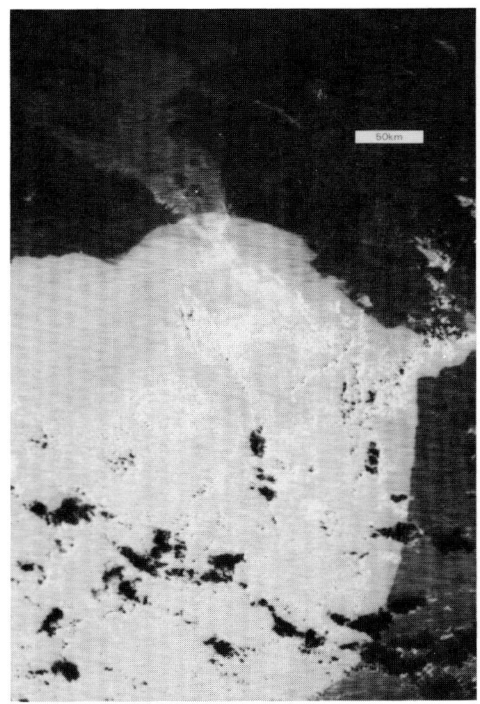

5.5.8 0829,26.11.87.3

5.5.9 0829,26.11.87.2

5.5.7, 5.5.8, 5.5.9
An interesting case of cloud illuminated by glint occurred where the valley of the Guadiana enters the Gulf of Cadiz. Some stratus had drained out to sea in almost calm conditions. In Ch 3 the cloud is clearly much brighter over the sea; the sea itself is bright with glint which is not detectable in Ch 2. At cloud level the glint is at a glancing angle, with the sun only just above the horizon, and the satellite sees the glint very brightly further east in the Mediterranean in Ch 2 (not seen here).

5.5.10 0850, 25.11.87,2

5.5.11 0850, 25.11.87,3

5.5.12 0850, 25.11.87,4

5.5.10, 5.5.11, 5.5.12
Nearly the full length of a pass is shown here, with the morning terminator down the middle so that the west half of the Ch 2 picture is black and is therefore omitted. Above is the Ch 3 picture in which the cut-off of sunshine is slightly to the east of that in Ch 2; to the west the scene is very much the negative of Ch 4.
To the east of the terminator the same is almost true, for the cirrus is black and the low cloud is bright in Ch 3, but there are big contrasts within the low cloud in Ch 3 which are not present in Ch 4.

5.6 STRATOCUMULUS AND STRATUS: PANCAKES AND CELLS

In many sea areas the convection is weak and there is little or no rain or drizzle. The cloud is very flat-topped, as if the cumulus was spreading out under an inversion not far above its base. Such clouds we call 'pancakes' because of their obvious similarity thereto.

When viewed in Ch 1 or 2 the cloud topography is revealed by the brighter illumination of the sunward sides of higher parts of the cloud top and corresponding shading on the other side. This is not the general rule for Ch 3; the centres and higher parts of the tops are usually darker because the average dropsize is greater. The updrafts are weak so that the tiniest droplets in the updrafts have evaporated before arriving at the top, but the nearer we are to the edge of the pancake the greater is the evaporation due to mixing with clear air at the same altitude and so all the droplets are smaller than in the middle. Possibly also the larger droplets are prevented from falling out in the middle of the pancake and are continually brought to the top by the updrafts, which are weaker around the edge. The pancake clouds which show this distribution of brightness are most commonly found in col areas where the wind is so light that no directional influence is evident in the cloud form (streets are not apparent and the cells are not neatly arranged in lines) and there is a tendency for the pancake size to grow to as much as 150 km in a layer as little as 1 km deep.

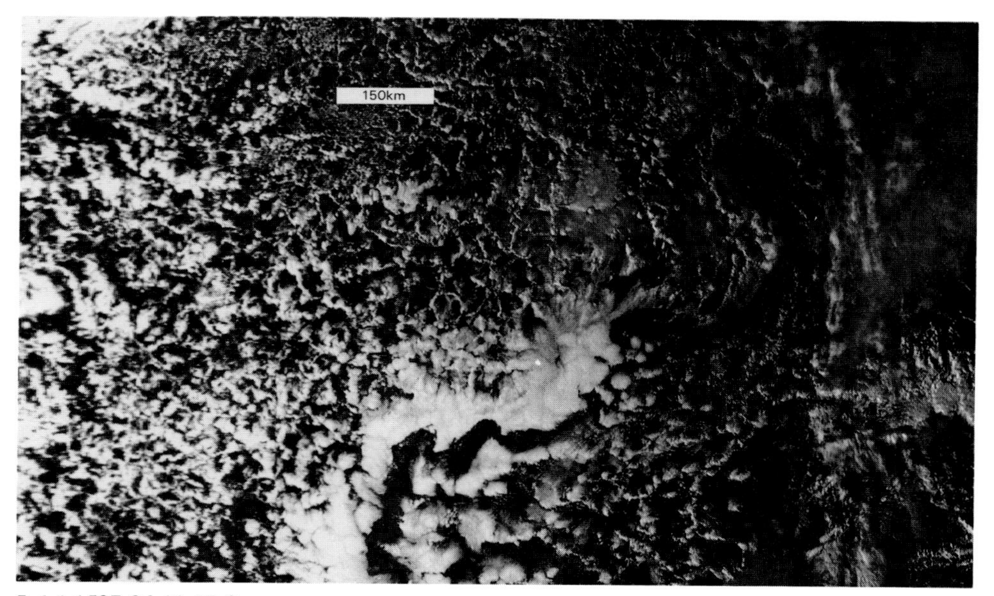

5.6.1 1537,26.11.87,2

5.6.1, 5.6.2, 5.6.3

When the air is subsiding, any inversion which limits the upward growth of convection cloud is strengthened. Cumulus spreads out into stratocumulus pancakes. In this example, in which we look at the cloud off the Portuguese coast, subsidence is concentrated between 200 and 500 km from the coast. Close to the coast are more isolated cumulus with very black tops in Ch 3, and the downdrafts induced by the showers there produce large gaps between the cloud groups (so-called 'open' cells). The sun is setting in the WSW so that in the north and east the cloud is

(*continued next page*)

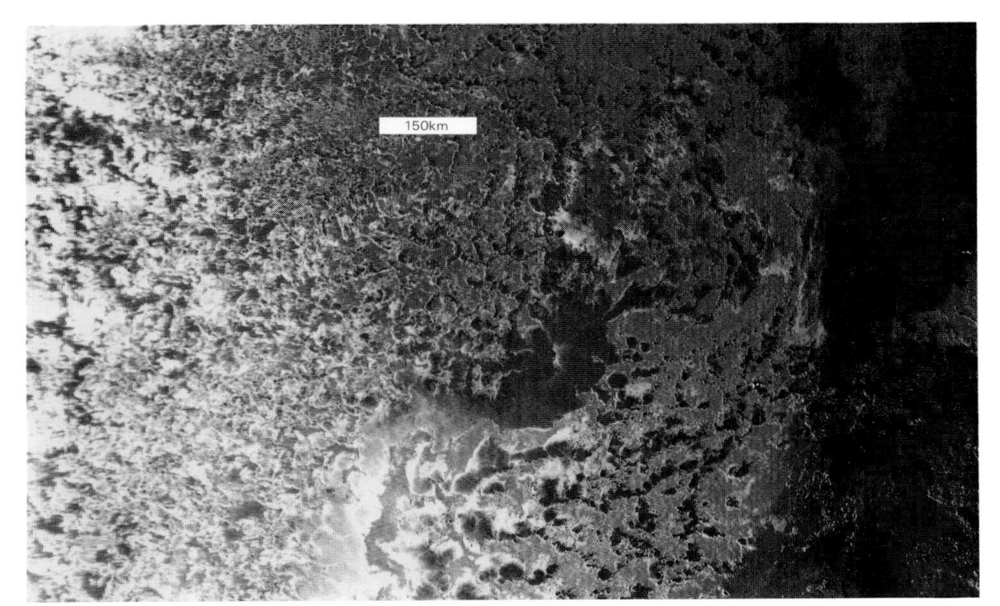

5.6.2 1537,26.11.87,3

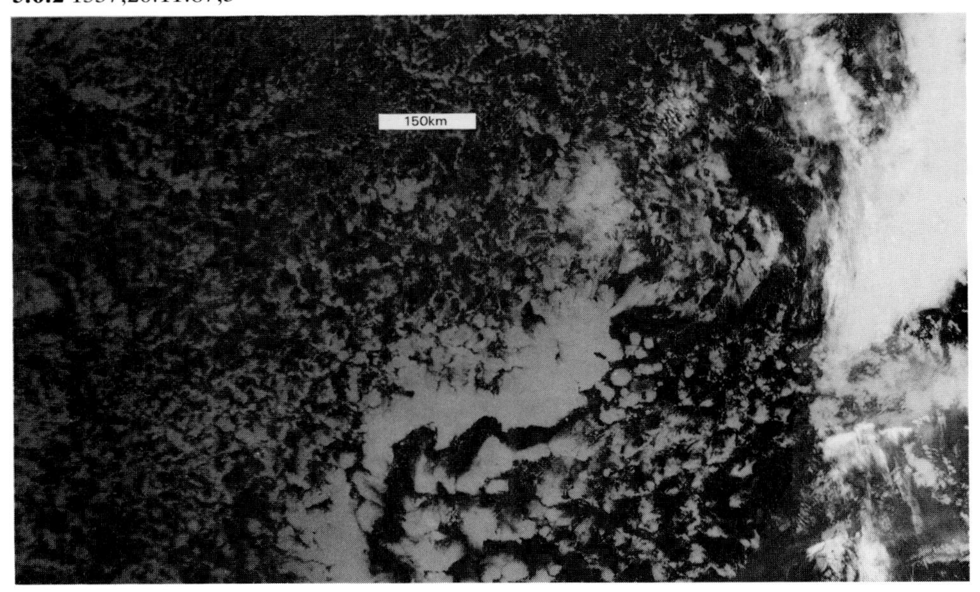

5.6.3 1537,26.11.87,4

5.6.1, 5.6.2, 5.6.3 (*continued*)
very dark in Ch 3.

Further west the pancakes are darkest in their centres while in Ch 2 they show the brightest illumination there (see also **13.1.10/11/12**).

There are wave clouds and short orographic cirrus streaks from the highest of them northeast of Lisbon and these are best seen in Ch 4, which also shows the uniform temperature of the rather flat tops of the pancakes.

There is no skylight to shine on the flat stratocumulus near the top of the picture in Ch 3.

5.6.4 1338,15.6.82,2

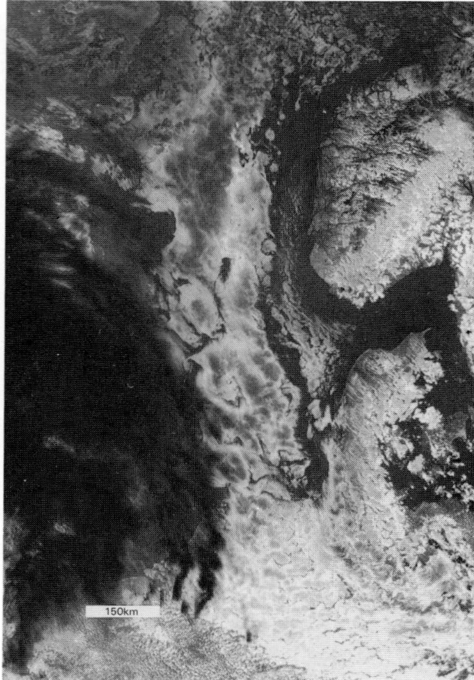

5.6.5 1338,15.6.82,3

5.6.6 1338,15.6.82,4

5.6.4, 5.5.5, 5.5.6

The subsidence in the col ahead of the warm front of a cyclone and behind a cyclone over western Russia produces typical pancakes over the North Sea. Ch 4 shows the flatness of their tops while Chs 3 and 2 give opposite shadings. The pancakes have grown large over the sea but remain smaller over land, where drying out is occurring over the warmer surface.

There is no sign of orographic cirrus over Scotland because of the opposing directions of the upper and lower winds ahead of this warm front.

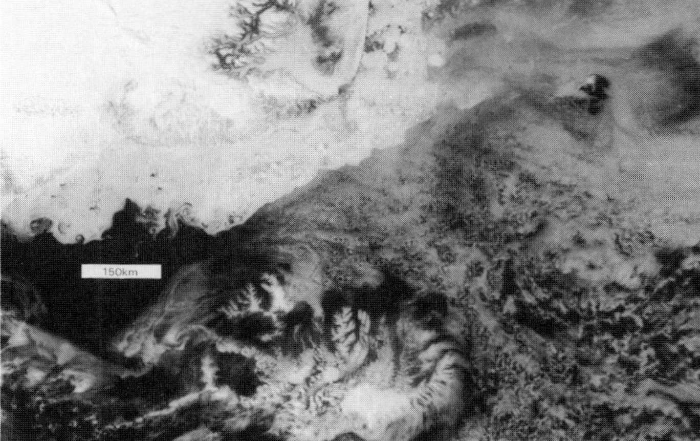
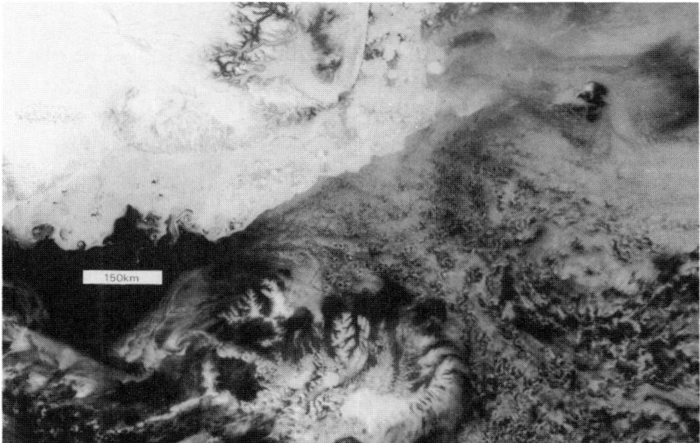

5.7.1 1521,15.6.82,1

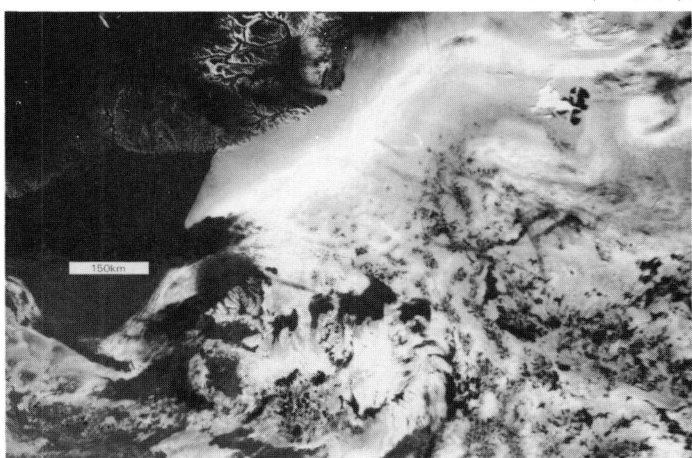

5.7.2 1521,15.6.82,3

5.7.3 1521,15.6.82,4

5.7.1, 5.7.2, 5.7.3

This is a remarkable case of a thin layer of low cloud being detected by Ch 3 but not by other wavelengths. Only in this channel is the cloud in bright contrast with the cold water and ice. This 'fog' has invaded the entrance of Scoresby Sound where the frozen sea is snow-covered.

The details of the ice are shown well in the visible, but the contrast between the cold fresh water from melted ice and the warm salty sea in Chs 3 and 4 overwhelms the difference between ice and snow, which are white, and the sea, which is dark, in Ch 1.

The midsummer sunshine in Iceland warms the land except where snow lies or cloud obscures and reduces it. Some of the areas in northeast Iceland which shine brightly in Ch 3 are areas of dry volcanic dust, which appears light brown to ordinary vision and is extensively windblown (but this day is calm). The bare rock on the sides of the glacier-eroded fjords of Greenland also shine brightly in the sunshine in Ch 3; but not in Ch 1, in which the snow on the higher ground is white. The summer sunshine also produces strong temperature contrasts across the snow line (see Chapter 4).

Ch 3 also shows the coastline of the northwest peninsula of Iceland very clearly. There is an interesting isolated fragment of water droplet cloud inland in Greenland.

5.7 SHARP DISTINCTIONS BY CH 3

Ch 3 exhibits several more extreme effects than the other two sunshine channels because of the powerful glint on water surfaces and the strong absorption in water and ice. Because Ch 3 has less than 1% of the intensity of the green component of sunshine, which is scarcely absorbed at all, the strong absorption has only a very small effect on the temperature of clouds. Absorption between 1.5 and 3.0 μm is much more significant, but it is sufficient to reduce to zero any radiation scattered within a cloud so that no sunshine penetrates clouds except through holes. Many objects such as ice and snow appear quite black, while much of the glint saturates the radiometer and appears pure white. Some fogs, particularly those that lie in valleys, are motionless and have cold ground below, appear very bright, particularly in the early morning while the ground is still very dark, there being no skylight. An exceptionally stark example is illustrated where low stratus was formed in a Greenland fjord and was much brighter than older stratus, which was widespread and could not be detected by Chs 2 or 4. Various mechanisms may be invoked to explain this observation: a manmade source of smoke might have polluted the fjord air so that it contained a much greater supply of good hygroscopic nuclei, with the consequence that this fog or stratus was composed of much smaller droplets than fogs in other fjords or stratus out to sea; or the mode of formation may have been lifting by a katabatic flow down the fjord of the air previously situated in the bottom of it.

In the sea area west of Greenland low stratus, ice, cirrus, snow and other cloud forms sometimes show unusual features, some of which are due to the cleanliness of the air. These include ship trails (see Chapter 6) which result from the trails of

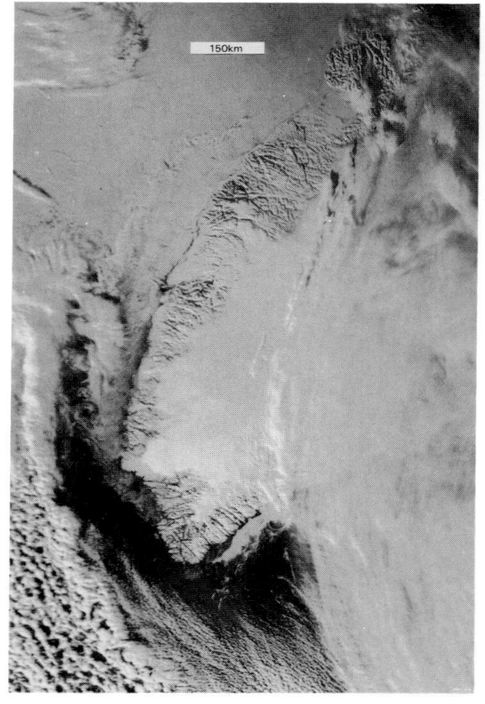

5.7.4 1514,8.2.82,2

5.7.5 1514,8.2.82,3

5.7.6 1514,8.2.82,4

5.7.4, 5.7.5, 5.7.6
The visible channel pictures of cloud edges illuminated by a low winter sun on the longitude of the eastern edge of the Greenland snow plateau. A narrow strip of coastal water is ice-free; cold air from the Davis Strait is moistened and warmed into convection south of Cape Farewell, to the east of which the familiar ice from the northeast is driven away from the coast. Further east the cirrus of a deep (935-mb) cyclone southwest of Iceland covers east Greenland. To the west is a mixture of clouds at various heights with some shadows revealing their relative heights which are confirmed by their temperatures seen in Ch 4. Some of the fjords remain unfrozen like the coastal strip of sea.

Ch 3 is unusually informative. Some of the unfrozen fjords contain fog, and two extensions out to sea at Godthab cast shadows in Ch 2. This calm fog, possibly due to katabatic flow down the fjords lifting the air resting on the warm water, has smaller particles than most other cloud except for small patches and edges in the cloud system of the cyclone and that further west in the Davis Strait. The fogs in Nordre Stromfjord and the smaller inlet just south of Disko Island only partially occupy them and there are suggestions of similar fog north of Disko, but we must be careful because of the cyclone cloud there.

The cumulus in the air which had been held 170 km off the Greenland coast by the katabatic flow from the plateau is more fully developed and is darkened by the larger water droplets in it.

The low intensity of the sunlight generally, particularly in Ch 3 in which there is little skylight or bright cloud and certainly no bright ice or snow, makes the warmth of the sea in the south and in the coastal strip look bright, but the very bright patches are fog on the coast north of Godthab.

The brightness of the fog may be due to pollution from human habitation which has reduced the average size of the droplets.

pollution. If a cloud allows no radiation to pass without encountering several droplets it will have achieved not far from its maximum albedo in Chs 1 and 2. But in Ch 3 the backscatter at the first droplet encounter is to be maximized because all multiple scatter is greatly reduced by absorption. Thus there are situations where the albedo may be increased by increasing the proportion of the very smallest droplets without significant effect on the albedo in other channels. To put it another way, pollution which achieves this particular kind of trail is detectable only by Ch 3.

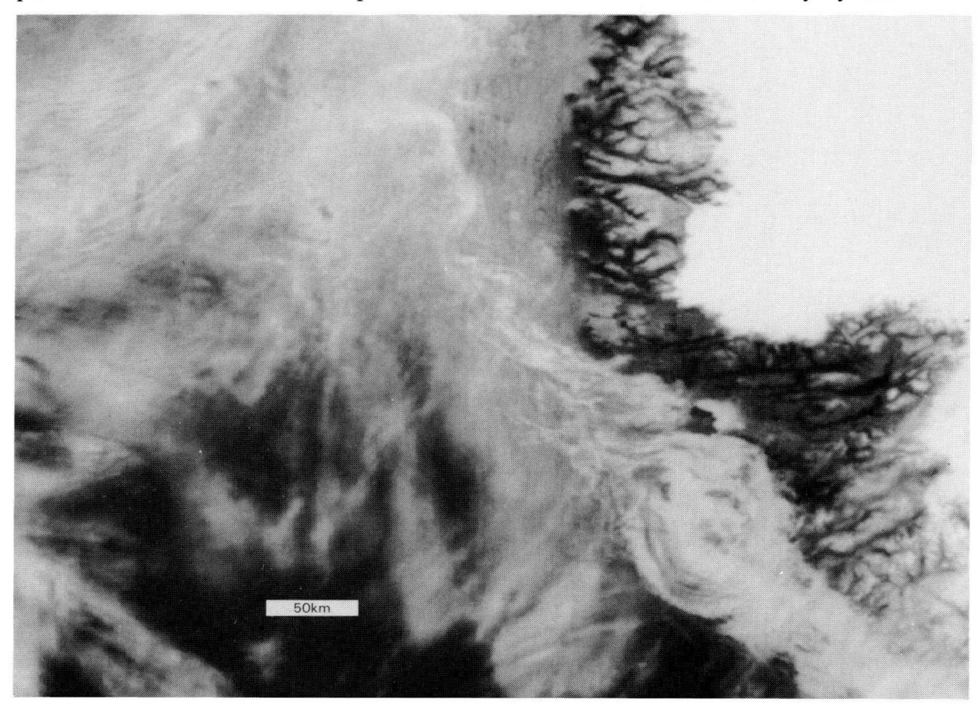

5.7.7 1609,11.6.82,2

5.7.7, 5.7.8, 5.7.9

If we look first at the Ch 3 picture we see an isolated cloud on the left and the edge of its visible shadow on its right side. On the left of it are two ship trails which appear to coalesce at their northern end. The southwest corner of Greenland appears on the east side of the pictures. Some of the other shadows on the lower cloud layer are of contrails to which fragments of cirrus lie roughly at right angles.

In the deep IR (Ch 4) picture the bare rock is warm in the midsummer sunshine and is correspondingly dark by comparison with the snow plateau.

We are accustomed to ship trails being visible in Ch 2 pictures, but in this case none appear, except faintly at the southern end of the eastern one. The isolated cloud first mentioned is also scarcely visible in Ch 2.

The floating ice close to the Greenland coast is very clear in Ch 2 because the image penetrates the stratus (see Chapter 3). The image is absorbed by the stratus in Ch 3, but in any case the ice is as black as the sea in Ch 3. In Ch 3 some contrail fragments are seen in white beside their dark shadows, and the dry rock of the mountains is brighter at the edges and has more contrasty detail than their Ch 2 image.

The ship trail is presumably composed of significantly smaller droplets than the stratus cloud, but this makes scarcely any difference to the reflectivity in Ch 2, for the cloud is already dense. The scene is shown on the following day in **3.1.1/2/3**.

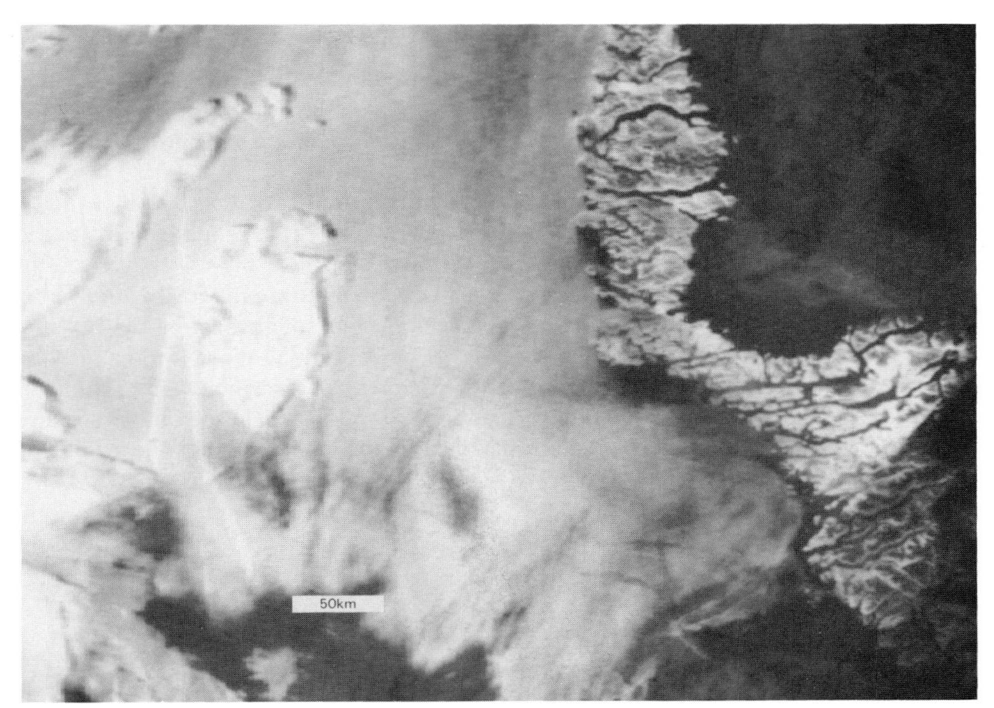

5.7.8 1609,11.6.82,3

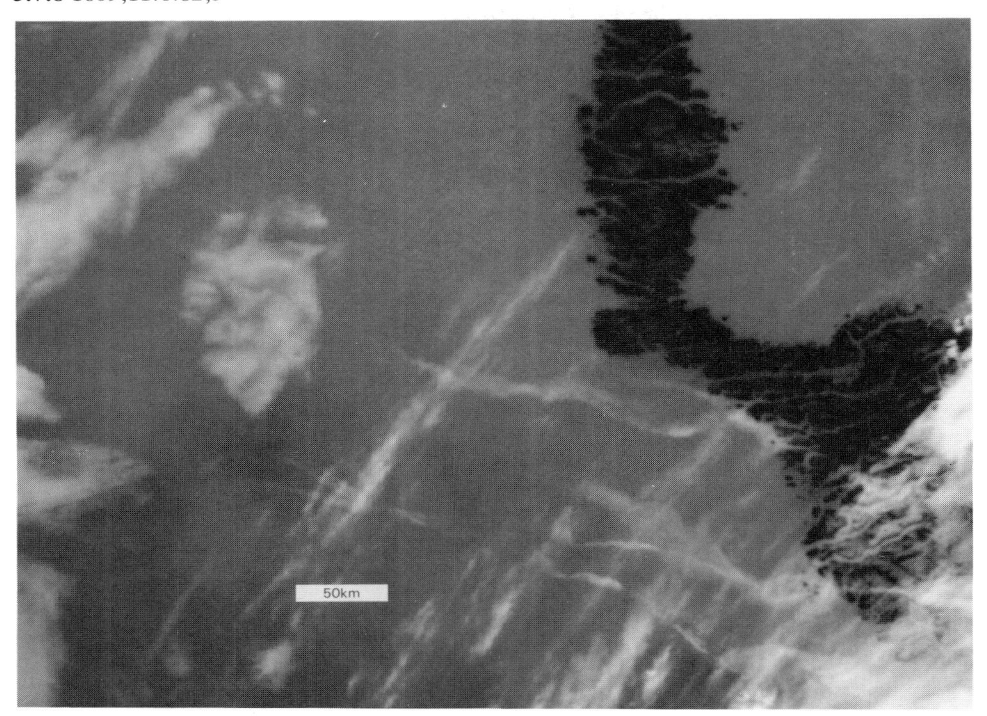

5.7.9 1609,11.6.82,4

5.8 HOT SPOTS

The emission spectrum of a hot body as represented by the Planck curve is shown in **5.8.1**. This shows the normalized intensity as a function of λT, where λ is the wavelength and T is the absolute temperature. The emission is also proportional to T^4 according to Stefan's law.

There are several horizontal scales, the bottom one being for $T = 6000°K$, which is roughly the temperature of the sun, and the curve then gives the relative intensities in the spectrum as emitted by the sun. At the earth's surface the UV wavelengths shorter than 0.3 μm are absent, having been absorbed by stratospheric ozone. Many of the wavelengths longer than 2.0 μm are absent because of absorption by CO_2 and other gases such as methane (CH_4) and the CFCs, but much more importantly by water vapour. Satellites view the earth using wavelengths to which water vapour is transparent (the water vapour windows), and Ch 3 is in one of these; Chs 4 and 5 are in another. There is still some absorption in passing through the atmosphere in these wavelengths.

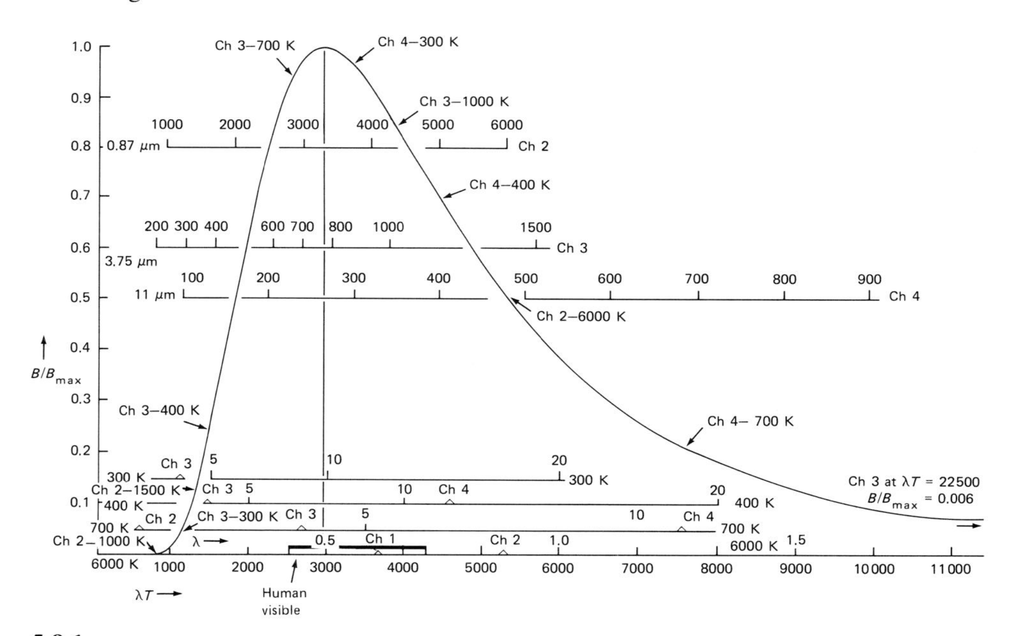

5.8.1
The Planck curve gives the black body emission intensity B as a proportion of the maximum as a function of λT. The maximum is found experimentally to be where $\lambda T = 29°K\mu m$. The values of λ are given at the bottom for four values of T corresponding to the temperature of the sun (6000°K), a body at dull red heat (700°K), an uncomfortably warm surface (400°K), and the an average temperature for the daytime earth (300°K).

In practice there is always some component of the medium present between the emitter and the recipient of the radiation which reduces the intensity for some selected wavebands. Thus the sunshine reaching the earth has all the radiation of wavelength less than 0.3 μm absorbed by the ozone and oxygen in the stratosphere (see sections 2.6 and 2.7 for more detail).

In this chapter the most important feature is that Ch 3 is at the maximum for emissions at dull red heat, and is therefore very sensitive to hot spots not detected by other AVHRR channels.

The next three horizontal scales are for $T=700°K$ (dull red heat), 400°K (steamy volcanic area), and 300°K (warm climate mean), and the positions of the middles of the satellite wavelength bands are shown on the various scales. The maximum is close to 0.5 μm (green) in the solar spectrum and close to Ch 3 for dull red heat. Ch 3 is at the maximum for a precise temperature of 770°K, and Ch 4 for 270°K.

As the temperature rises from 300°K to 770°K the relative intensity of Ch 3 rises from 0.05 to 1.0. At the same time the intensity of Ch 4 decreases from close to 1.0 to just under 0.2. For this temperature rise T^4 increases by a factor of about 44. Consequently the intensity of a spot is increased by a factor of 880 $\left(=44 \times \dfrac{1}{0.05}\right)$ in Ch 3 and a factor of only 8.8 $\left(=44 \times \dfrac{0.2}{1}\right)$ in Ch 4.

By this simple calculation we see that for the detection of a hot spot at dull red heat Ch 3 is 100 times as sensitive as Ch 4. The hot spot need not occupy a whole pixel to be detected; for if the spot is to make the pixel that contains it four times as bright as the neighbouring pixels only 4/880 of its area needs to be at 770°K. A pixel near the

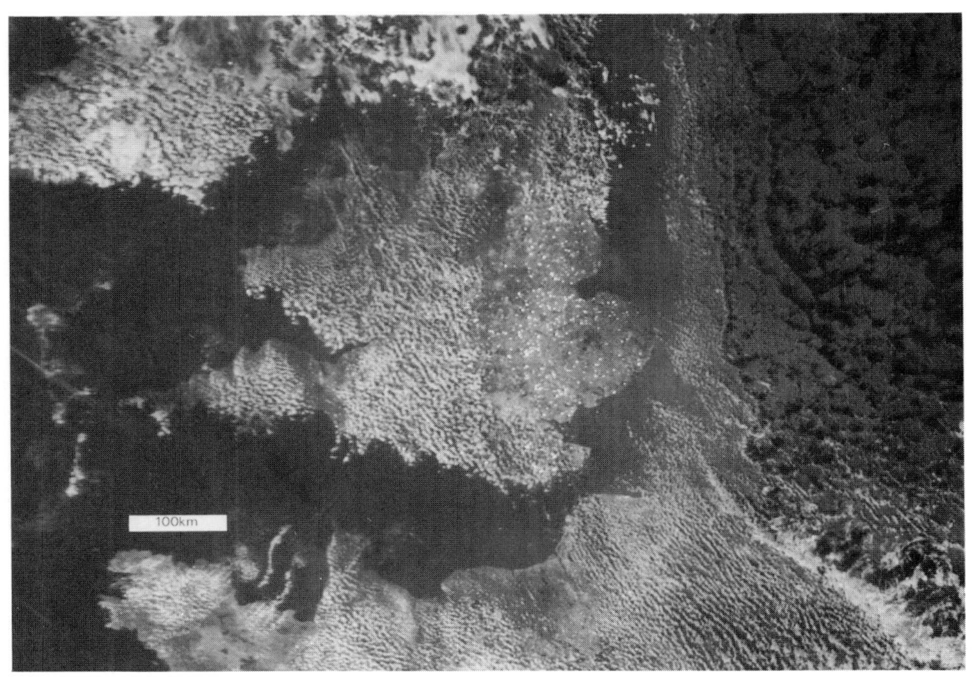

5.8.2 1310,6.9.85,3

5.8.2
Cumulus cloud under an inversion covers the south and west of Britain and the adjacent continent. Hot spots caused by harvest straw burning can be seen all over eastern England and some can be seen shining through among the clouds. A few may be identified as steel works (e.g. Dunkirk) which can be seen almost every day. (See also 9.5.3.)

In the North Sea there is a sharp dividing line to the east of which the cumulus are producing showers. They have black tops and the downdrafts produce large cloud-free areas between the showers.

centre of the satellite's swathe has an area of 1.2 km^2 and the hot area needs to be at least the area of a square of side 73 m.

To be four times as bright as surrounding pixels a flame at 1500°K within it would have to occupy the area of a square of side 10 m.

A warm pixel at 400°K in surroundings at 300 deg would be about 2.2 times as bright in Ch 4, but 15 times as bright in Ch 3. It is clearly not possible to determine the temperature of a hot spot without knowing the area of it and even then it may not be possible if the radiometer is saturated by the pixel in question. Ch3 is certainly very useful in locating them, and the detection of straw fires on farms at harvest time is a particularly spectacular example. Others are shown in **6.3.1/2**, **9.5.3** and **11.2.3**.

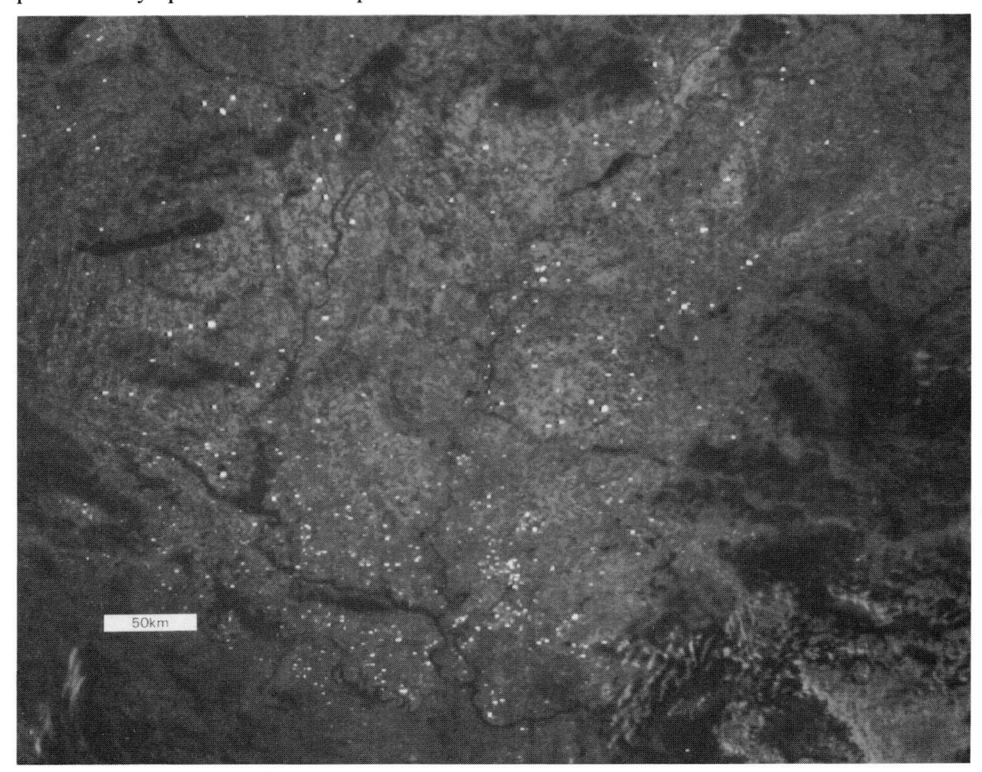

5.8.3 1315, 15.10.86,3

5.8.3
Some sort of waste burning also occurs in Hungary and the surrounding farms of Yugoslavia and Rumania, but much later in the year than in Britain. Straw-burning accompanies the earlier harvest in France, while in Denmark it takes place at about the same time as in Britain.

It must be remembered that a hot or warm spot may have the same appearance as a small cumulus cloud in Ch 3, but not in Ch 4 or Ch 2, and in this way can be identified as a hot spot.

6
Ocean plumes and trails

In this chapter we describe effects which are detectable far downwind of a variety of influences, some of which were well-known before satellites operated, others which have been discovered by satellites (see refs. [7] and [iv]).

6.1 PACIFIC OCEAN SHIP TRAILS

These trails were quickly observed in the early days of regular meteorological satellites. They occur when a ship leaves a trail of engine exhaust containing an abundance of hygroscopic or other condensation nuclei which reduce the average size of droplets in a cloud and increase its albedo (whiteness).

Suitable clouds are found in anticyclones in which a thin layer of stratus or stratocumulus is trapped below a strong inversion. If the air containing this cloud has been over an 'ocean desert' for several days the natural condensation nuclei are precipitated out and not replaced. A 'desert' is a region deficient in nutrients where algae do not proliferate and do not generate new condensation nuclei. The air is deficient in nuclei and may be described as very 'clean'. It does not appear possible for it to become clean enough for thermodynamically significant supersaturations to occur, but the number of droplets in any cloud that is formed is reduced and this means that the cloud water is condensed into larger droplets. This increases the visibility within the cloud and may even make it semi-transparent. We call this a 'lace curtain' cloud.

The updrafts in such a cloud are rather weak because the convection is determined by the radiative heat loss from the cloud top to space; a steady state is reached when the cloud and the air below it are cooled down until it is cool enough for convection and long wave radiation from the sea to maintain the temperature. Such a cloud may appear warmer to Ch 4 because some of the radiation received is directly from the sea through the gaps in the cloud. It also appears less bright in sunshine in Chs 1 and 2, but the greatest effect is in Ch 3 because of the dearth of droplets smaller than the wavelength which would scatter by diffraction (see Chapter 5).

After the passage of a ship the abundance of nuclei is spread throughout the convection cells into which it is released, and they become much whiter as the number of droplets is increased. Although the engine exhaust usually contains water vapour which increases the humidity initially, the quantity supplied, when diluted into a convection cell, is negligible compared with that already present in air over the ocean.

In these circumstances the convection appears not only rather slow but also to take place with very little transport of the pollution across the cell boundaries into the neighbouring cells. The trail, which is the plume from a moving chimney in a light wind, is spread laterally very slowly, and this is important information about the dispersion of any pollution under cloud showing a cellular convection pattern.

In strong winds convection cells are aligned along the wind shear in streets. In such winds ship trails are not seen, probably because too many nuclei are thrown up from the sea surface for the cloud to be deficient to the point of having a reduced albedo. We cannot be certain that in streets pollution would be transferred across street boundaries, although it is certainly spread along the streets as a result of the wind shear.

Typically, when a ship sails across the lines of cells and leaves a trail, it appears to be zigzagged. This is because if any part of a cell is traversed the albedo of the whole cell is soon increased.

The observations in the Pacific Ocean caused it to be said that ship trails are a phenomenon of summer ocean anticyclones in low latitudes [7].

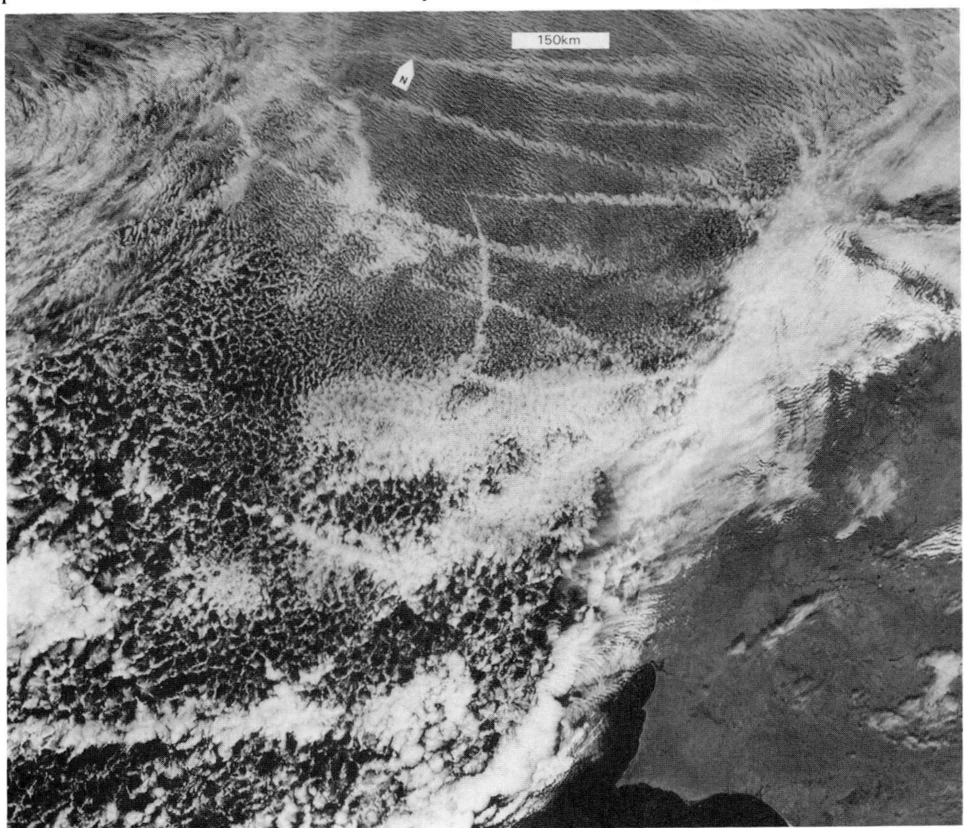

6.1.1 1208,15.12.82,CZ5

6.1.1
The cellular structure is arranged like streets, and we see how the ship's pollution is spread along the direction of the wind, giving a saw-toothed, or zigzag, appearance to the ship trail as it pollutes each cell through which it passes.

Further south the polluted cells are easily identified, and it is noticeable that the cloud which has passed southwards over Portugal is much whiter than further out in the ocean.

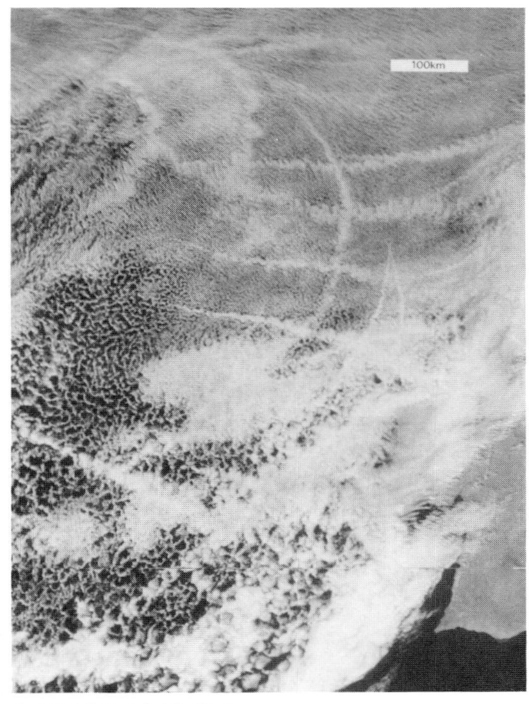

6.1.2
The same scene as **6.1.1** but $3\frac{1}{2}$ h later contains several new trails in the busy sea lanes in this area, but many of the old trails are still present and not much changed.

6.1.2 1542,15.12.82,2

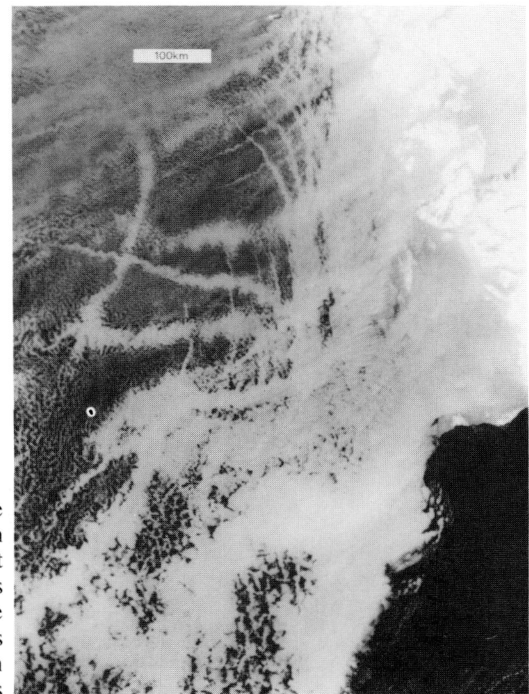

6.1.3
With suitable processing the temperature difference between the sea and cloud can be emphasized, but cloud top detail is not obtained in layer cloud in Ch 4. Ten hours later than **6.1.2** the scene is much the same but only some of the dense cloud patches can be identified. With so many ships in the same area the persistence of trails cannot be checked with certainty.

6.1.3 0354,16.12.82,4

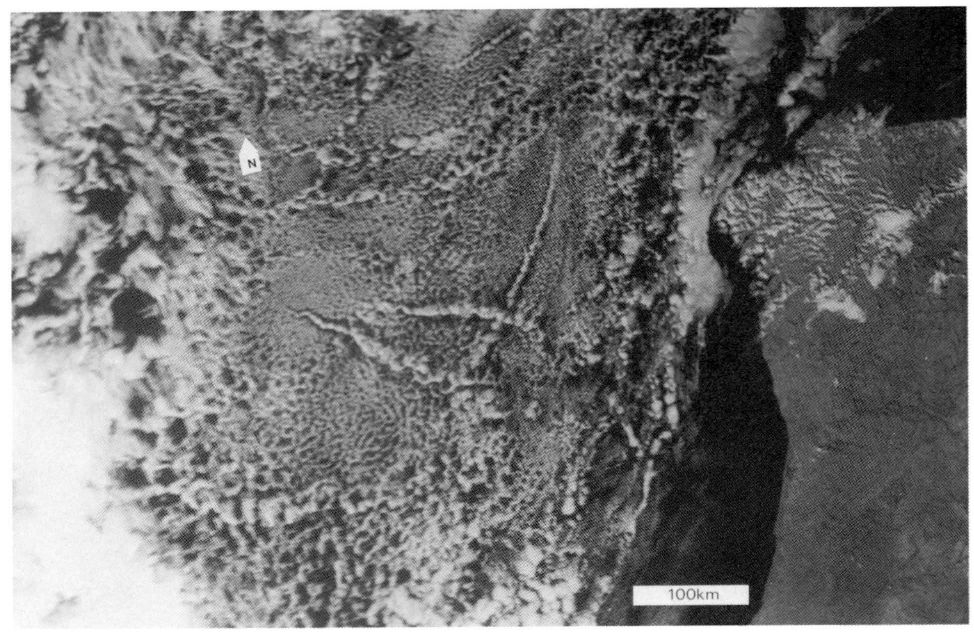

6.1.4 1436,17.11.82,2

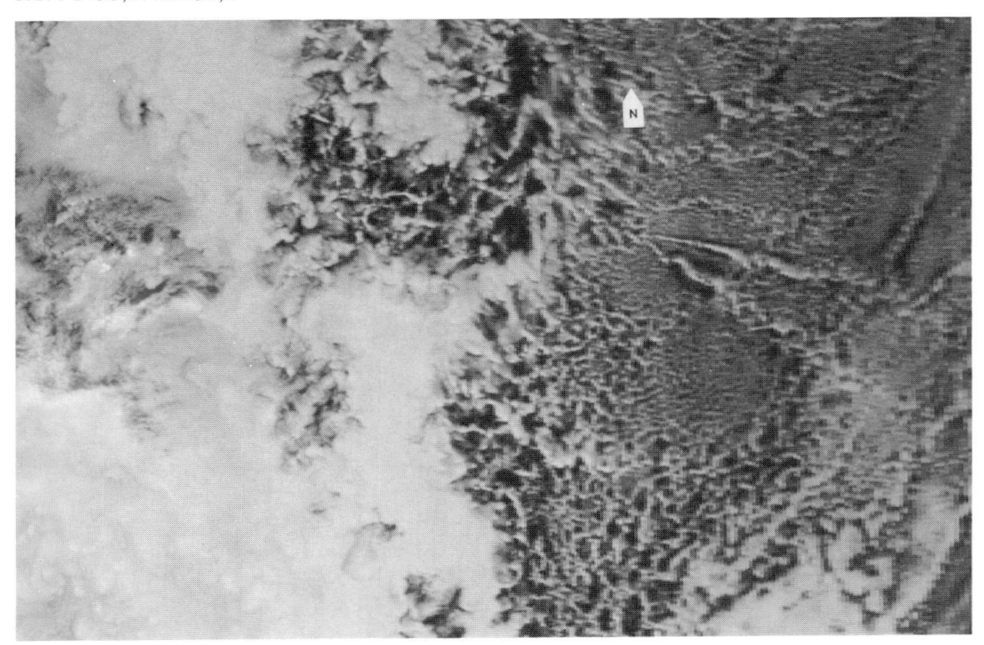

6.1.5 1619,17.11.82,2

6.1.4, 6.1.5

The second of these two pictures was close to the edge of the swath so that the definition is greatly reduced. This case illustrates the creation of a cloud-free strip on both sides of the trail in this case where the cells in the trails appear to be larger than in the surrounding cloud. One of the trails has created some ringed cells, i.e. large holes surrounded by narrow dense cloud.

6.1.6 0900,23.7.82,2

6.1.6

This patch of ship trails is in the centre of an anticyclone west of Ireland. It is seen that the parts of the cellular cloud area which have acquired pollution from the land are much whiter. The absence of cloud off the west coast of Ireland indicates offshore wind, with cells appearing 100 km or so out to sea and growing downstream.

The lee waves in the north are from the Faroe Islands.

6.2 ATLANTIC OCEAN SHIP TRAILS

There are no 'ocean deserts' in the Atlantic, but there are frequent incursions of arctic air from regions where there is no source of condensation nuclei. From time to time fairly long periods of light winds occur which transport the air large distances over smooth water from which large nuclei are not thrown up. Typically such air enters the Atlantic from the Davis Strait or the Greenland or Norwegian Seas and is transported southwards and usually eastwards by an anticyclone over the Azores or further to the west. Sometimes it is carried behind a cold front which travels southeastwards through the middle of an anticyclone; such air masses may be found anywhere from Cape Farewell to Ireland and southwards to Madeira.

Under more common situations, air of arctic origin is subject to strong convection over relatively warm sea. This not only generates whitecaps which release salt nuclei into the air, but also mixes the clean low level air with layers above which are of continental origin and plentifully endowed with nuclei.

Consequently, air in which ships would form trails does appear, but not very frequently. A study made for the period 1979–1985 [iv] showed that the average frequency of appearance of ship trails in the northeast Atlantic was about once in six weeks, with great variations in the intervals. On many occasions trails were seen in the same air mass for two or three days on end, and this was termed one occasion.

The main features of interest are illustrated in the pictures. As in the Pacific they show the zigzag form as if the ship were tacking. Sometimes the cloud is evaporated each side of the trail as if the trail were causing subsidence there, and the mechanism of this is elusive if only because it is not observed in every case.

6.2.1 Occurrences of ship trails during 1980–85 in the North Atlantic area.

	1980	1981	1982	1983	1984	1985	Total
Jan				15–16	9–10		2
Feb			28	10–12	8		3
Mar			1	26–27			3
			11				
Apr	19						1
May	31					21	2
Jun	21	20		21	15		5
					22–24		
Jul		15–18	22–23	15	13–15		5
				26			
Aug	21	10	7–9	10	24–25		9
			18	17	31		
			22–24				
Sep			1	6			2
Oct				4		15–16	4
				19			
				23			
Nov	11	2	17–18	12			5
		30					
Dec	4–5	6	15–16	8–9	22		6
Total	7	6	10	14	8	2	47

With the aid of the Dundee archive a study of the occurrence of ship trails in the north-east Atlantic during the six years 1980–1985 was made in 1986. The IR pictures of the Atlantic were examined for every day, and the Ch 2 pictures also where they were available.

The final year was almost barren, although it provided one of the very rare cases of trails in the North Sea which is well illustrated in **6.4.9–16** later in this chapter. The other case is reported in **6.3.1–3**. A few fragments did occur on other occasions, but these two cases are the only ones in which plumes were seen from flares on oil platforms and the whole area of the North sea was involved.

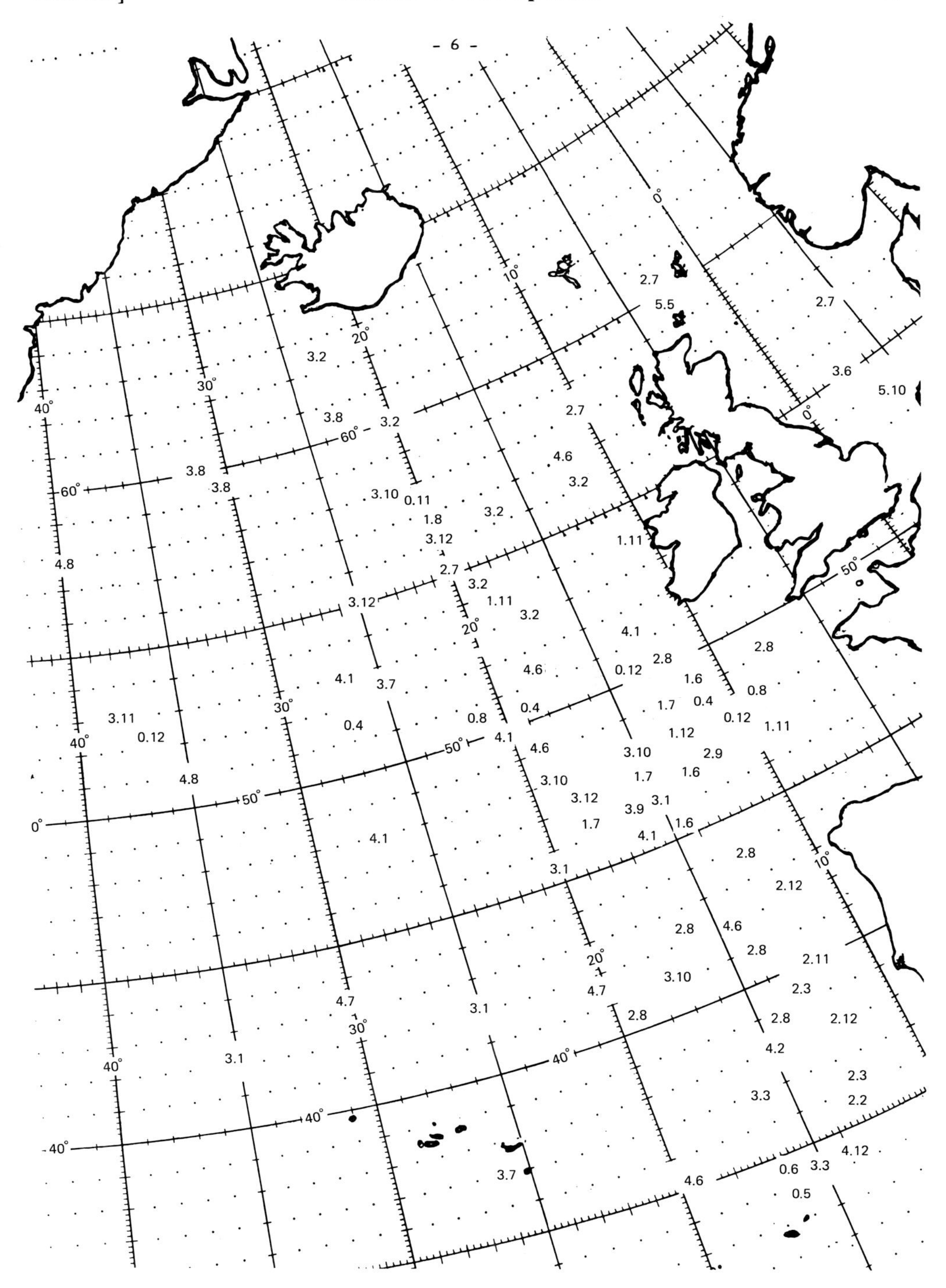

6.2.2
Map of occurrences given in **6.2.1**: the first number gives the last digit of the year and the second gives the month. The numbers are placed in the general area occupied by the trails.

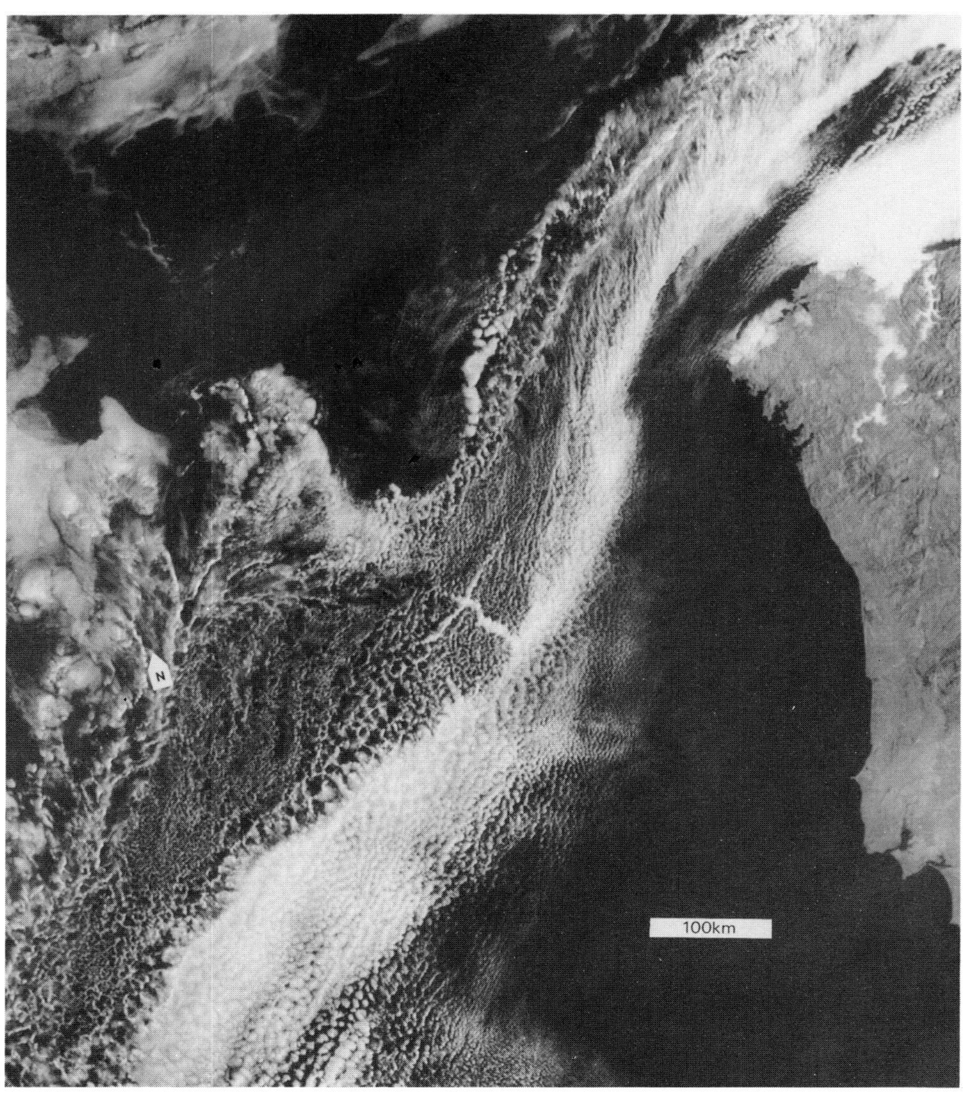

6.2.3 0855,19.10.83,2

6.2.3, 6.2.4

The terminator is well-defined in the cloud top in these pictures, particularly in Ch 3, where the ship trail crossing it is detected on the light and dark sides. To the west the cloud tops in Ch 2 are illuminated by skylight, but in Ch 3 they are dark and not as bright as the warm sea between the cloud and the coast of Portugal, which is glowing with its own Ch 3 emission. The water is seen to be cold close to the coast.

Beyond the main cloud belt two features stand out in Ch 3. First, the newest part (85 km) of the ship trail running parallel to the terminator is illuminated, presumably because it is higher (due to buoyancy) while it is still less than a cell size wide and not fully diluted. Secondly, to the west of the main cloud belt, some fragments of cloud are brighter than the sea beneath them, although they are not in direct sunshine and are not warmer (for otherwise all the low cloud would be of the same shade because we would be looking at emissions, not at scattered sunshine, as was the case with the pancakes shown in section 5.6). (*continued next page*)

6.2.4 0855,19.10.83,3

6.2.3, 6.2.4 (*continued*)

A full analysis of this second phenomenon would require a comparative statement about the dependence of transmission and emission of tenuous cloud or haze upon the ratio a/λ (a particle size; λ, wavelength). Certainly transmission decreases as particle size and emission increase and we switch from measuring mainly the emission of the sea below to the emission of the cloud particles, at a given wavelength. If the two changes did not exactly compensate for one another we would see different pictures in Ch 3 and Ch 4 at the intermediate particle sizes.

But in the north (top left) of the pictures the south-facing cloud edge is clearly illuminated in both Chs 2 and 3. This is explicable in terms of the wave-guide provided by the forward scattering by the very tiny cloud droplets close to the base of the clouds and the very strong reflection by the sea surface in Ch 3. The same mechanism would operate in Ch 2 but relative to illumination by skylight would be much weaker than in Ch 3.

6.3 NORTH SEA TRAILS

In most cases of ship trails in the northeast Atlantic the air can be traced back to the Davis Strait, where trails also occur in light wind situations. It is most unlikely that air of that origin will enter the North Sea without crossing the British Isles. Trails are therefore relatively rare in the North Sea.

Trails have not been detected in the Mediterranean Sea because the air must have crossed land to arrive there. Any land except that which is covered with snow, even if no human activity is taking place there, will supply condensation nuclei in plenty, and there are so many different kinds possible that no air over land may be regarded as clean. Antarctica is a windy and mountainous area which supplies ice crystals in plenty to the air, but the ice of the Arctic Ocean provides wide flat expanses over which calm air may become very clean.

The North Sea is occasionally invaded by air from the region of Spitzbergen and beyond. Usually such incursions have fairly strong winds and frequent showers, and it requires some precision even in these cold outbursts for low level air to enter the North Sea without crossing Scotland or Scandinavia and without picking up plenty of nuclei. Only two such occasions have been found in the Dundee archive between the beginning of 1979 and the end of 1988: these were 19–21.6.83 and 15–16.10.86. Trail fragments have been seen in the Shetland area on other occasions and they were associated with trails seen further west, indicating that the air was of Davis Strait origin.

On the two notable occasions just mentioned the most prominent trails were plumes from oil platforms on which the flares were observed as hot spots by Ch 3 (section 5.8). In the 1983 case the sea fog invaded the east coast of England during the night of 20–21 June, and plumes from many coastal sources of pollution could be traced many miles inland. The convective structure can be seen as streets along the wind. Lateral dispersion was very slow, even over land, the cold air being very shallow.

6.3.1, 6.3.2

This event has many points of interest. The cloud entered the north sea between Scotland and Norway, and spread down the east coast of Britain, invading England during the night. Behind it came a gust of air from the Skagerrak, and this is well shown up by the darkening of the glint on the sea. The sea around the Danish islands is much darker because it is almost a flat calm and produces a narrow-angle reflected beam which did not impinge on the satellite.

The rivers and lakes of northwest Europe are bright with glint, the wet surface of Holland more so than the rest.

In the North Sea, close to the northern cloud boundary, are two hot spots on oil platforms, and each has given rise to a plume, the more westerly one being difficult to detect in this print, although the hot spot itself can be seen through the cloud. Trails of several ships travelling from east to west can be seen, but the ships must have a southerly component of velocity about equal to that of the wind.

At several places small patches of dark high cloud, including some contrails, cast dark shadows on the low stratus or glint below them (note the patch just off southwest Norway).

The cloud over England is shown in enlargement because of the special interest of the cloud streets which, at first sight and with the lesson of ship trails in mind, appear to indicate slow dispersion across the wind. But the cloud streets have many breaks and amalgamations along their lengths, in spite of which the rather ragged dark line on the east cloud edge is of remarkably persistent width. Note also the clean cloud edge parting from the east coast of East Anglia, and the lee waves over the southeast corner of England.

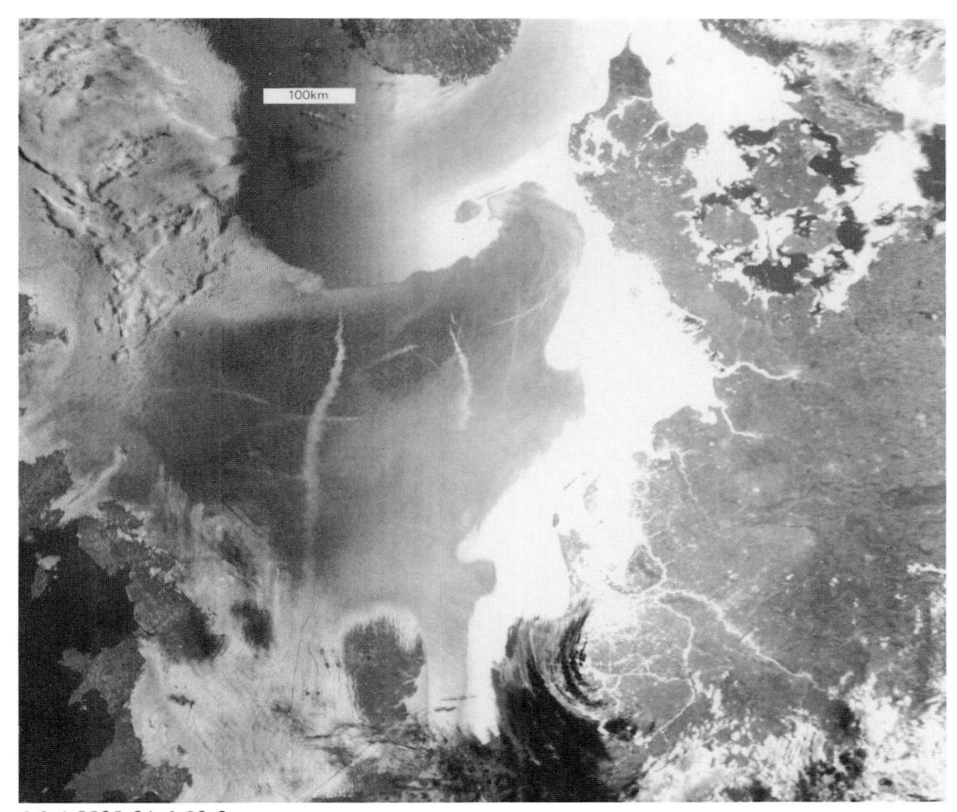

6.3.1 0838,21.6.83,3

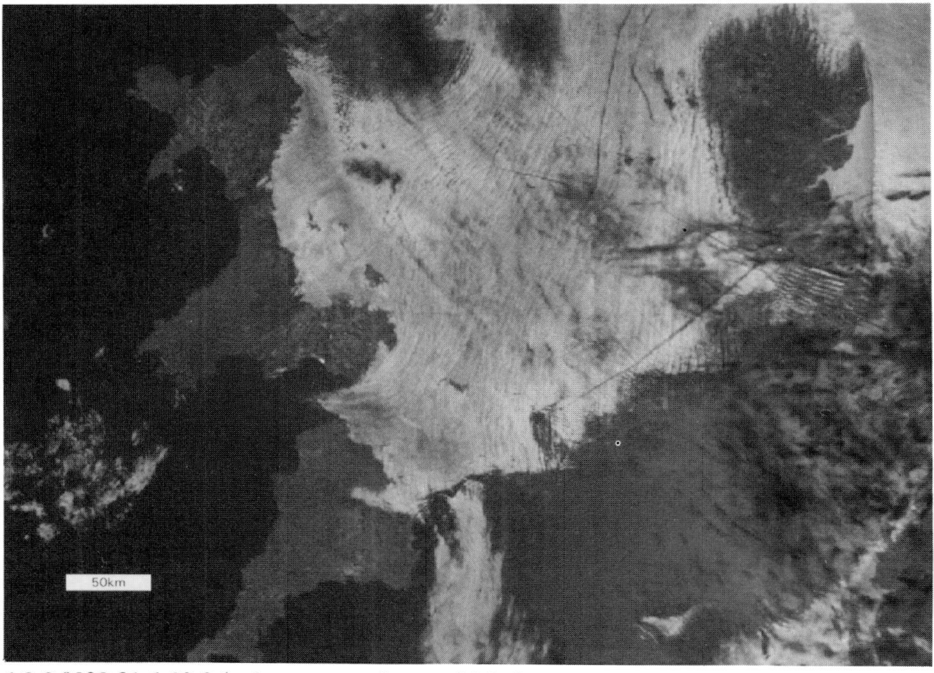

6.3.2 0838,21.6.83,3 (enlargement of part of 6.3.1)

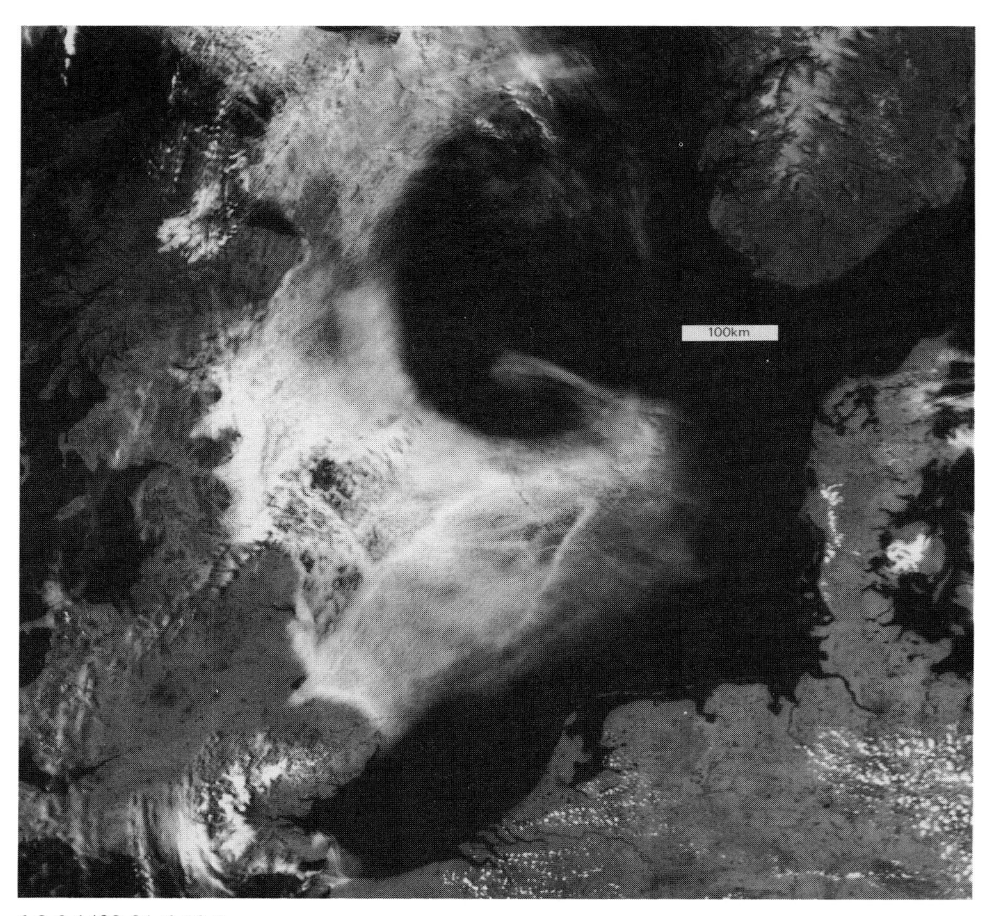

6.3.3 1432,21.6.83,2

6.3.3
The main high pressure centre over Britain and the North Sea was moving eastwards but was weak. The entry of the cold air into the North Sea at sea level was far from geostrophic and was largely isallobaric.

The afternoon picture shows how the stratus was burnt up inland while the coast continued to shine through the cloud in Ch s 2 and 3. This raises the question of how much was pollution responsible for the change in the cloud' s brightness across the coast (see 3.3). Lee waves and stratus reveal some of the character of the airflow through the Straits of Dover.

6.4 INCINERATOR PLUMES AT SEA AND LAND PLUMES

On several occasions the plume from an incinerator ship, usually stationed at about 54°N 4°E in the North Sea, has been clearly identified, although these are only a small minority of occasions on which the incineration of chemical waste has been carried out. The operation is carried out at sea because of objections to this particular operation on land, but since the dispersal of plumes has been observed to be very slow over the Atlantic and the North Sea it is possible that at landfall the concentration in the incinerated plume may still be objectionable. The principle of

6.4.1

This is not a typical example of a plume from the North Sea incinerator ships, but it is not unprecedented. It is a trail extending northwards from the ship (at about 54°N 4°E) a distance of about 300 km, at which point it disappears into cloud. This is CZCS channel 4 which is very sensitive, and the other markings are a mixture of cloud, haze, and sea colour variations.

The main plume draws attention to a second, weaker plume parallel to it about 20 km to the east. In both cases there appears to be a short white plume at the ship.

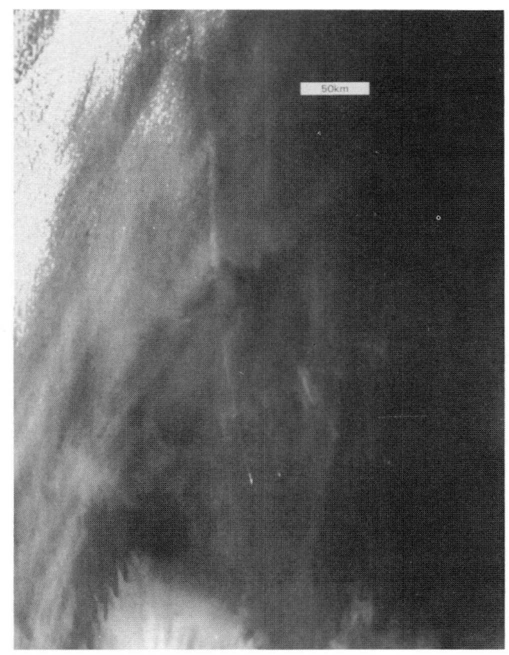

6.4.1 1048,21.2.82,CZ4

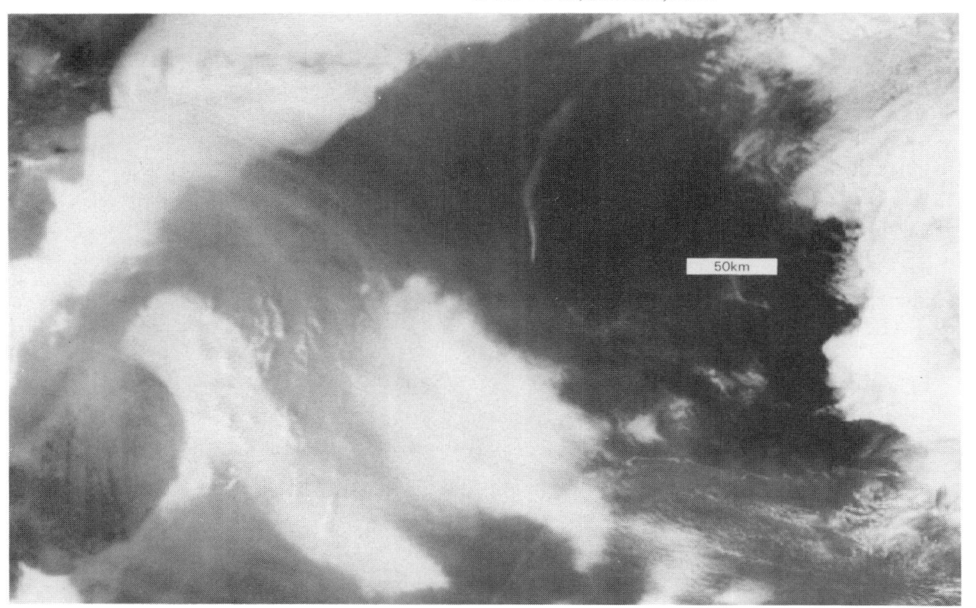

6.4.2 1342,17.9.82,1

6.4.2

Very short plumes (see later pictures) are common, and this case of an incinerator ship plume resembles more than usually a typical industrial or power station plume which disappears by dilution rather than by evaporation. It is clearly visible for about 100 km, which is very much longer than an industrial plume showing this sort of shape over land remains detectable.

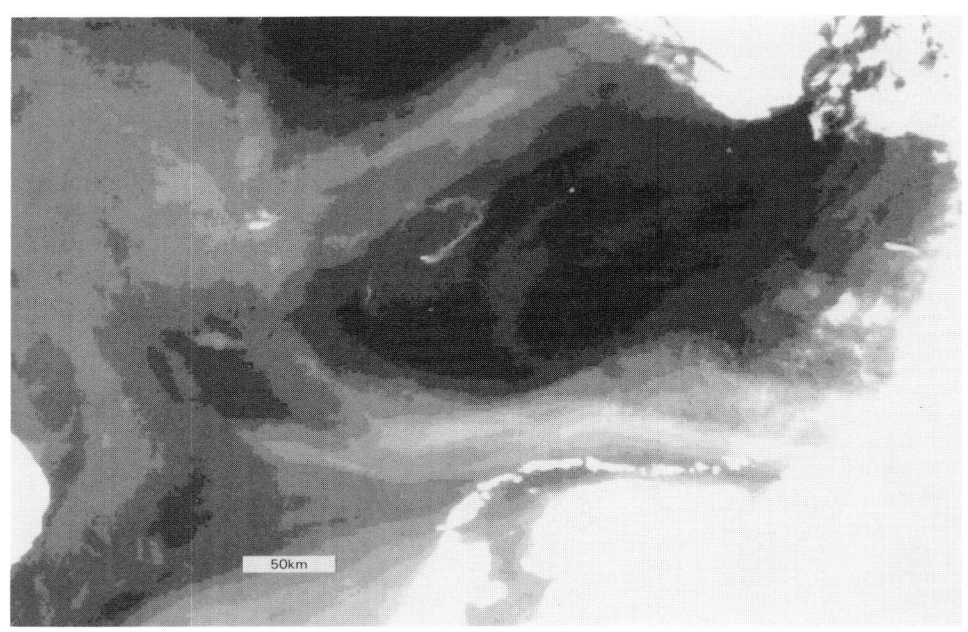

6.4.3 1450,19.8.84,2

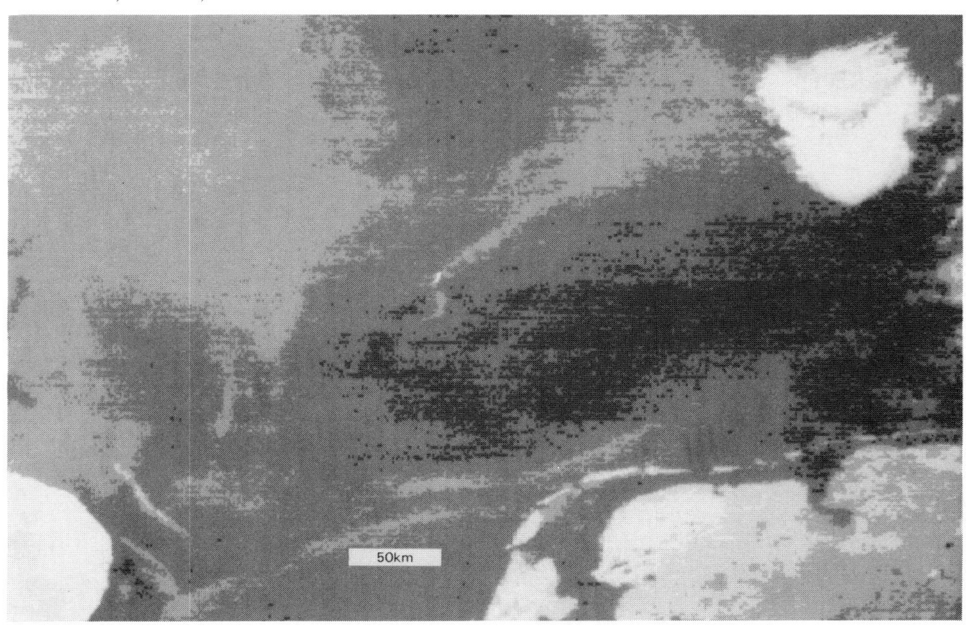

6.4.4 1843,19.8.84,2

6.4.3, 6.4.4

The processing has been carried almost to extremes to emphasize the small range of shades which appear as contours. This enables the plume from the bright spot (which was obvious in the ordinarily developed picture) to be shown up.

Along the north coast of Holland is a belt of pollution which was soon carried southwards. The evening picture shows the same general layout with less contrast over the sea.

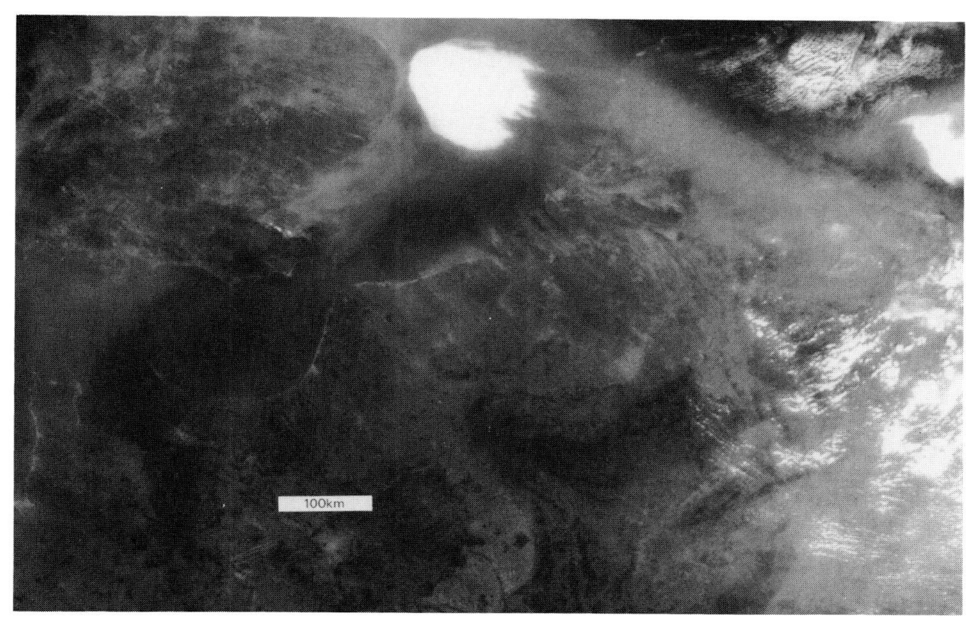

6.4.5 1110,20.10.84,CZ1

6.4.5, 6.4.6
On the following morning the belt of haze had been driven into Holland by a surge of northeast wind which produced cloud streets. The patch of cloud over the sea soon impinged on the English coast (see 9.5.3).

Enlargements of the CZCS pictures showed the incinerators at work and **6.4.6** shows the haze belt as dense white on its southern side, with cloud and haze over the sea.

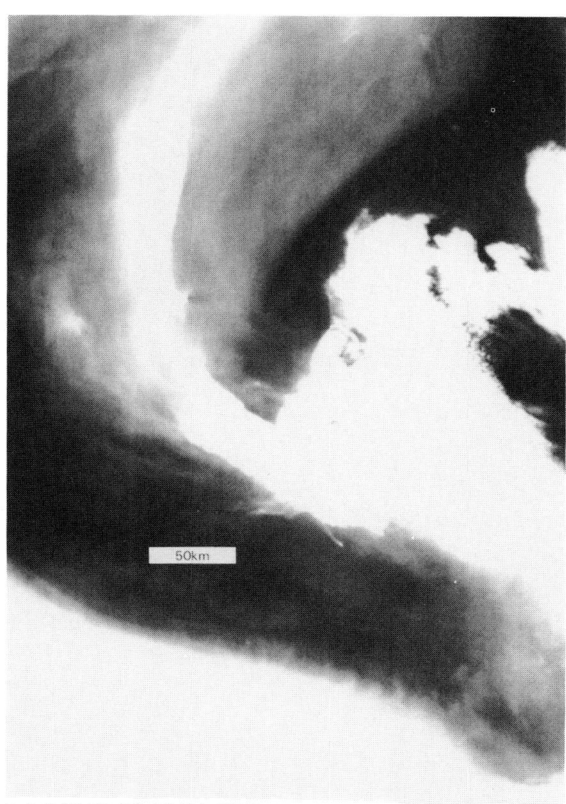

6.4.6 1110,20.10.84,CZ3

doing dirty jobs at sea, where regular supervision is less easy to carry out, raises many questions of environmental pollution control which have not yet been answered.

The plume has been observed only on occasions when the wind was from a southerly point, indicating that an early landfall was being avoided. In winds from the south the air is coming from industrial areas, which ensures that it is already well-polluted, and it is surprising that the ship's plume is such as to make it observable at all, but the microphysics is quite different from the formation of the ship trails previously discussed.

The second occasion of North Sea ship trails mentioned above is illustrated because it showed, better than the previous case, how land sources may also be detected in clean air sometimes.

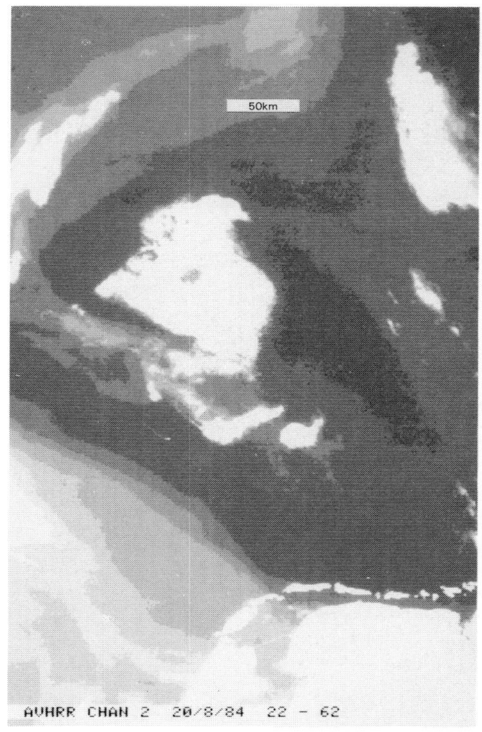

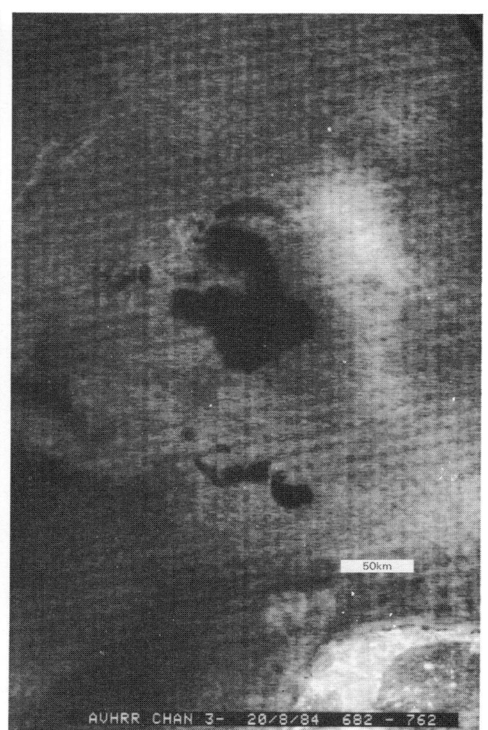

6.4.7 1437,20.8.84,2

6.4.8 1819,20.8.84,3

6.4.7
Ch 2 sees through the haze better than CZ 3. Two ship plumes are detected.

6.4.8
In the weak evening sun the cloud appears black in Ch 3 against the warmer sea. The ships show hot spots like oil platforms.

In Ch 2 at this time only the northeasterly of the two ships had a white spot (a condensation cloud which quickly evaporated).

6.4.9 0302,15.10.85,3

6.4.9

This series shows features of a small mass of clean air which entered the North Sea from the north. Four plumes are evident, the two northerly ones being from oil platforms and the one originating inland near the Dutch coast being from the Haarlem steel works. The plumes are black because they are cold compared with the sea, which shines through the lace curtain cloud and is responsible for the white areas where there is no cloud. Interference (stripes) spoils some details.

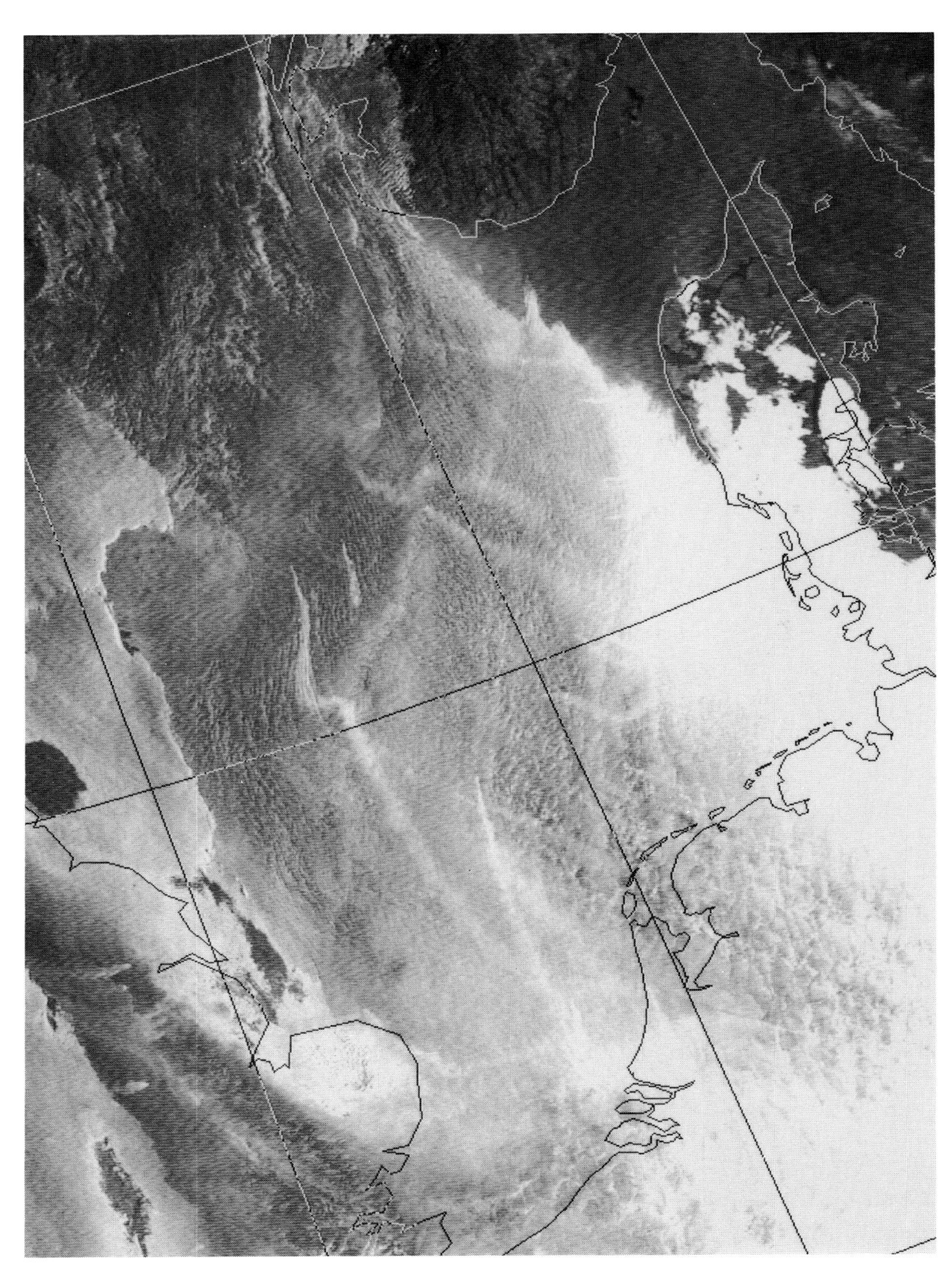

6.4.10 0836,15.10.85,3

6.4.10

During the next $5\frac{1}{2}$ h the plume sources remained fixed but the eastmost of the northerly ones has started again after a period without a plume and presumably without a flame. Many ship trails advancing from the east are plumes from moving sources. The plumes are now in bright sunshine and some of the structure of the lace curtain cloud is evident.

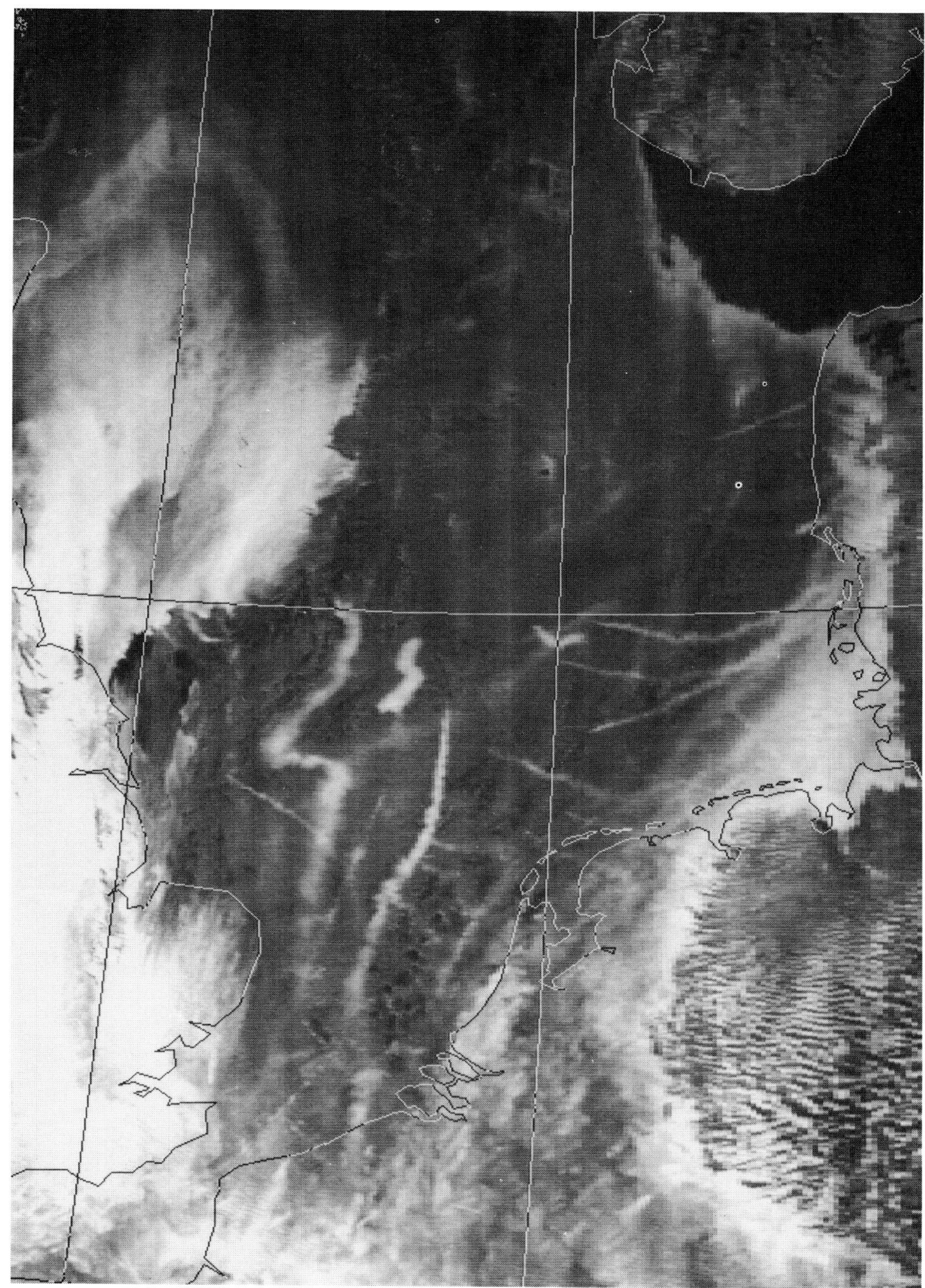

6.4.11 1437,15.10.85,3

6.4.11
The plumes, with their details of wind changes, which can be seen to have moved southwards, can be identified from the previous picture, but the ship trails are not as easily tracked.

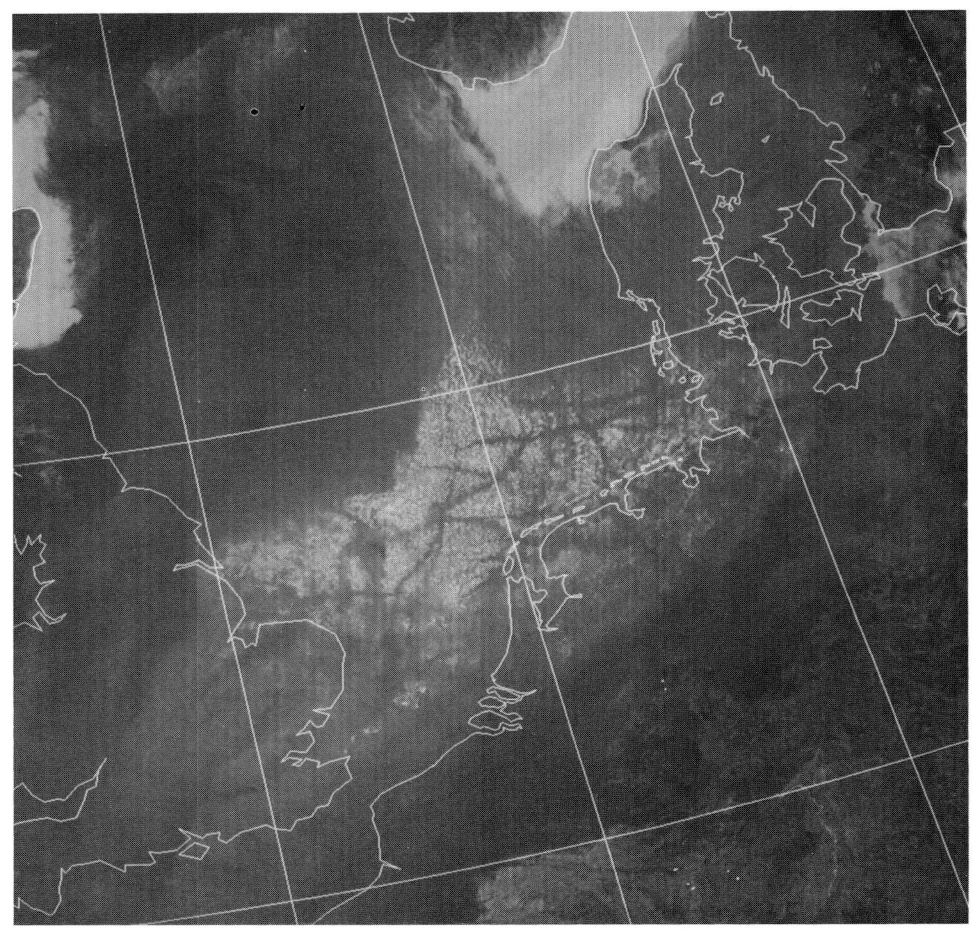

6.4.12 0251,16.10.85,3

6.4.12
The following night it was clear that the clean air with its trails had moved into Holland and Germany. The night was clear so that land pollution was not stirred up into it from under the radiation inversion.

6.4.13, 6.4.14
By this time the clean air had reached the Ruhr where several dense plume heads can be seen. Others are in the Gent–Antwerp–Brussels area. It is noticeable in **6.4.13** that white clouds of all kinds are seen as brighter where the scatter is more nearly forward, while in Ch 2 (14) the illumination is less contrasted.

The westerly of the two prominent plumes in the Ruhr industrial neighbourhood is from Wesel, and the dense ones further south are from Neuss and, to the south west, from Julich.

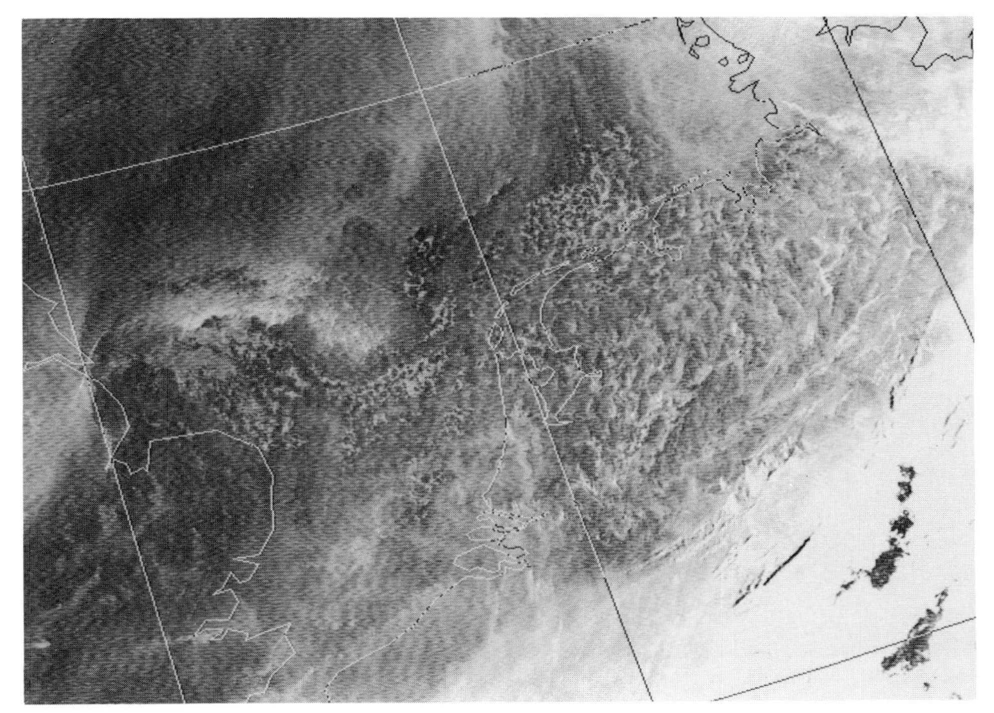

6.4.13 0815,16.10.85,3

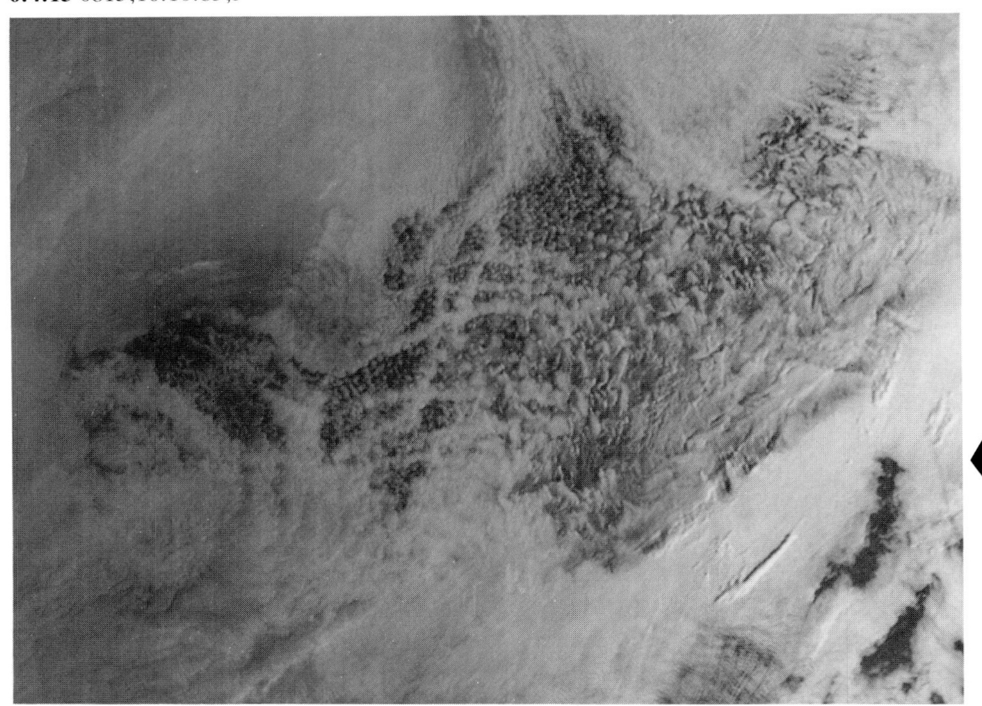

6.4.14 0815,16.10.85,2

6.4.15 1246,16.10.85,3

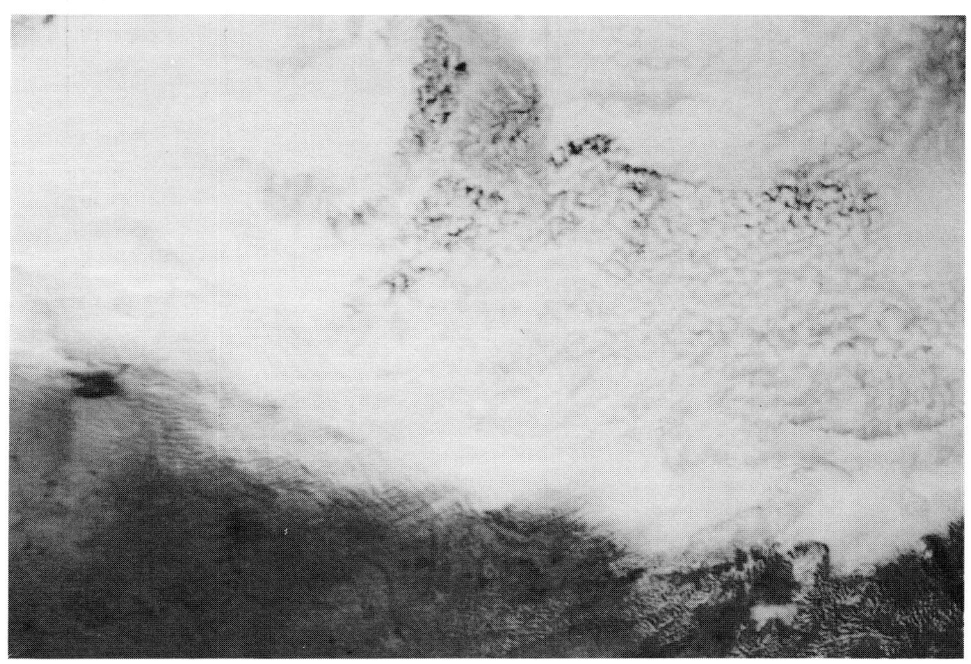

6.4.16 1246,16.10.85,2

6.4.15, 6.4.16
The German industrial plumes are now very clear. There are also many from the Rotterdam area southwards. Lee waves in Picardy have a wavelength just over 4 km.

6.5 OTHER POSSIBLE SHIP TRAILS

At the time of writing I have seen no reports of ship trails in the Western Pacific, the Indian Ocean, the South Atlantic, or the southern ocean surrounding Antarctica. Trails are not to be expected in those oceans because there is no adjacent source of air deficient in condensation nuclei. Antarctica itself is well known for the frequent displays of optical phenomena in almost cloudless skies, and these are probably caused by ice crystals raised by the wind from the surface. The usually fairly stormy seas indicate a considerable wind shear with vertical mixing, which means that any air which may have been cleaned by residence over the snow or ice would quickly be mixed with air from above of different origin and well-supplied with nuclei.

6.6 CLOUD LINES

Many cases have been observed of a cloud line stretching hundreds of miles downwind from an island. The persistence of the line for such large distances indicates the persistence of a motion pattern temporarily imposed on a layer of cloud. The lines are most common in cellular convection patterns, and this indicates that the radiative heat loss from the cloud top is important in driving the convection. Over land the position of streets may be maintained by the reduction of ground heating in their shadows, but we are not here concerned with streets which occur over a large area, several together. Solitary lines retain their identity for several hours among cellular patterns and even in street patterns which are not parallel to the line. The lines are plumes in that they indicate the track of the source in the moving air mass; they are sometimes very obviously not trajectories of the particles composing them.

Convection plumes are regions of convection cloud growing where the air has moved to warmer sea. The starting point of such a plume often generates cloud lines extending for hundreds of miles; plumes are seen where air moves from the cold sea near Newfoundland to the Gulf stream, and from the cold southeast European land mass across the Adriatic and Tyrrhenian Seas in winter. As would be expected if they move over progressively warmer sea, such plumes grow wider and the clouds taller. The cloud lines we are concerned with here do not grow wider. Nor is there any suggestion in their appearance that they are island pollution trails because the clouds in them do not differ in albedo from the surrounding clouds as ship trails do.

6.6.1

At any moment a plume, which is the ' line' containing all the particles which passed over the source point (or emerged from it) will not, in general, coincide with the track taken by those particles. In this picture there are several plumes which do not run parallel with the streets. Streets lie along the direction of shear, which, near the surface, is close to the direction of the wind. Some source points seem to be in eastern Sweden and the wind has varied during the times of travel of the particles to the positions indicated by the plume. Thus at the Finland end of the plume the motion of the air relative to the ground has lost the component from the north which it possessed in Sweden.

The pictures **6.4.9–16** should be viewed with the same considerations in mind.

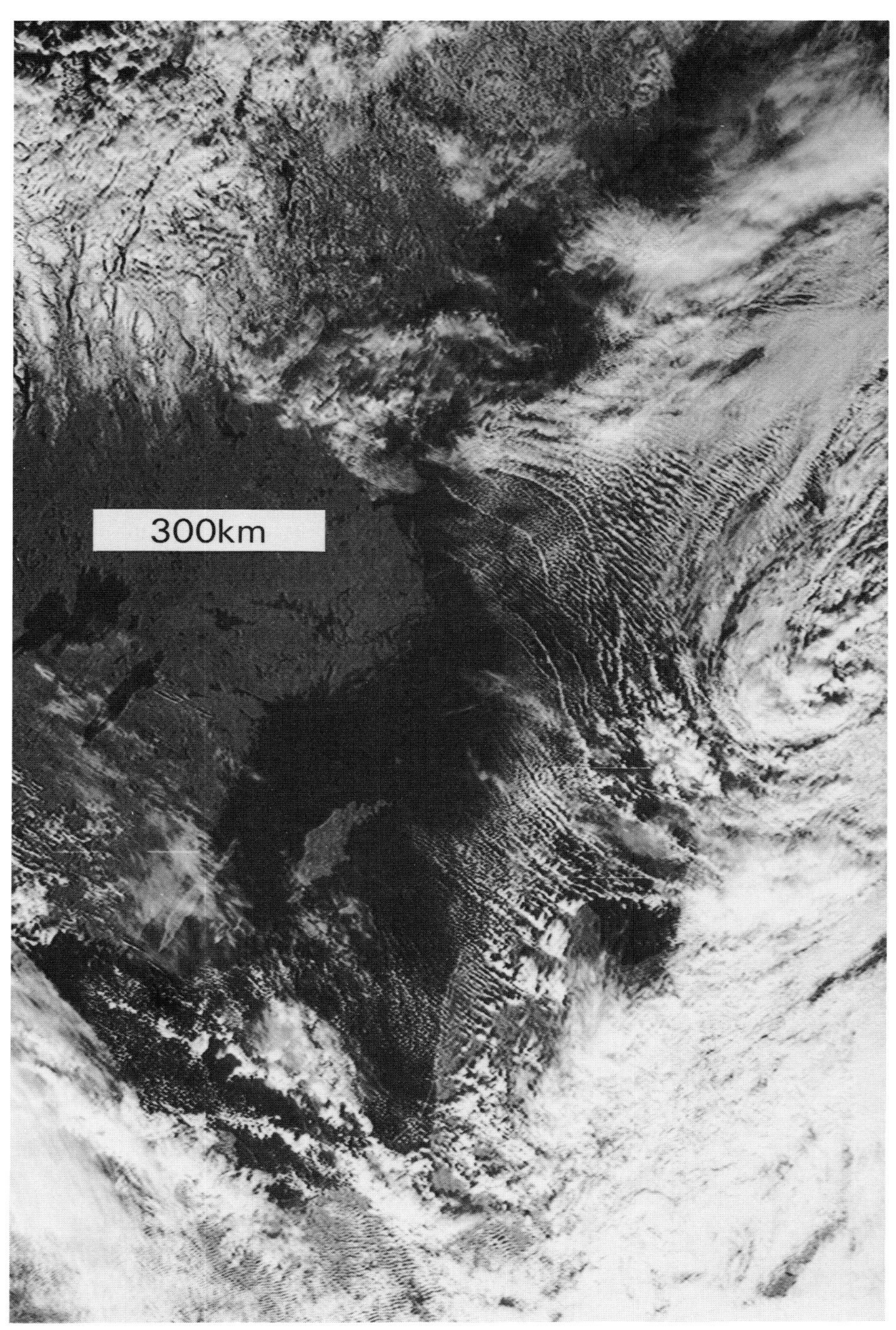

6.6.1 0958,21.10.80,CZ5

6.6.2 1144,15.1.73,IR

6.6.2

This early (NOAA-2) picture shows cloud lines in the Bering sea, one prominent one being of denser cloud, while two others are dark lines where the cloud is absent, and these are in the lee of the Probilof Islands. The longest streak is more than 600 km in length.

The ice off the coast of southwest Alaska is clearly significantly colder than the sea. The satellite track is northeast–southwest so that north is in the top left corner. The islands are 75 km apart.

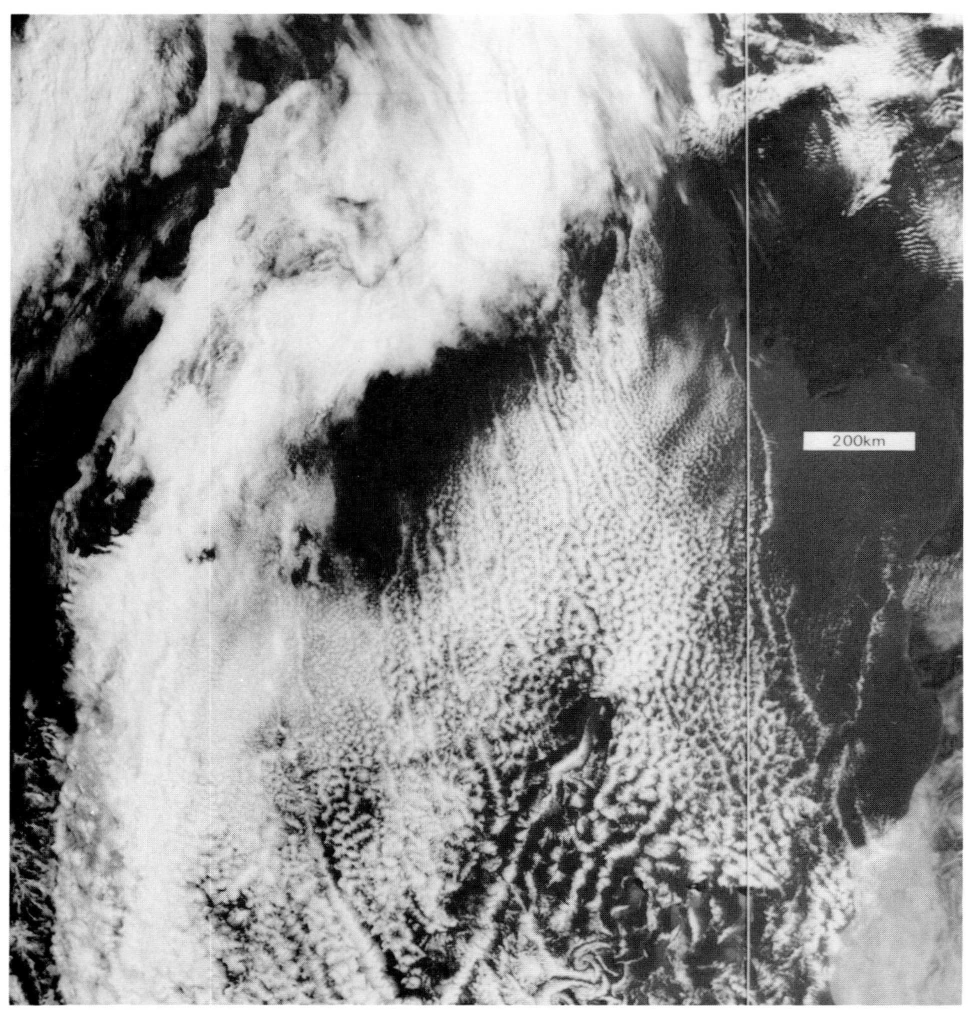

6.6.3 1039,30.3.77,VIS

6.6.3

The dark lines (cloud absent) are quite commonly seen in the lee of Jan Mayen, but equally common are single strong cloud lines. In this picture we see one being formed by Madeira while the Canary Islands, in the south-east, produce vortex streets. Note also the bright sea in the lee caused by an increase in glint due to the sea being less rough in the lee of the islands (see [1], 9.29). It is relevant to note here that under illumination from a different direction the bright areas of sea would mean that the sea was rougher in those areas.

In the present context interest lies in the cloud lines inclined at about 45° to the wind, which is clearly defined by the Madeira trail. Some of these appear further north out of nothing where the cloud is forming and this may be because an island has impressed a trail in the humidity distribution further upstream. The northwest end of the long trail in the west of the picture is very close to Santa Maria, the most southeasterly of the Azores, and the other trails might have undergone wind changes since being formed in the lee of the other islands, for the wind had undergone great changes with the movements of the high pressure centres in the last two days, while their strength remained light. Within the area shown there is a change close to the African coast to northeasterly.

7

Plumes from land sources

7.1 THE WESTERN ATLANTIC AND NORTH SEA

Several cases of plumes from industrial areas in the coastal states of northeast USA extending far out into the western Atlantic have been reported in the American literature. Some of these are cloud lines, which indicates that the water vapour content of the industrial effluent was an important factor in rendering them visible. Plumes of haze are frequently seen from Germany and the Low Countries extending far out across the North Sea.

These are measures of the industrial activity and may be used to test computer models of plumes and trajectories, but are not seen at present as introducing any new idea into the microphysics of clouds and haze. It is interesting that such pollution streaks are not seen coming from Britain into the North Sea, but this cannot be taken

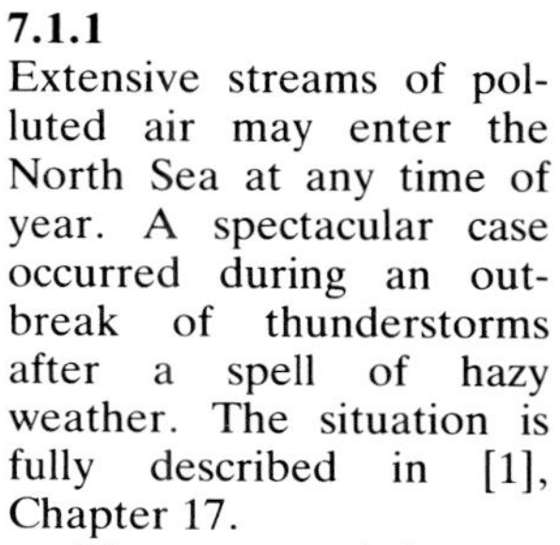

7.1.1
Extensive streams of polluted air may enter the North Sea at any time of year. A spectacular case occurred during an outbreak of thunderstorms after a spell of hazy weather. The situation is fully described in [1], Chapter 17.

The storms left very damp air near the surface and a few cirrus fragments high up. The stratus over the sea north of Holland is the remains of one which drew dense haze out of Europe in a whirl over Denmark. The snow on the Norwegian mountains is seen to the north of it.

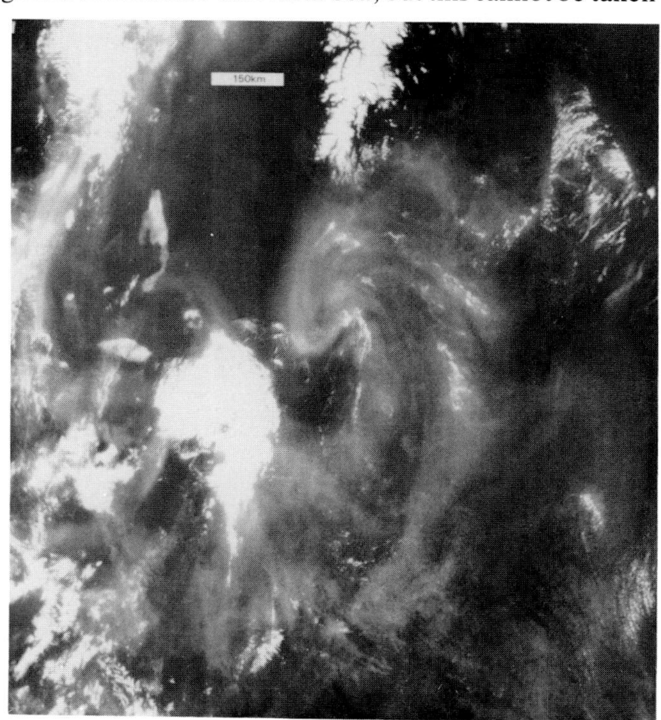

7.1.1 1046,2.6.82,CZ1

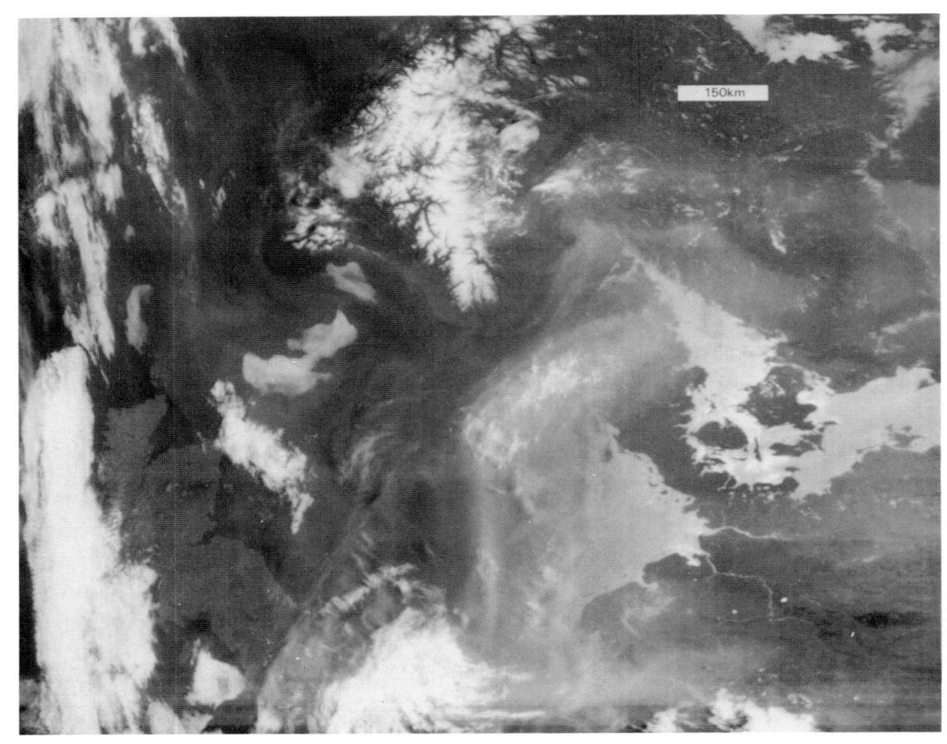

7.1.2 0831,3.6.82,1

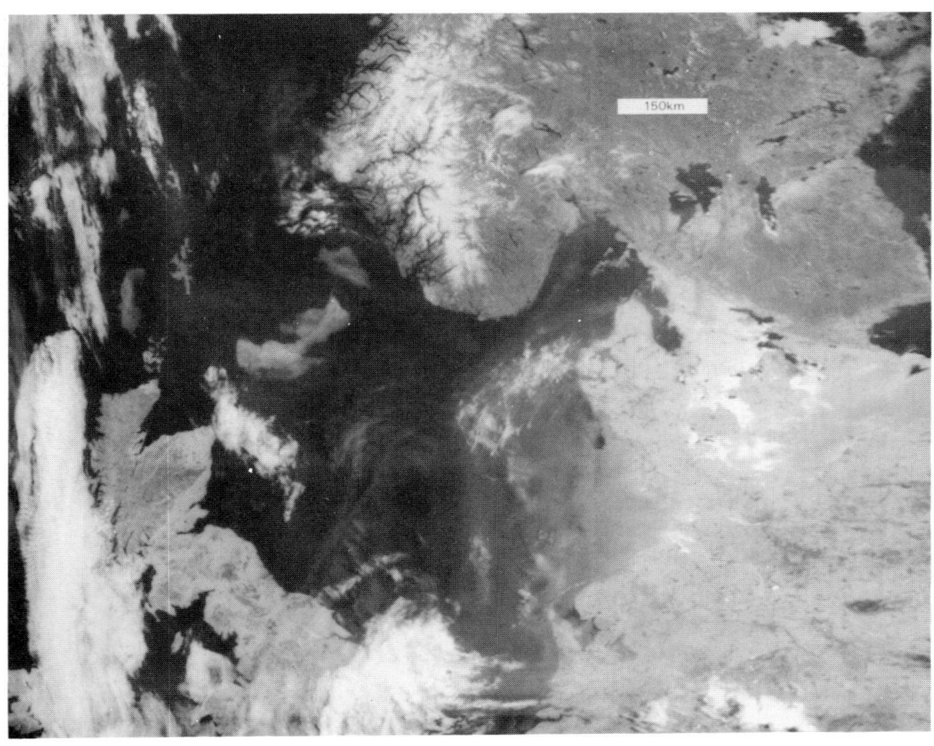

7.1.3 0831,3.6.82,2

to indicate for certain that Britain produces less pollution than Holland, Belgium and the Ruhr, although that may be the case, because almost invariably a wind from the west possesses a much deeper mixing layer than one from the southeast, and there is usually much more shear within the mixing layer. At the same time it must be remarked that often the air entering the North Sea has come from the very large pollution sources in East Germany, and Poland.

7.1.4 0831,3.6.82,4

7.1.2, 7.1.3, 7.1.4

The cloud over southeast England is the remains of a storm which raged the previous night over the Belgian coast. The haze and glint are shown by Chs 1 and 2 and it is seen how the brightness of the land affects the details presented. Ch 4 shows that cirrus over the German Bight is mainly shown in Ch 1 or 2 by its shadows on the haze below, and that the very fibrous cloud off the Jutland coast is not cirrus but low stratus!

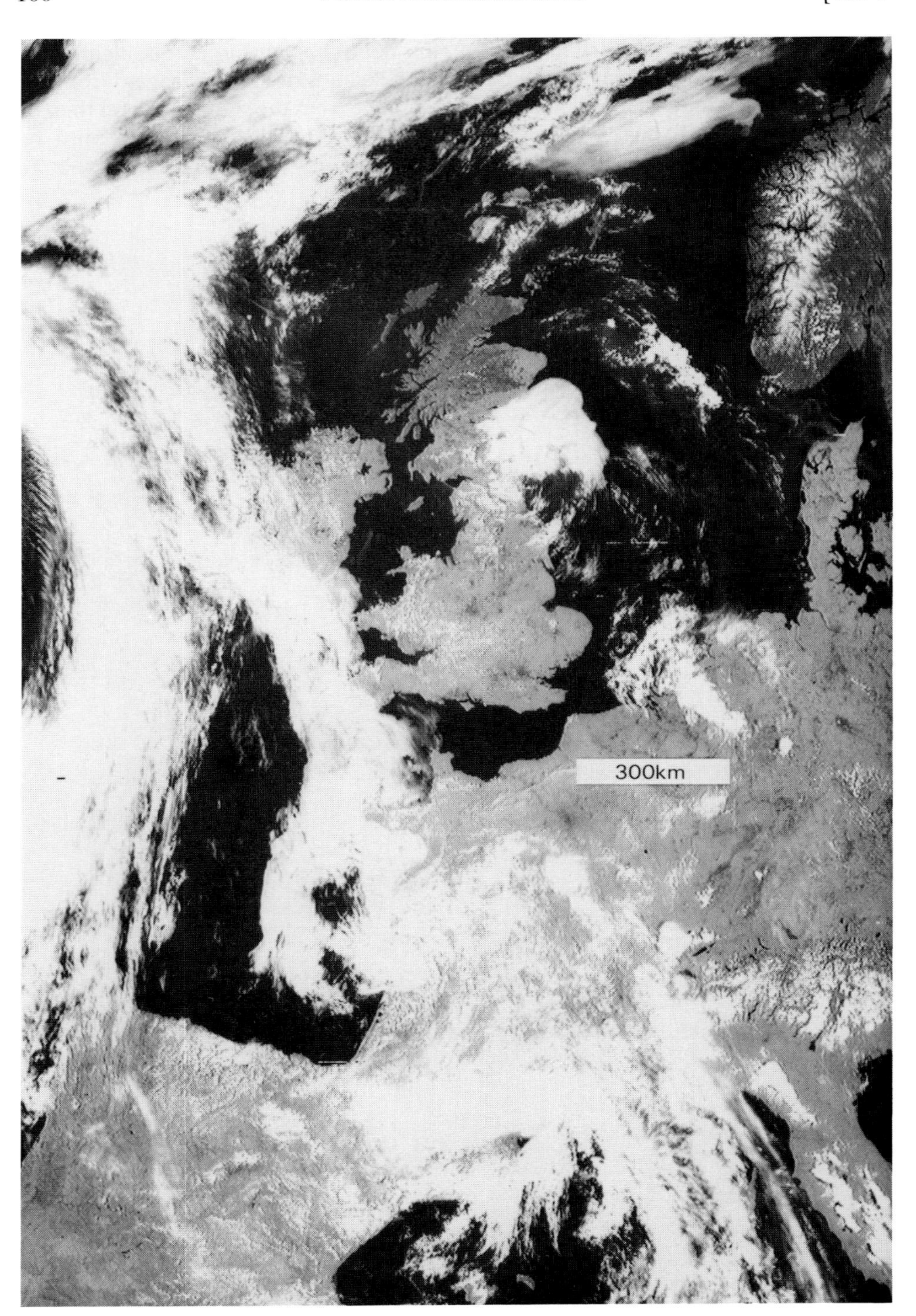

7.1.5 1122,4.6.82,CZ5

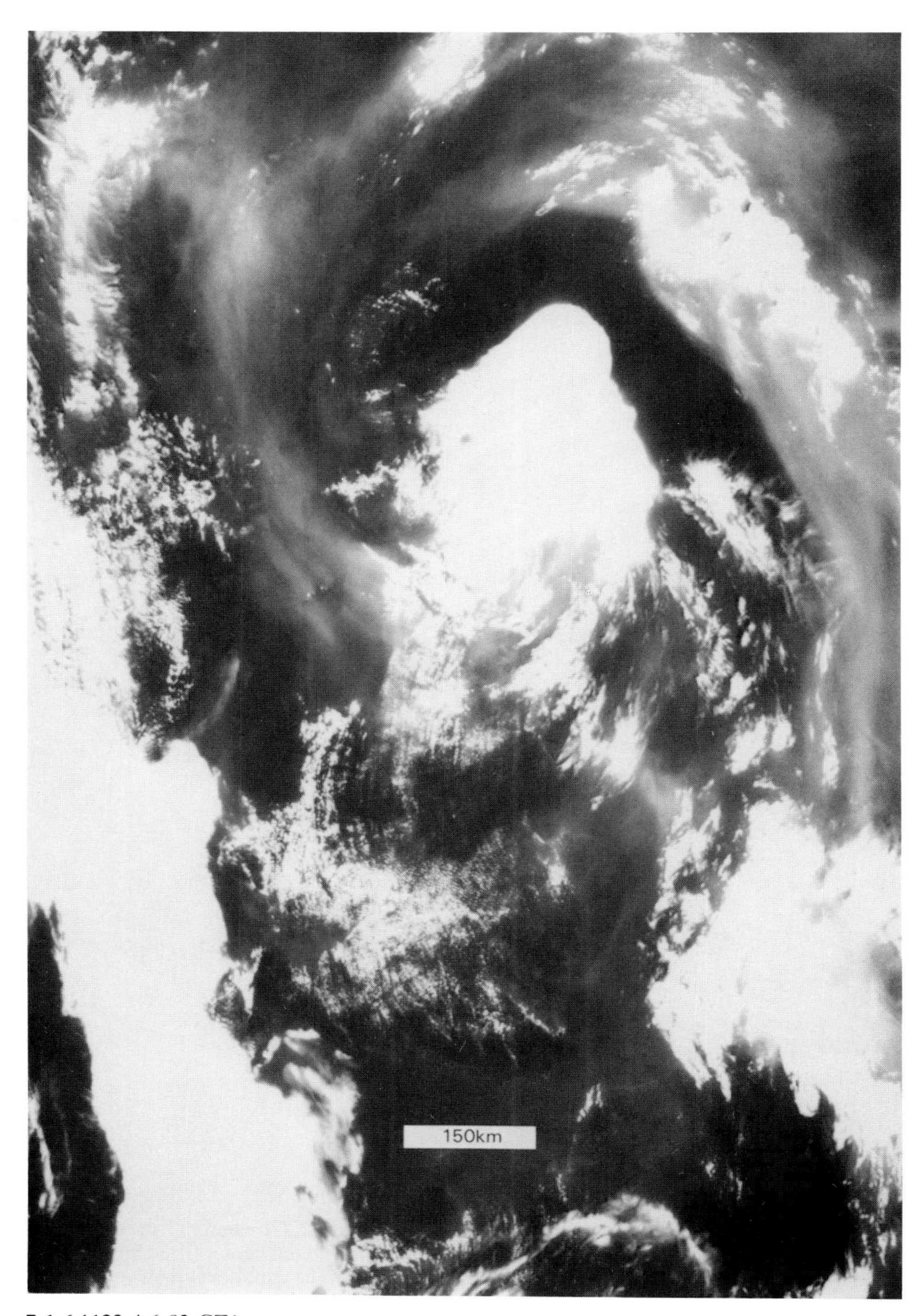

7.1.6 1122,4.6.82,CZ1

7.1.7 1404,4.6.82,2

7.1.5, 7.1.6, 7.1.7, 7.1.8
The small-scale picture, CZ 5, shows the scene the day after **7.1.2** when the stratus lies off eastern Scotland. But CZ 1 shows the great whirl of polluted air has been carried into Norway and Denmark; it also shows the wave and street structure of the cloud over England.

The wave structure has vanished by the afternoon (**7.1.7**) with some cumulus appearing. The stratus has moved up the Scottish coast: it and the haze were not visible by Ch 4, but Ch 3 (7.1.8) separates cloud from haze and distinguishes various cloud types.

7.1.8 1404,4.6.82,3

7.2 SOLITARY CHIMNEY PLUMES

While it is not difficult to find many hot spots in industrial areas or in intense farming areas at harvest time, only a few of these produce plumes which can be seen by satellite. In any case the variegated reflectivity of typical land areas makes it difficult to recognize a plume of dust particles. A plume of water droplet cloud would be seen best by Ch 1 because the land is darker than in Ch 2. Cloud is generally brighter relative to the rest of the scene in Ch 1 than in Ch 3, with the possible exception of a plume heavily laden with SO_3, which keeps small droplets in existence.

It is generally difficult to distinguish any particulate material in the air from surface detail when the particle size is typical of haze except over the sea, which is usually dark. However, when the ground is covered with snow it appears black in Ch 3. Thus the plumes from four large power stations in East Germany have been seen very clearly on several occasions.

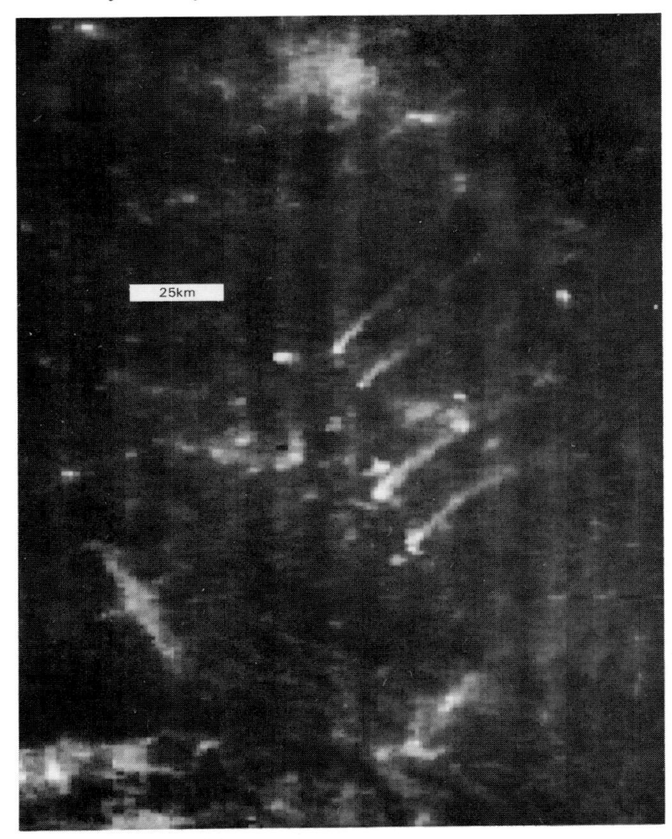

7.2.1 1225,27.2.86,3

7.2.1
The four large power stations in the DDR south of Berlin (the white area at the top) northeast of Dresden (the long area to their southwest) have plumes from 25 to 50 km in length. In the surroundings there are several other hot spots, typical of central Europe.

7.2.2, 7.2.3
The same power stations as in **7.2.1** were identified during examination of pictures of a severe smog incident in the Ruhr area. They are not all operating equally. By Ch 1 they were seen with widening plumes about 22 km in length. Because of the snow backgrounds they were seen more clearly by Ch 3 together with other lesser plume sources nearby.

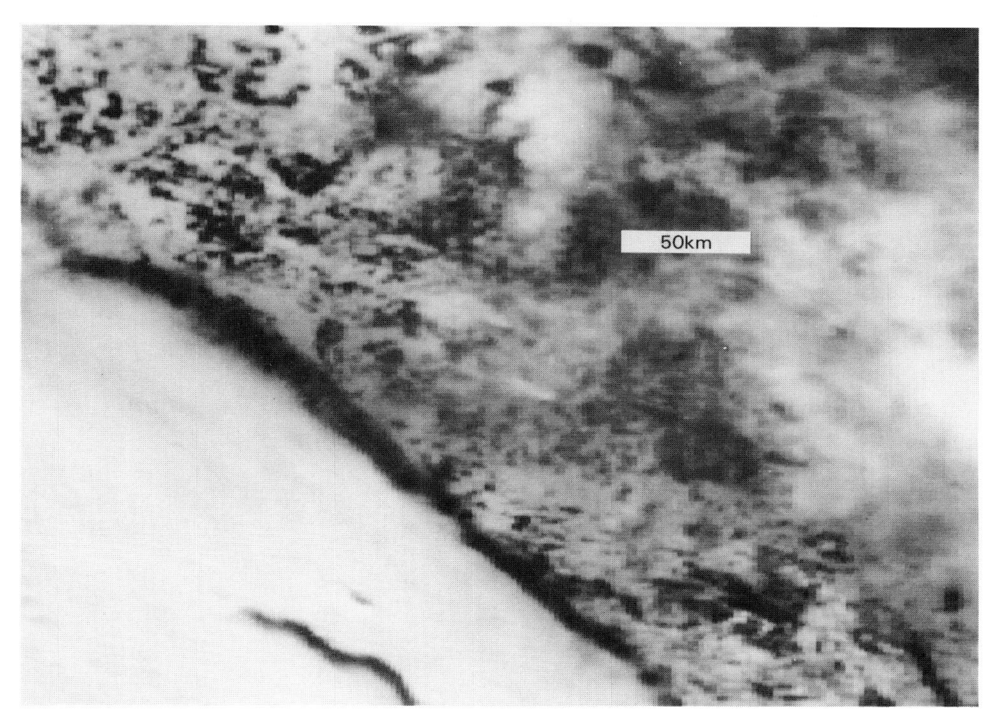

7.2.2 1334,20.1.85,1

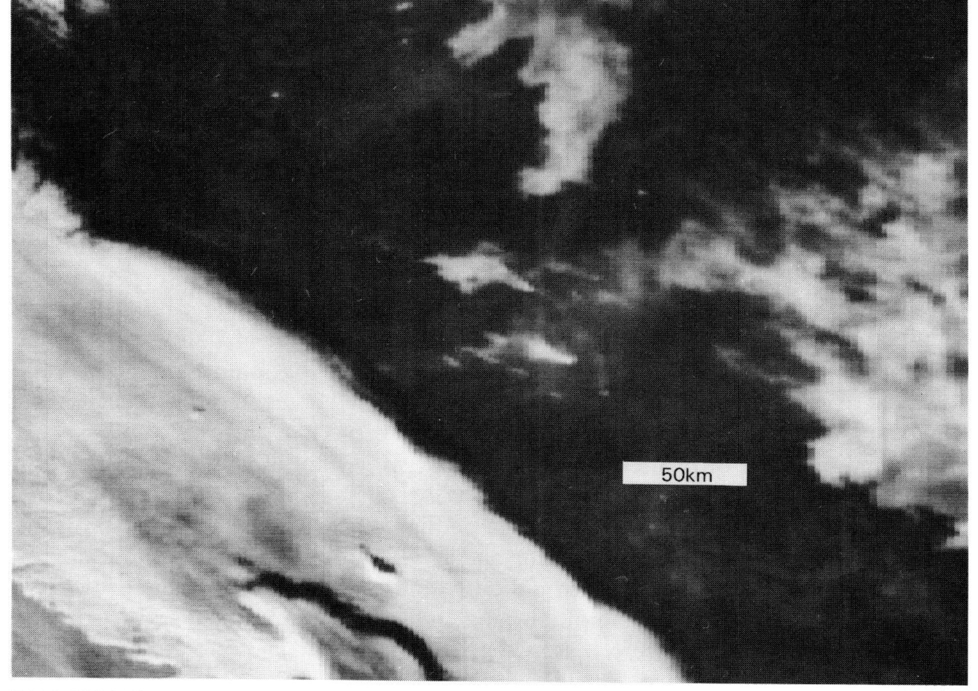

7.2.3 1334,20.1.85,3

7.2.4 0721,20.1.85,4

7.2.4
Early in the morning of the same day as **7.2.3** the plumes were seen by Ch 4 because of the warmth of effluent compared with the cold ground below. The air was almost stagnant.

It is not immediately obvious why the plumes of these power stations are so clearly visible only in Ch 3. The following are the relevant considerations:

— If the plumes had contained water droplets due to the condensation of the product of the combustion of hydrogen in the fuel they would probably have been visible by Ch 1 or 2, even against a snow background, and possibly with the aid of contrast with their own shadows. Dust cloud, by contrast, is much less bright because it lacks scatter by refraction, and would probably not be seen against a well-illuminated background. A dense water droplet cloud might be just detectable by Ch 4 because of its higher temperature than that of a snow surface. A dust-laden plume, like any other, is subject to dispersion, but this could be very slight above a snow-covered surface in very stable air lacking turbulence. The DDR plumes were detected up to a distance of over 35 km, and one, dubiously, to 50 km. A plume of condensed water vapour would not escape evaporation for anywhere near that distance. In this context it should be noted that the evaporation of water droplets causes cooling, which in turn causes buoyant motion which leads to further mixing and evaporation. Thus a wet plume cannot escape evaporation if it is surrounded by unsaturated air. A plume of non-evaporating particles, on the other hand, reflects as much sunshine by primary scatter after dilution as before, if by the dilution the particles are not carried out of the pixel. A dust plume might remain detectable even when spread over two or three pixel widths when the background is dark, like snow in Ch 3.

— The persistence for such a long distance as 30 km implies that the scatter was primary and not much decreased in total by the dispersion. If the particles were not on the average larger than about 2 μm the scatter would scarcely be detectable by Ch 3, and anything much larger would be removed from the plume without much difficulty by electrostatic precipitators. Since all four of these plumes have often been seen together, it seems unlikely that a malfunction of the precipitators was the cause of them being visible by Ch 3. There is the possibility that the particles may be flaky when they are visible in the circumstances described. They would then present a much larger area for the reflection of sunshine but would have a very small fallspeed. The diffuse reflection of Ch 3 by flakes would be much greater than by droplets of the same cross section area because of the strong absorption in water droplets.

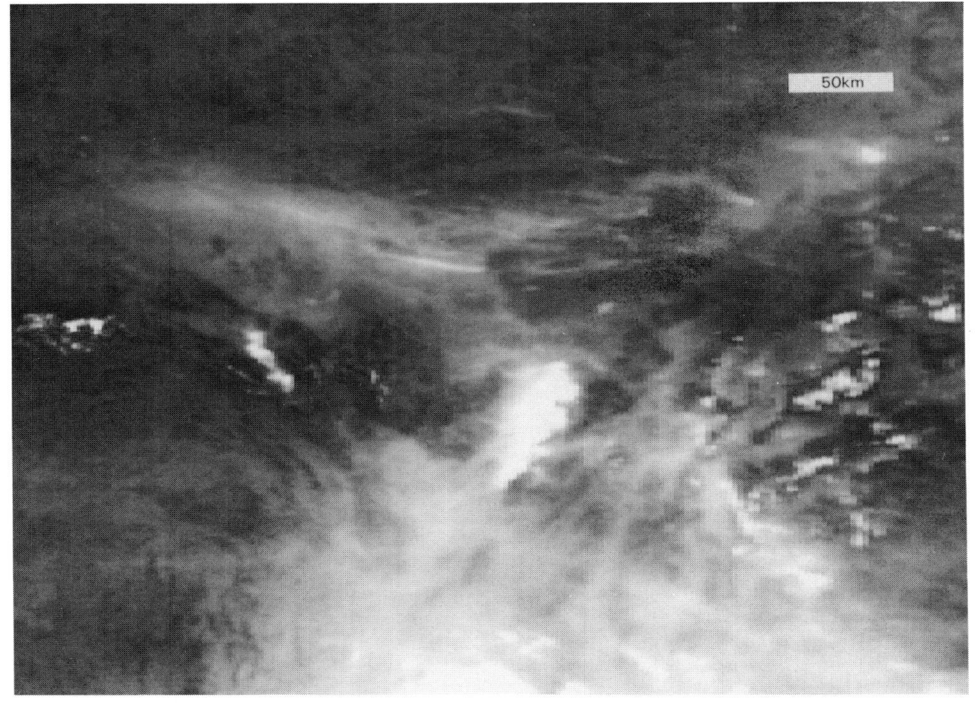

7.2.5 0803,1.4.82,1

7.2.5

Very often individual plumes can be seen in a highly polluted industrial area; but the illumination, together with its composition, will determine which channel is best to observe it. Most often the brightest plume is one containing water droplets, because that gives the best reflection of sunshine. It also depends on the angles of illumination and view, and the droplet size and background illumination. This is a picture looking towards the early morning sun, which is a good situation for detecting haze. Later in the day it would be difficult to see the plume against the detail of the ground in sunshine, and the solar heating may often evaporate such plumes.

The clouds in the centre and left of this picture are a common occurrence associated with industry in Bohemia. On the right is the industrial region of Upper Silesia, which is a major source of visible pollution. The smoke is drifting north-west wards and is being reinforced by sources which produce condensation and visible plumes around 50 km long.

— The power stations whose plumes are observed burn lignite, which has a much higher ash and sulphur content than black coal. It is likely, therefore, that the plumes have a much higher SO_3 content than the plumes from British power stations, from which no plumes have been observed. The control of the production of SO_3 in the furnaces and flues of large power stations has only been achieved in Britain in quite recent years, and so there may remain many stations where quite important amounts of SO_3 are produced. If these plumes were composed of SO_3 particles or droplets of sulphuric acid which do not evaporate, and which are large and numerous enough to scatter Ch 3 by diffraction, we have an alternative explanation of their brightness.

Some bright plumes are seen from time to time within the general haze, and these can be attributed to industries which emit a large amount of water vapour. Their sudden disappearance by evaporation rather than a fading by dilution confirms that they are probably composed of condensed water droplets.

Many of the plumes observed by Ch 3 have the appearance of coming from hot spots, but this cannot be inferred with certainty because a cloud formed above wet

7.2.6 0801,21.1.80,2

7.2.6, 7.2.7, 7.2.8
In the Ch 2 picture the cloud in the centre is over Katowice in Upper Silesia and is probably caused by the concentration of industry. That it lies both below an anticyclonic inversion and above a strong winter ground radiation inversion is shown by the Ch 4 picture which shows it to be warmer than the ground. Thus Ch 3 has the advantage, and in a magnified version we find several plumes of water droplet cloud in the neighbourhood, even near the edge of the pass where the pixel size is enlarged.

7.2.7 0801,21.1.80,4

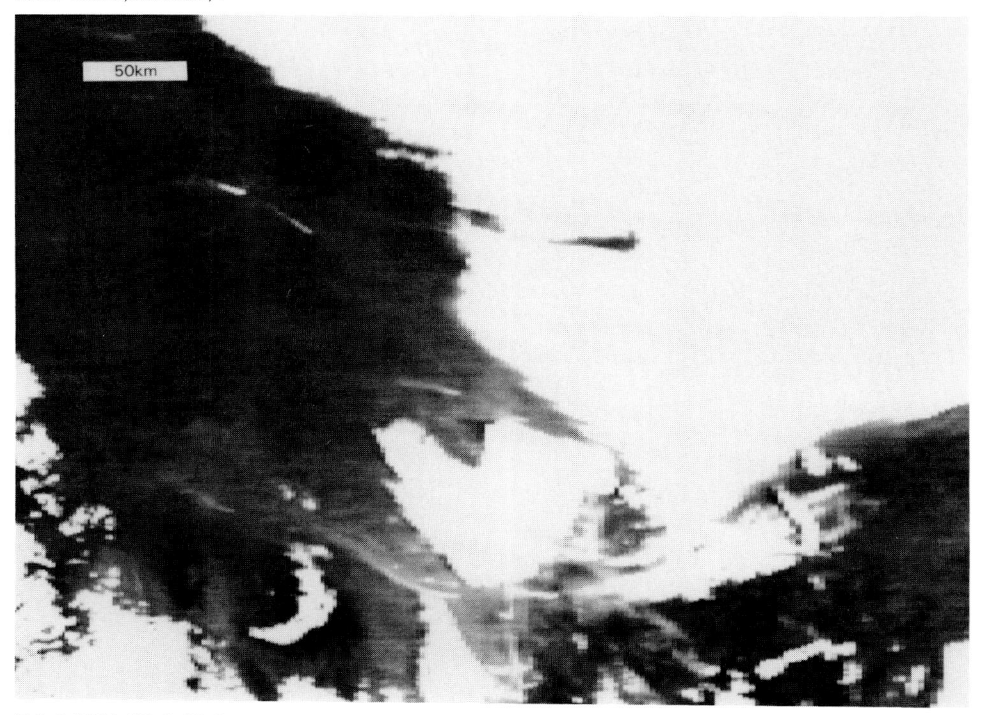

7.2.8 0801,21.1.80,3

cooling towers could produce as bright a spot in sunshine. Inspection by Ch 1 or 2 would usually reveal a white cloud if there was one.

No examples of plumes from solitary chimneys seen by satellite have been found in Western Europe, where the ash and sulphur content of the coal is much less than in East Germany and Czechoslovakia in particular. (But see **6.4.13-17** where very clean air invaded Holland and northwest Germany and plumes were clearly seen. These, however, were probably caused by the large input of condensation nuclei rather than ash or moisture.) The four plumes in the first pictures of this chapter have occasionally been observed at sunrise, when the plumes, well above the ground which received only glancing sunshine, were brightly illuminated and easily distinguished from the darker background. This, like snow on the ground, is a chance circumstance which cannot be relied upon for routine purposes.

Many streaks of haze are produced by industrial complexes, and observations of these are described in section 9.5.

8

High-altitude trails

As already explained in section 4.3, in most circumstances ice clouds are created by the freezing of clouds of supercooled water droplets. This usually produces fallstreaks, either from supercooled layer clouds, in which case fallstreak holes are produced, typically when the passage of an aircraft initiates the freezing; or from supercooled castellatus clouds, in which case the tower may or may not disappear. In all cases the ice crystals fall to below the condensation level.

The size of the ice crystals is determined by the circumstances immediately following the freezing, and this makes an important difference to their fallout properties and the appearance of the subsequent ice cloud.

8.1 OROGRAPHIC CIRRUS

This section describes the natural mechanism which creates an ice cloud trail from above a mountain. It is quite a common feature of jet streams, and the cloud produced is not easily distinguished from the usual frontal clouds, which have lines along the direction of the wind shear (thermal wind). This direction is usually close to the line of the front, but if there is significant ageostrophic outflow at the top, fallstreaks appear to slope down the frontal surface. Sometimes the front has a significant velocity transporting it forwards, in which case orographic cirrus trails lie along the wind direction at their level, which would be outwards from the warm air of an advancing warm front. The cloud lines formed by the thermal wind lie along the front; the orographic cirrus lies at an angle to it.

In a jet stream running around a stationary (warm) anticyclone the orographic cirrus has an anticyclonic curvature which indicates that the anticyclone is not moving.

In the case of a long mountain barrier such as the Alps, high-level flow across it is usually accompanied by blocked low level-flow, so that at middle levels we see cloud now far from mountain top height evaporating on the lee side in a series of short wavelength lee waves. As it descends the lee slopes it is decelerated as it moves into higher pressure; this is made possible by the ascent of air at higher levels. The cloud in this high-level ascending air is orographic cirrus, and because of the blockage the air that does traverse the mountain range does not usually (and certainly not quickly) return to the level from which it originated upwind. How the flow achieves this flow pattern requires quite complicated mathematical analysis to explain; the simple description given here is merely intended to make the mathematical description of the steady flow which contains these features appear plausibly to represent actuality.

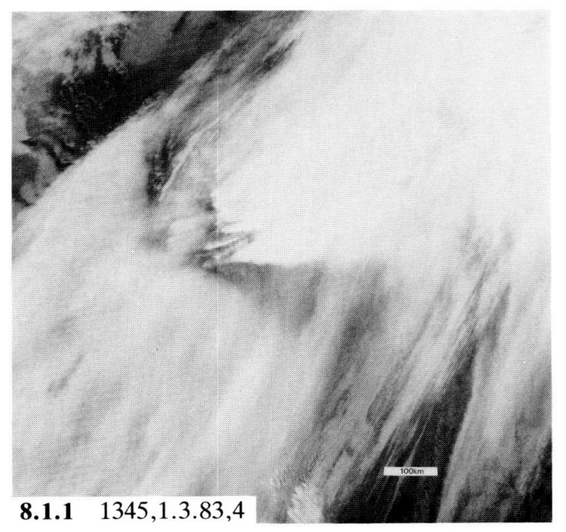

8.1.1 1345,1.3.83,4

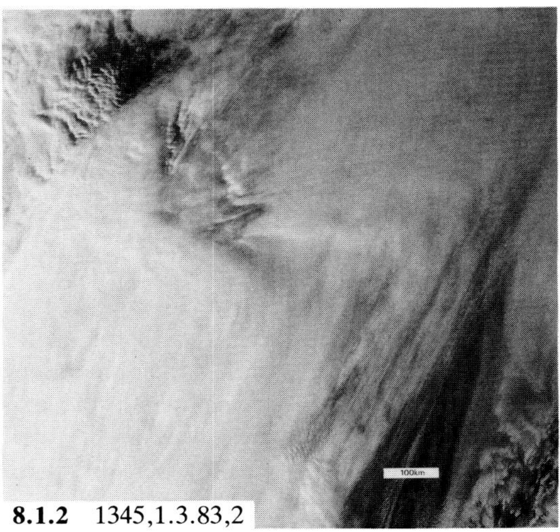

8.1.2 1345,1.3.83,2

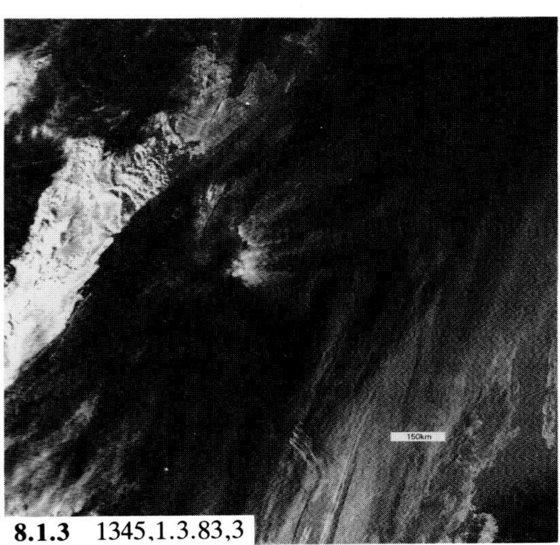

8.1.3 1345,1.3.83,3

8.1.1, 8.1.2, 8.1.3

The thermal wind is along the jetstream, which is from southwest to northeast across Iceland in this case. Iceland is at a point of recent occlusion with a narrow warm sector to the south advancing eastwards. The cloud lines display the effect of many hours of thermal wind from the southwest. A small wave begins to be carried across Iceland and thin orographic cirrus is carried across the thermal wind streaks. Notice also the dip and evaporation at cirrus level as the air at the surface makes its ascent of the mountains.

The Faroes have their own display of lee waves indicating a WSW wind at low cloud level and orographic cirrus indicating a NNW wind relative to the ground.

This orographic cirrus is thin and does not completely obscure the structure of the cloud below. It does so much less in Ch 2 because the cirrus is darker than the cloud below, and also because there is strong absorption of the message from below in Chs 3 and 4. The jetstream lines seem to be at a different level from, but even so fairly close to, the level of the orographic cirrus and its lines, and are therefore not fully obscured (see **3.1**, **3.2**).

Although much spoiled by interference, the Ch 3 picture nevertheless contains remarkable detail.

8.1.4 1235,13.4.86,CZ5

8.1.4
Orographic cirrus shows anticyclonic curvature on the northeast and east side of a warm anticyclone. It is particularly noticeable in this case that the cloud first appears where the surface air descends from the Greenland plateau. This is a typical springtime picture, with the ice carried from the northeast down the coast to Cape Farewell becoming very mobile.

Not shown in this picture, but detected at 1609, were some contrails at the southern end of the trail. In Ch 3 they were white (with dark shadows) while the cirrus trail was dark.

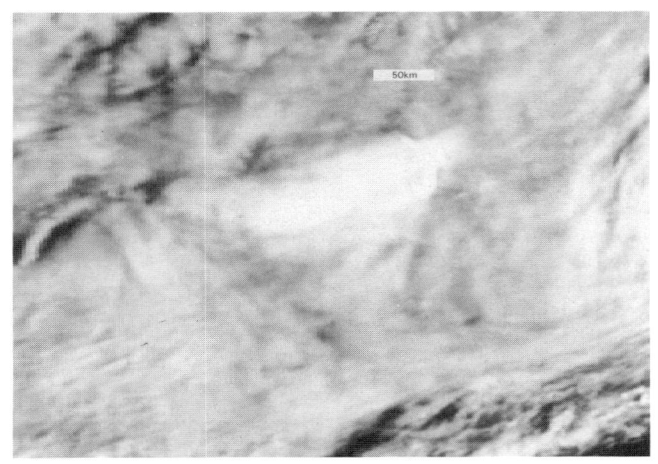

8.1.5 0816,14.4.84,2

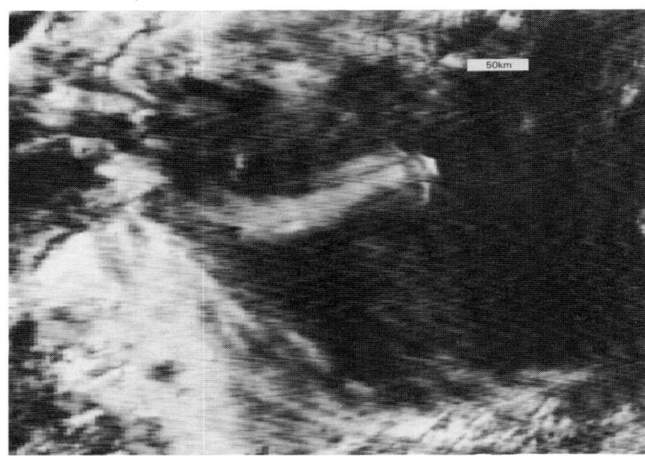

8.1.6 0816,14.4.84,3

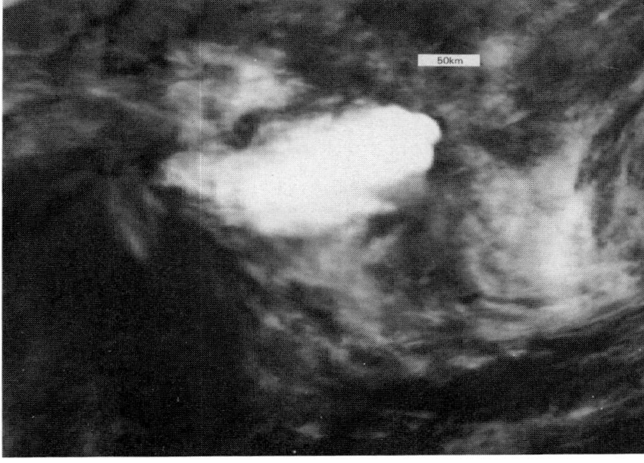

8.1.7 0816,14.4.84,4

8.1.5, 8.1.6, 8.1.7, 8.1.8, 8.1.9, 8.1.10

The first sign of the growth of orographic cirrus over Jan Mayen on this occasion was at 0413, when the arc cloud was present without a plume. By 0816 the plume had extended about 200 km downwind, and by 1408 was over 500 km in length and almost 200 km at its widest. Jan Mayen is 55 km in length but in this case was lying almost along the wind with a width at sea level of only 20 km. Mountain and lee waves are quite commonly of the shape indicated here, but the size of the plume is exceptional. This was because a front was lying almost along the wind in this neighbourhood with a low just east of Iceland and south of Jan Mayen.

The Ch 3 pictures are specially interesting because they reveal smaller particles near the middle of the plume where the cloud is white, than at the outer reaches where it is black. Ch 2 pictures show slightly greater transparency of the cirrus than in Ch 4.

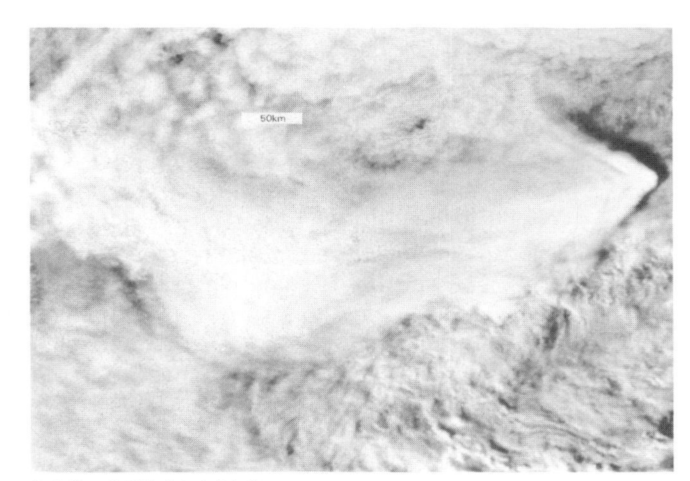

8.1.8 1407,14.4.84,2

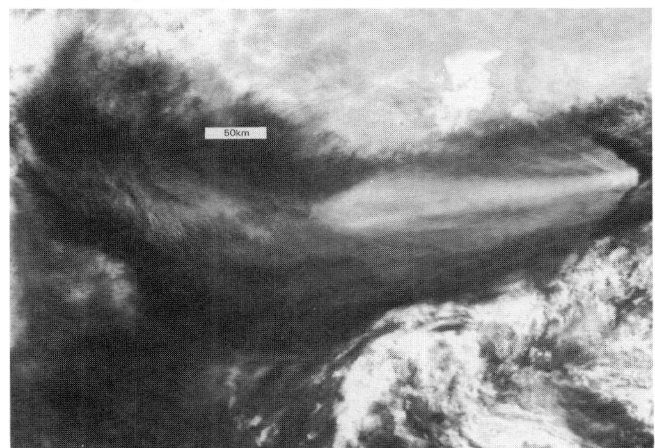

8.1.9 1407,14.4.84,3

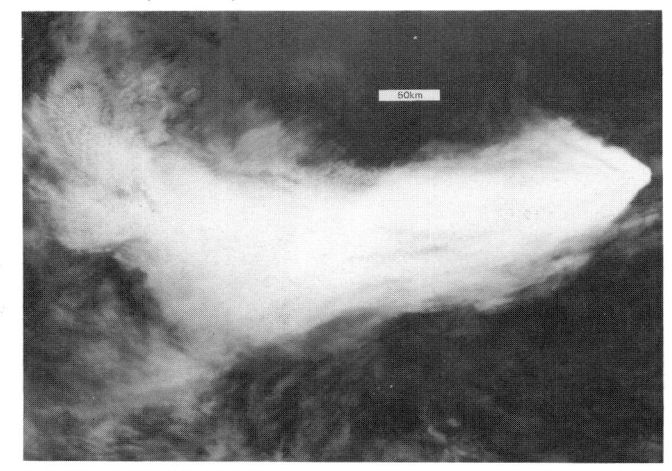

8.1.10 1407,14.4.84,4

That circumlocution is to avoid the logical error of stating that a mathematical solution to the steady-state problem must necessarily represent the actual flow, even when it possesses many of the desired features. It could be said, however, that this solution is an 'attractor', and this may justify the claim that it is the solution in the minds of those who rather like this modern terminology. Because of its respectable history this attractor does not deserve the attribute 'strange'.

In cases such as that just described the air was not saturated with respect to water when it was on the upwind side of the mountain, and the altitude to which it ultimately descends on the lee side plays a part in determining the particle size. Cirrus does not often take on a grey appearance in Ch 3 pictures of these circumstances and when it does it indicates a smaller than usual particle size. Pictures of such cases have to be examined carefully, because if it acquired about the same brightness as the cloud below it, it could become virtually invisible and could seem to be more transparent than it actually is in Ch 3. The situation is complicated by the fact that the brightness depends very much on the angle of view in relation to the sunshine.

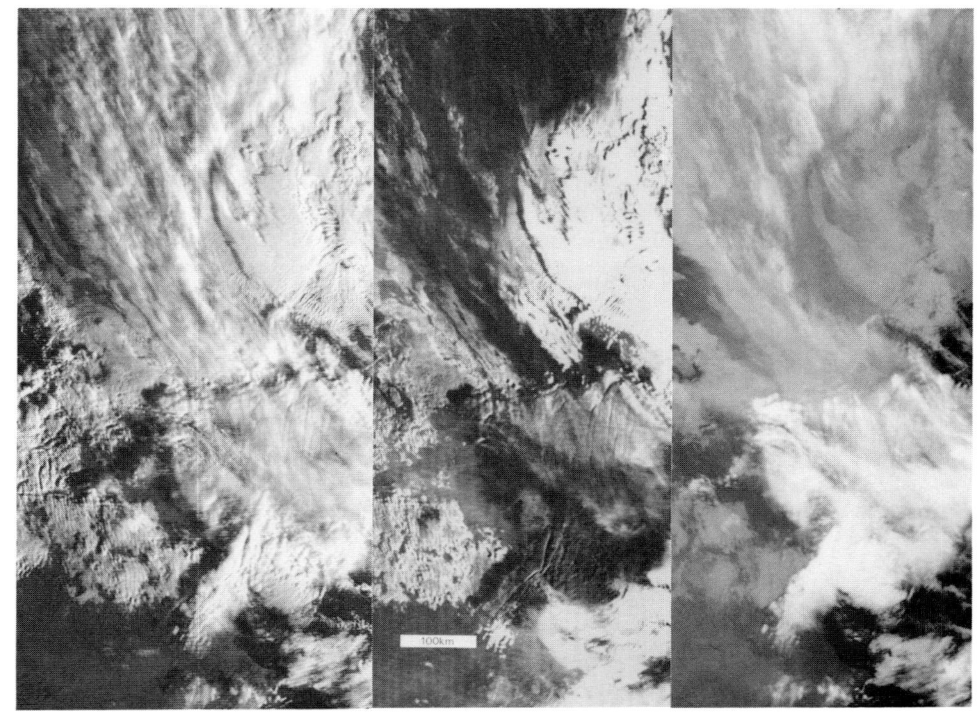

8.1.11 0829,26.11.87,2 **8.1.12** 0829,26.11.87,3 **8.1.13** 0829,26.11.87,4

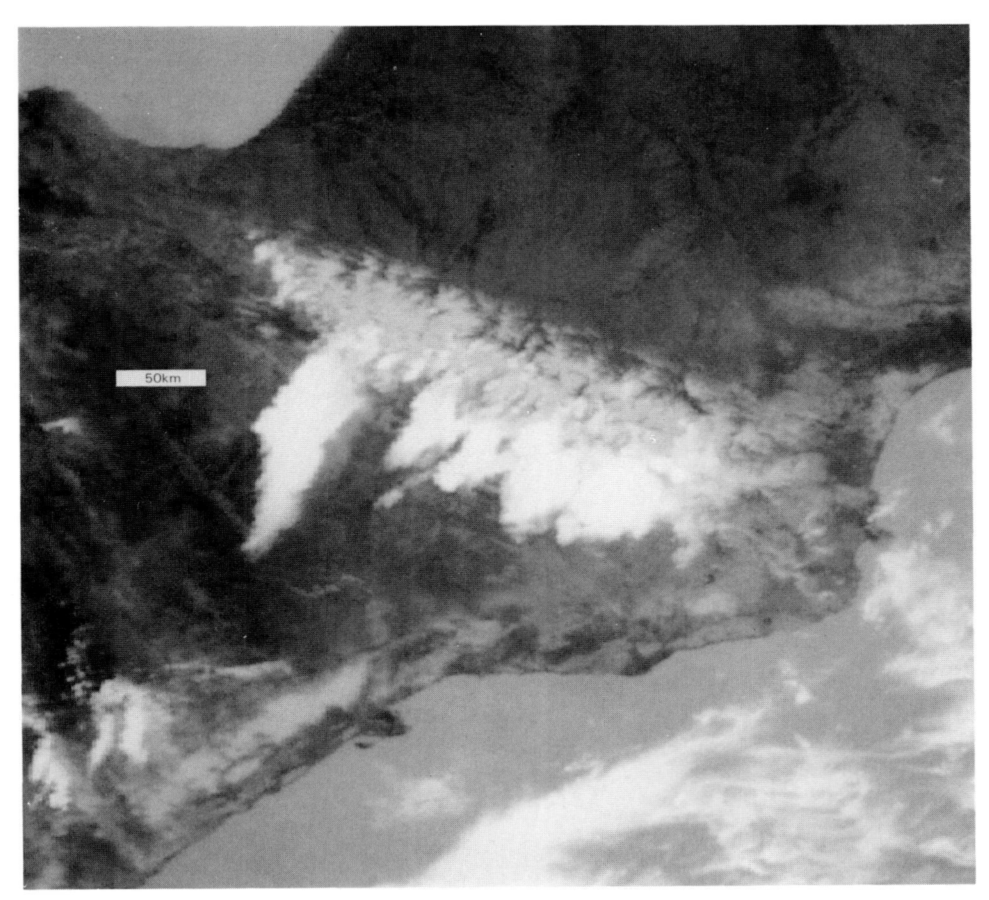

8.1.14 1434,20.4.84,4

8.1.11, 8.1.12, 8.1.13, 8.1.14
There is a great deal of detail in this group of pictures of cloud in the lee of the Pyrenees. Cloud at three or more levels is revealed by inspection of the three main channels, with streamers of cloud in at least three different directions. Unfortunately Ch 3 is spoiled by two systems of interference lines close on either side of the horizontal (left to right, i.e. parallel to top).

There are lee waves in the lowest clouds; plumes are beginning on peaks; the bottom boundary of the picture crosses the coast of Spain at Cape de la Nao (15′E 38° 45′N) and the sides of the picture are parallel to the line Valencia–Toulouse; the black area in the northeast in Ch 4 is sea and a piece of coast at Perpignan clearly visible.

In Ch 4 the western end of the high cloud over the Pyrenees is a few miles west of the Pic du Midi d'Ossau (2855 m at about 30′W). The cloud just north of Valencia, where the Spanish coast is obscured by it, is generated where the surface air descends to the Costa del Azahar from heights around 1500 m and is denser than the clouds in the lee of the Pyrenees, from which there must be separation of the airflow and upwind of which the ground is at a much lower level.

This case is typical of what may be seen in the lee of the Austrian Alps (see **4.3.1/2/3**), with almost transparent cirrus indicating at least a 45° variation of wind direction with height. The region is illustrated in a quite different context by **8.1.14**: the pressure gradient is slack and anabatic winds feed large cumulus on the Pyrenees and, with a sea breeze, upslope stratus on the Castellon mountains north of Valencia.

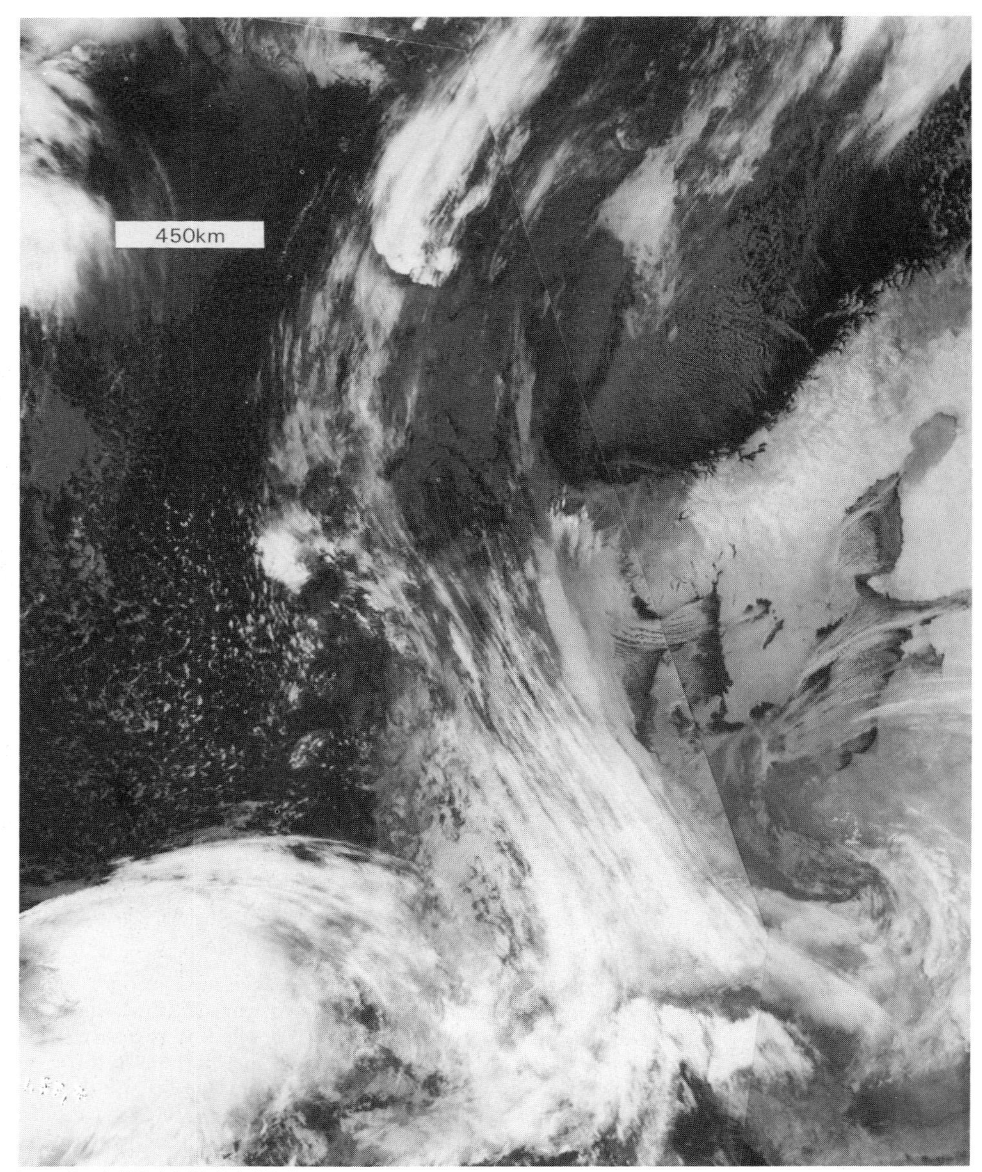

8.1.15 1437-1257,9.1.87,4

8.1.15
Orographic cirrus often throws unexpected light on jetstream cirrus. In this case the cirrus plume from Iceland indicates a southerly wind while that over the Alps a wind from the NNW. The jet is at a standstill somewhere in between! Although joined in the middle, these two portions of jetstream cirrus originated in fronts which had been widely separated, but both blocked by the Scandinavian high. As the fronts become almost stationary the wind becomes parallel with the jetstream's thermal wind.

A quite separate feature of this episode was the 'jet' of air containing cumulus coming out of the Skagerrak. It deposited a very heavy snowfall on a rather short stretch of the coast midway between Newcastle and Edinburgh, with much lighter falls elsewhere.

8.2 CONTRAILS

When they are first formed behind an aircraft, condensation trails are composed of a very large number of very small particles because the exhaust contains a great abundance of condensation nuclei. That it is initially unfrozen can sometimes be confirmed by the appearance of iridescent colours if the angle of view is appropriate to the direction of sunshine. But iridescence is not seen except at the very front of the trail because the spectrum of particle sizes is rapidly widening. After being frozen they often produce hexagonal prisms or plates with vertical axes in which brilliant mock suns are seen, again if the angle of sunshine is right. That they do not contribute significantly to the 22° halo indicates that the axes are predominantly vertical. They have been shown, in several published pictures, close to circumzenithal arcs, which are produced through the horizontal and vertical faces of vertical prisms, but rarely contributing to the arcs. It will be noticed that these remarks are hedged about with qualifications such as 'sometimes', 'occasionally', 'usually', and so on; this is intended to imply a corresponding degree of uncertainty in the expectations resulting from many years of wide-eyed observing. The implication concerning the circumzenithal arc is that the ends of the prisms are not horizontal, but may be pyramidal inwards or outwards, in which case they might contribute to the upper or lower tangent arcs of the rare smaller haloes of 8°, 17°, 19°, etc. (See ref. [14].)

The shape of crystals is influenced by the temperature and rate at which they grow. Thus we find dendritic crystals forming snow flakes in freezing wet clouds. Contrails are unusual in having initially a superabundance of very small droplets, which may all freeze very quickly if the temperature is much below −40°C. Alternatively they may freeze slowly, in which case some have time to grow much larger among those remaining unfrozen. The time available for the transformation of the particle size spectrum depends on the rate of dilution by surrounding air and the ambient humidity, for this determines how long the crystals remain in air supersaturated for ice. If the exhaust from outboard engines (near the wing tips) is entrained into the tip vortices the ambient pressure is lowered and the mixing delayed; thus vortex tube trails may persist for nearly a minute even when unfrozen, and frozen trails may disappear rapidly if the engines are central so that the exhaust is more quickly mixed, and the ambient vapour pressure is below ice saturation.

The consequence of this is that contrails may present a variety of different appearances in Ch 3 pictures. In Ch 2 pictures they are normally white, but less bright than layers of cloud below and therefore difficult to identify; but their shadows are often very obvious as dark lines on the cloud below. In Ch 4 pictures they are usually very clearly identified as being at least as cold as any cirrus that may be present, and they are opaque to Ch 4 radiation and do not let through emissions from the cloud below, which is almost certain to be darker. They are therefore as white as or whiter than any cirrus that may be present. In Ch 3 pictures they appear black, like cirrus, if the particles are old large crystals of at least 10 μm, but they may appear almost white if the crystals are more numerous and small, in the region of 1 μm or less; at sizes in between the trail may be invisible or not according to the brightness of the cloud or surface below. Usually, however, their shadows are very well defined. Many contrails are less than one pixel in width. To be observed they must be significantly brighter (in Ch 2 or 4) or darker (in Ch 3) than cirrus has to be to be noticed.

In Ch 3 contrails may be either dark, like cirrus, or white, like clouds with small particles. In my experience dark trails appeared to be more common, and I searched for white ones which seemed to be a scarcity. But in recent years white ones have appeared to be more common. In addition to the possible unrepresentativeness of the earlier studies it is possible that aircraft design changes may have played a part in altering the rate of dilution. For example the by-pass engine certainly causes a more rapid dilution in the early stages by applying a greater volume of air directly to the exhaust. The newer three-engined aircraft place one-third of the exhaust in the centre of the wake and this is immediately spread around the *outside* of the accompanying air of the vortex pair which the aircraft leaves behind. This air is more rapidly mixed with the exterior, non-accompanying, air and enters the curtain of exhaust eroded from the accompanying fluid and left above it ([10] sections 9.10, 11.3, 11.4). Both of these effects decrease the importance of water vapour in the exhaust relative to that in the ambient air in the early seconds or minute or two after the creation of the trail, so that crystal growth after freezing would be reduced. The trail would then be more likely to appear white because of scatter rather than dark because of absorption of the Ch 3 component of sunshine.

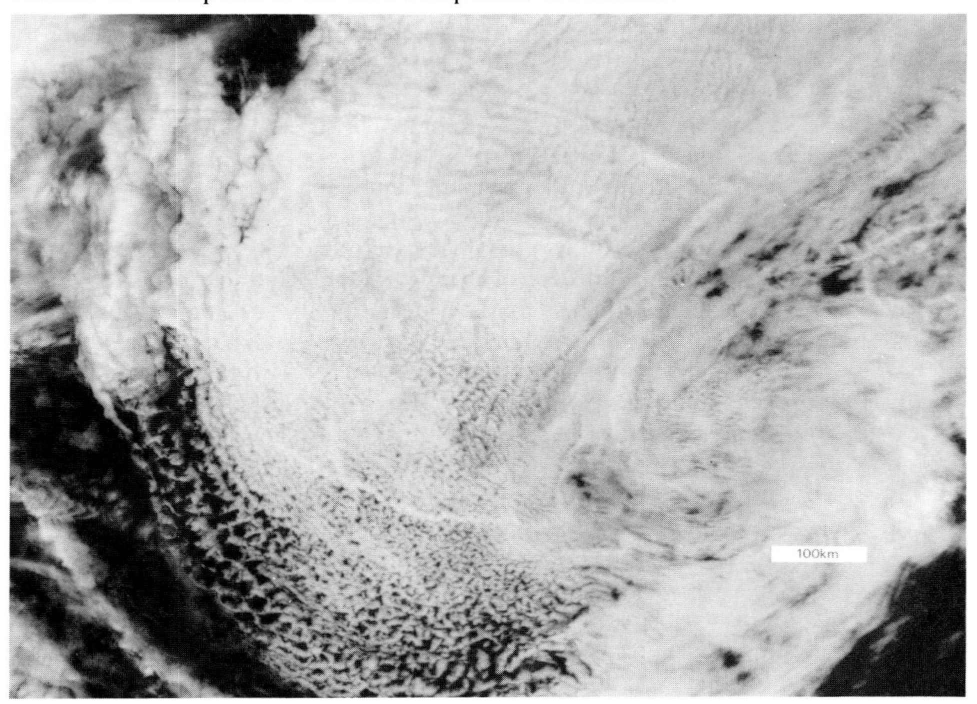

8.2.1 1707,30.3.87,2

8.2.1, 8.2.2, 8.2.3

These pictures are centered at about 50°N 36°W. They show an area of clean air with contrails, which are white in all three channels, in the north and some ship trails, which are scarcely visible by Ch 4, in the south. The newest parts of the ship trails are brighter than the rest in Ch 3, but not in Ch 2. There is an old trail and a few other trail fragments at the southern extremity of the area which are detected by being whiter only in Ch 3. Likewise the cirrus among the contrails is distinguished from them by being dark only in Ch 3. The low cloud in the north which is part of the same air mass is distinguished from the clean air by its whiteness (it has been polluted) only in Ch 3.

8.2.2 1707,30.3.87,3

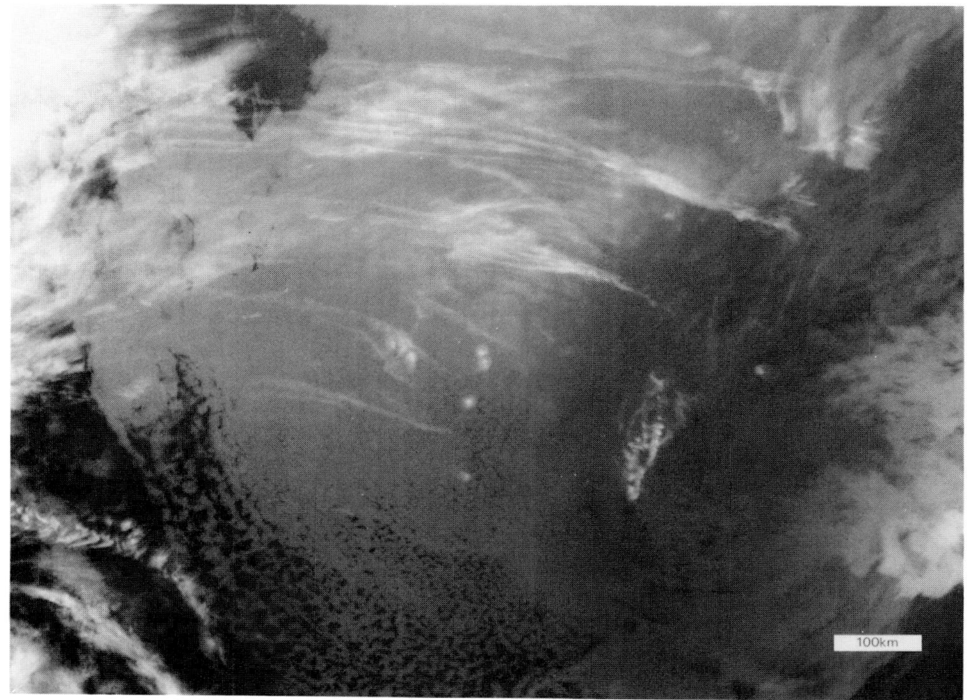

8.2.3 1707,30.3.87,4

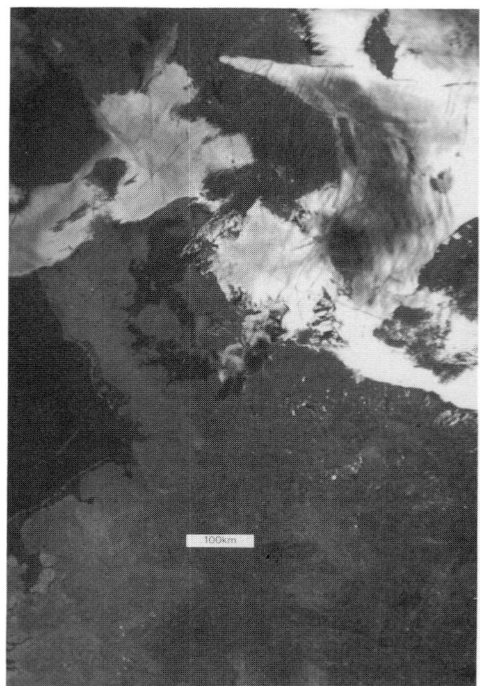

8.2.4 0817,24.4.87,3

8.2.5 0817,24.4.87,4

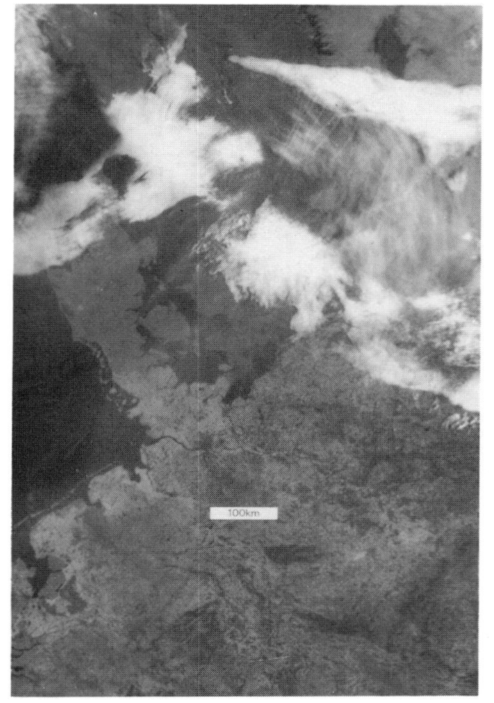

8.2.6 0817,24.4.87,2

8.2.4, 8.2.5, 8.2.6

When a contrail is above a dark surface, such as the ground over southern Sweden, we may see it in Ch 3 even if it is rather dark, but over glint or bright low cloud we see the shadow, not the trail. Thus over a dark surface the Ch 3 trails are in the same place as the Ch 4 trails, but over a bright surface their shadows appear about 17 km further west in these early morning pictures; the distance depends on the orientation. The cirrus is dark in Ch 3.

In Ch 2 we see the white trails over a dark surface but shadows on a bright surface, and these are weaker than the Ch 3 shadows. The cirrus is white in Ch 2, and it obscures vision of the island of Oland in Chs 2 and 4. The lake Vattern (into which Oland fits, according to Swedish mythology) is detected in all three channels, but much of it appears to be covered by fog or weak glint.

Several hot spots are seen in the DDR. These cannot be clouds, such as those produced by cooling towers, because they do not appear in Ch 2.

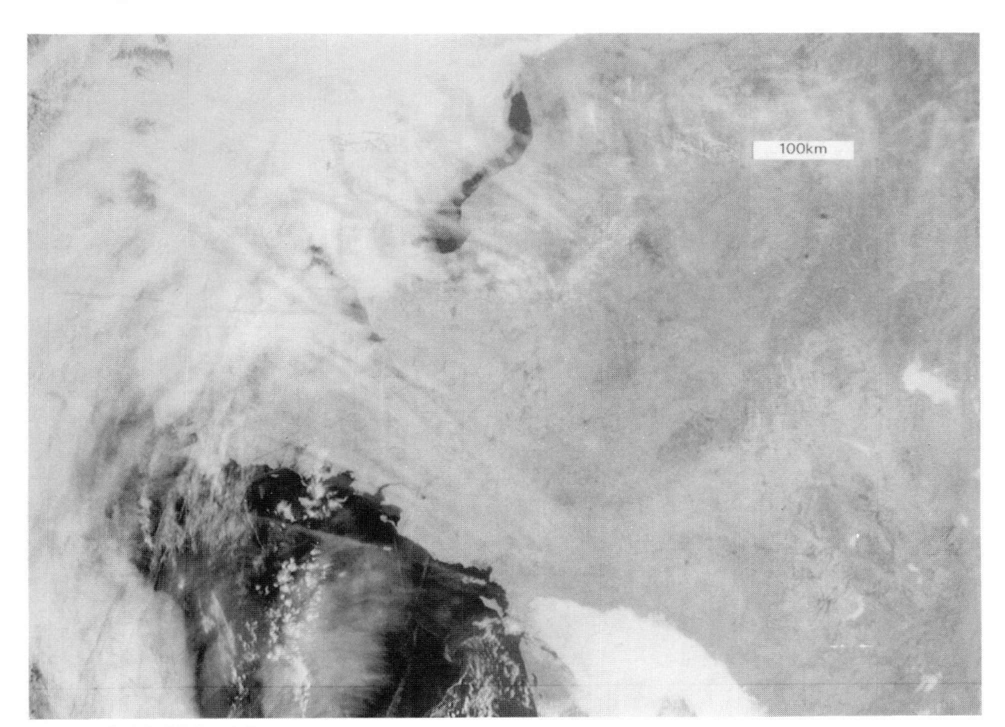

8.2.7 0829,27.9.83,2

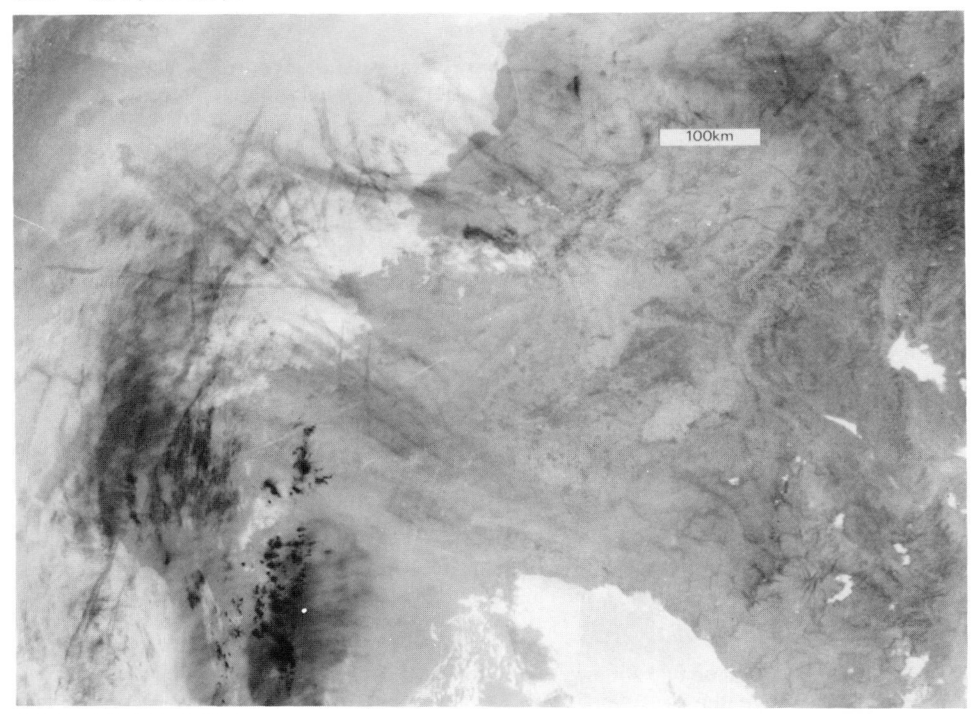

8.2.8 0829,27.9.83,3

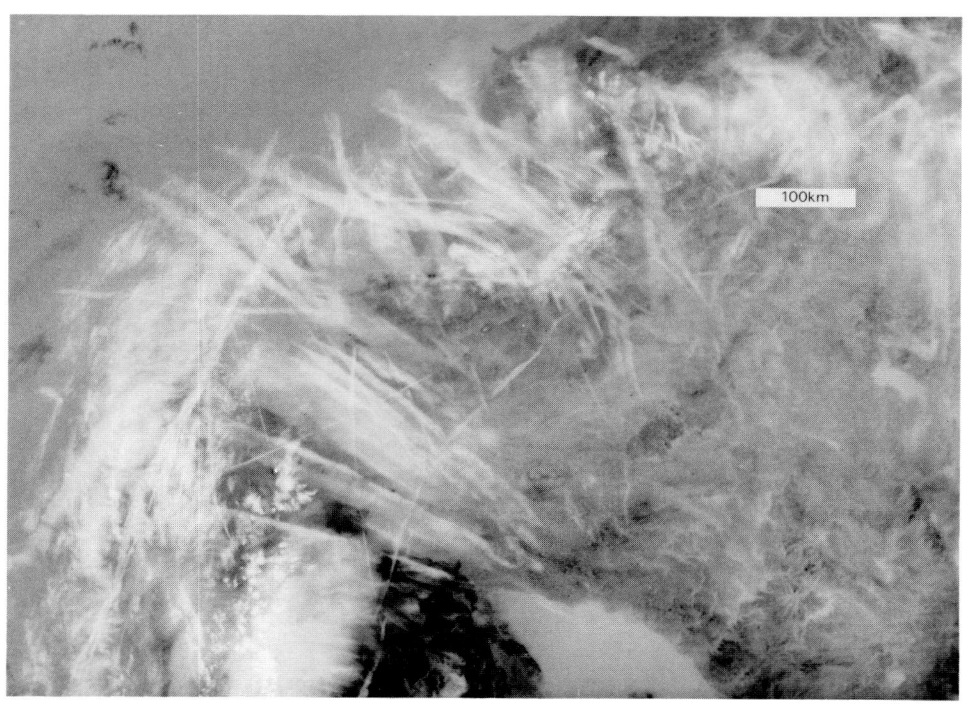

8.2.9 0829,27.9.83,4

8.2.7, 8.2.8, 8.2.9

Successive flights along the same route tend to produce parallel contrails, which might be mistaken for a contrail and its shadow. Occasionally a busy area may contain a very complicated collection of trails, as in this example centered close to the island of Jersey.

Most of the French coast is buried in sea fog except for a stretch from Boulogne to just beyond the mouth of the Seine, a small part of the Cotentin peninsula, and the southern coast of Brittany.

In Ch 2 we see a few white trails over France with dark shadows 25 km distant in the low morning sunshine where their orientation is north–south, but much less where the trails are more nearly along the direction of sunshine.

In Ch 3 some trails are dark and some white. The castellatus is black in the south, very black in fact, indicating large particles.

Ch 4 makes no shadows and therefore only one representation of each trail, which can be seen to be displaced relative to the shadows which are more prominent than the trails in Ch 2. Ch 4 does not separate the sea and the sea fog.

8.3.1, 8.3.2

The sunshine is roughly from the SSW in this picture, and the separation of the line of a trail from that of its shadow may be up to 40 km. The picture is over western France. The looped trails, which have been seen several times, but always white in Ch 3, are off the south coast of Brittany. Several other white trails which have obvious shadows are seen to the east. The type of trail that is looped in this picture is probably from single-engined military aircraft on reconnaissance.

The fog over Aquitaine had some very odd shadows cast on it, the trails themselves being almost invisible above the white cloud.

These trails were not present when the area was seen by CZCS two hours earlier.

8.3.1 1328,17.12.85,2

8.3.2 1328,17.12.85,3

8.3 LOOPED CONTRAILS

Occasionally trail loops are seen which are somewhat thinner than typical airline contrails and are probably made by much smaller aircraft. See 8.3.1, 8.3.2.

There are many occasions when they are very clear in Ch 3 but can only be identified clearly in Chs 1, 2, 4 and 5 after detection in the Ch 3 picture. This is particularly true when there is a great deal of thin streaky cirrus present and the trails and cirrus are both bright in Chs 2 and 4, whereas the trails may be white and the cirrus black in Ch 3. When there is no cirrus but a good cover of low stratiform cloud and trails are black in Ch 3, the shadows are also seen as black duplicates, while in Ch 2 only the shadows are seen, and in Ch 4 only the trails are white against a white background. On the other hand if the trails are neither white nor dark in Ch 3 they may be less easily seen in that channel than in the others.

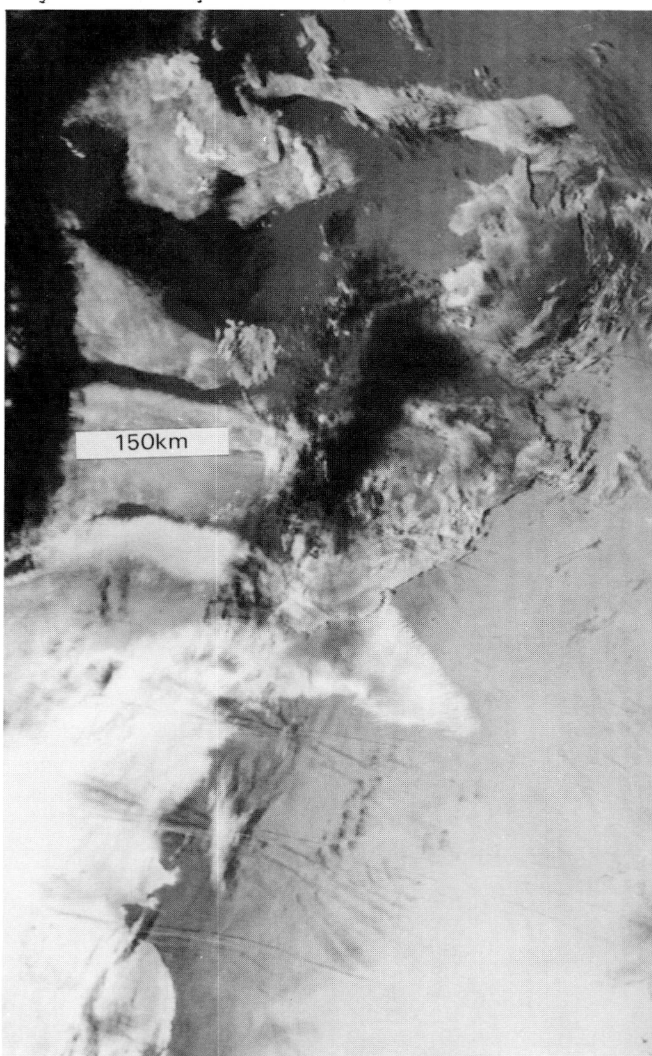

8.3.3 1659,7.5.87,3

8.3.3
This is a good example of duplication of contrails by their shadows in Ch 3. There are also a few fragments of cirrus with the same effect. The sea is bright, this being on the edge of a large area of glint; it is at about 45°N 55°W.

9

Haze, dust, glint and smoke

9.1 GRIT, DUST AND BLUE SMOKE: THE USE OF IR

Haze is usually transparent to IR radiation; this, of course, is one of the reasons for using the red rather than the blue end of the spectrum for routine meteorological observations of cloud systems. The emissions from industrial, urban and agricultural sources have increased since the industrial revolution got under way two centuries ago. The use of fuel has become excessive, the heat or electricity often being very inefficiently employed, much of it frivolously, even where the combustion itself is efficiently performed. Haze and dust have been transformed from a mere obstacle to meteorological and oceanographic studies into an object of great concern.

Sometimes clouds of dust can be seen by IR, and a brief note of this fact is relevant, if only because it makes some useful night observations possible for tracking their movement. The most outstanding examples are provided by the great deserts. Although much better observations are obtained by shorter wavelengths, the fact that dust clouds are not always transparent to IR means that some carry a much greater load and larger particles than others.

Dust is convected up to 3000 m or more in the Sahara Desert; this has been exemplified by the hiding of the highest peak of the Canary Islands in the dust cloud. The appearance of lee waves on the dust cloud top on one (perhaps rare) occasion ([1], 9.30) indicated that there was a very strong inversion of potential temperature at the top of the cloud. It was the upper limit of the convection over Africa and continued to be the top of the layer which was being cooled by radiation from the dust.

Most of the dense industrial hazes frequently seen from the air are much shallower than these examples and they are a much smaller barrier to the direct passage of rays of all wavelengths. The particles raised by the wind are generally larger than industrial haze particles, which are mostly smoke and photochemical haze, together with a variable amount of ash from the combustion of pulverized coal and also a small amount from oil.

Haze particles often grow by aggregation if the haze is very dense. Smoke is a condensation of vapours distilled from the fuel which have not been fully combusted, and the particles are so small that the smoke often has a bluish colour in sunlight when viewed with a dark background. Aggregation into larger particles may be very slow when the smoke cloud is diluted with cleaner air.

The same amount of pollution, but in the form of larger particles, reduces the optical barrier and if the particles are very large they will fall out. It is an old saying

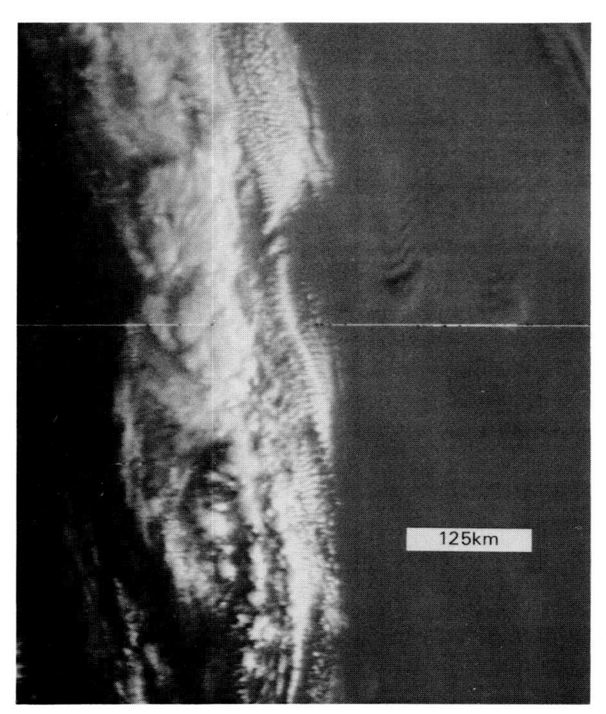

9.1.1 0909,3.3.77,VIS

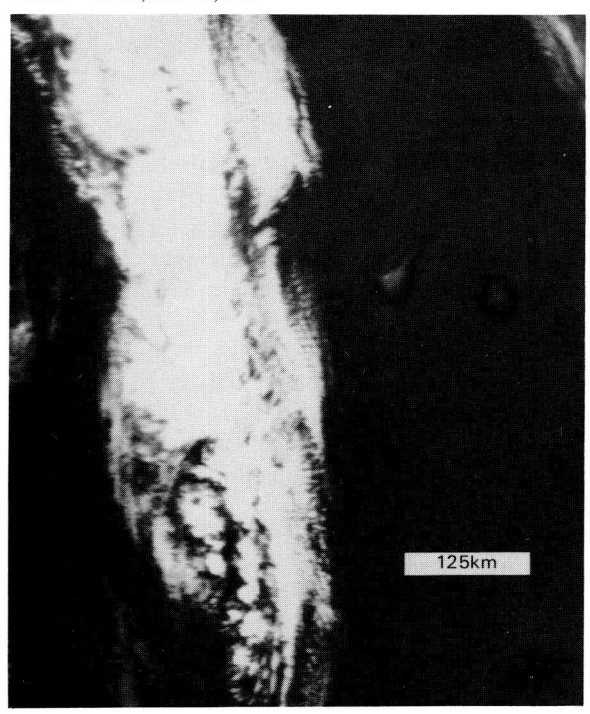

9.1.2 0909,3.3.77,IR

9.1.1, 9.1.2
The dust from the Sahara covered the top of Tenerife where there are two peaks close together (3070 m and 3718 m) in the west, another (2887 m) about 18 km to the ENE, and a 40 km peninsula to the northeast with a ridge averaging about 1000 m high. The neighbouring island of Gran Canaria has a peak at 1949 m. This outpouring of dust lasted about 4 days. The IR picture shows the degree of detection of the waves possible by IR, which is very surprising considering the clarity with which the haze is penetrated. It may be noted that lee waves keep the same phase down to the ground so that the whole vertical air column is warmed in the troughs and cooled in the crests. There is no cloud in the island's lee waves, and in IR waves are usually seen by the gaps between wave clouds, like the billows in the nearby alto-cloud in this case, rather than by temperature differences between crest and trough.

9.1.3 1515,23.8.80,2

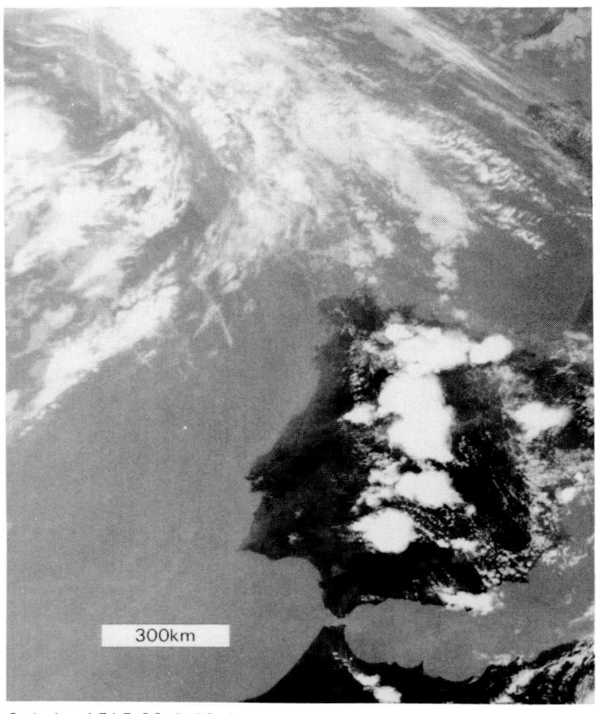

9.1.4 1515,23.8.80,4

9.1.3, 9.1.4
These pictures lie in the western side of the swath and the sea is therefore likely to show glint. Several days of Sahara dust in this general area certainly prompt the thought that southern Portugal is under dust at the time of this picture. On the other hand the dark areas off the lee coast of north Portugal and Mediterranean Morocco indicate calm areas of sea not showing glint, and it happens that the brightness of the glint is close to that of the land in areas where the coast is indistinct. The Ch 4 picture shows no indication of dust' merely large temperature contrasts in sunny Iberia where there is or has been fog and are or have been showers. Pictures below show that deep desert dust clouds show well in some cases in the deep IR channel.

that 'The haze you can see will not trouble your vision but the grit you cannot see may hit you in the eye'. Smoke does not get in your eyes!

When a dust cloud travels over a large distance without the continual convection which raised it initially, fallout reduces the load of larger particles and the cloud becomes less easily detected by IR. A 100-μm mineral particle has a fallspeed of about 0.5 m s^{-1}, which will cause it to reach the surface from 3000 m in 100 min in calm air; a 10-μm particle will take 4 days. Dry organic material which has been pulverized into fragments by blown dust at the ground may have much smaller fallspeeds, and the Harmattan type of haze which is frequently seen on the edges of the North African and some other deserts remains in suspension for many days and would for many weeks if the organic lightweight particles were in the 1-μm range. Although some ash particles are in this range their total amount is small compared with other haze particles.

Black smoke in the form of carbon particles may be created by the reduction of CO or CO_2 in an oxygen-deficient region of a furnace, and it is not easily oxidized if the gases are cooled by the addition of more air. To avoid this, enough warm air must be added in the right place. The particles of both black and ordinary smoke are usually too small to deplete the dust cloud by fallout and must await capture in a wet cloud to be removed by rainout. This is discussed in section 10.2.

Blue smoke is sometimes so small that it is not even captured in the lungs of a cigarette smoker. The smoke that smokers breathe out is bluer than that which they inhaled because all the larger particles are removed by the naturally operating air cleaning mechanisms of the bronchial system; but if they have been damaged by previous smoking, the smoke penetrates to the alveoli of the lung with the danger of causing emphysema. In the flue-gas washing chambers of the old Battersea power station the very small particles of SO_3 present were not captured by the copious rain of water drops because they followed the airflow around them, although 96% of the sulphur in the form of SO_2 gas was absorbed and removed. The SO_3 emerged from the washing chambers as a blue smoke.

A surprising feature of clouds of Sahara dust seen over Iberia and the neighbouring ocean when viewed by Ch 4 is not that the dust is distinguishable over the sea, which is usually warmer than the air above 1000 m, but that it may be seen to be warmer than the ground and is of a shade between the land and sea on those occasions (e.g. **9.1.4**). The boundary of the dust-laden air is often very clearly defined, but tends to become less well-defined as soon as shear, which includes a change of direction as much as speed with height, develops in the environment of latitudes north of 35°N which are much more baroclinic than in the Sahara of its origin.

With particles small enough to remain airborne for a few days a dust cloud which

9.1.5, 9.1.6, 9.1.7, 9.1.8
Sahara dust arrives in France rather less frequently from the southeast. In this case it is very clearly depicted in Ch 4 over the land but less well over the sea. Ch 1 is preferable to Ch 2 in this case because the brightness of the land in Ch 2 obscures the dust. To the west of Iberia the glint is very bright, particularly in Ch 3, but over France and the sea Ch 3 scarcely detects dust, although in this case it is sensitive to the small arc of warm surface water just west of Majorca. Dry inland Iberia is very bright in Ch 3, and over the hot land the high cloud is darker than its shadows, but over the cool but glint-bright sea the shadows are brighter than the cirrus!

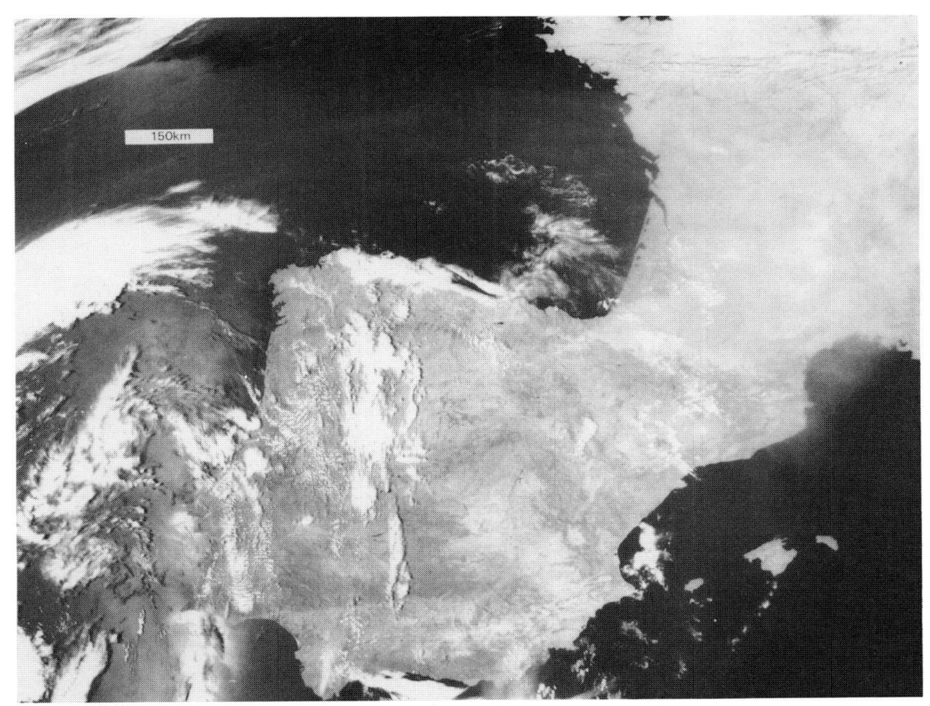

9.1.5 1442,5.7.82,2

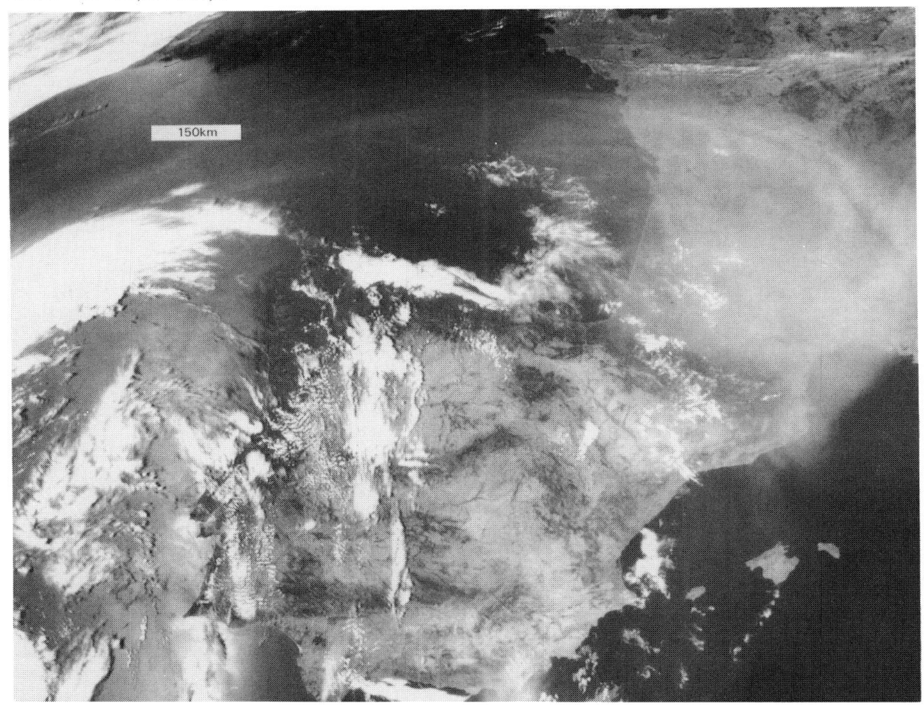

9.1.6 1442,5.7.82,1

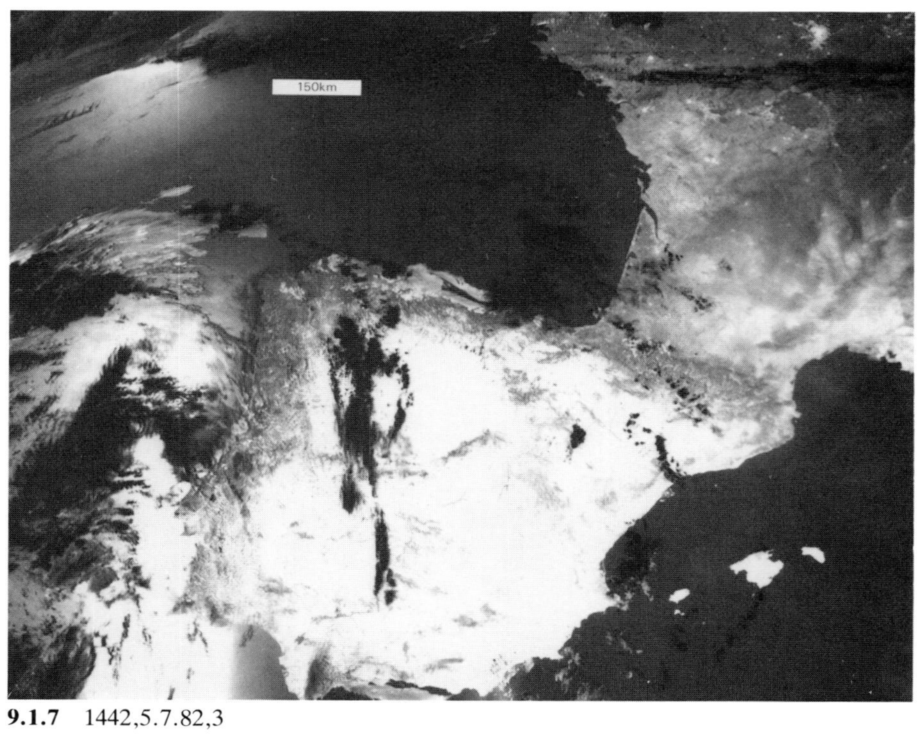

9.1.7 1442,5.7.82,3

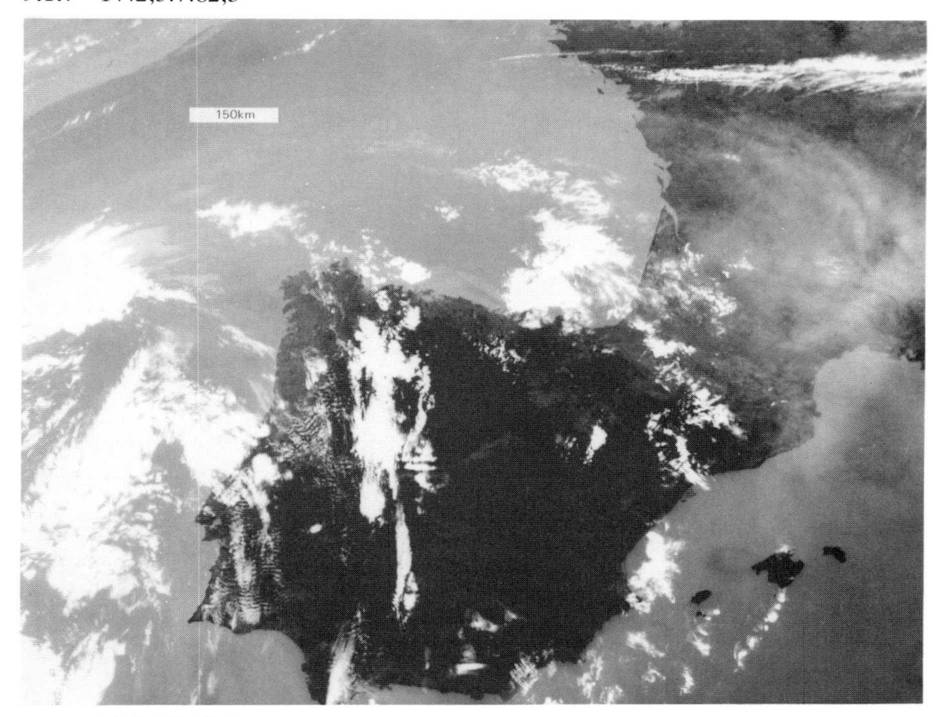

9.1.8 1442,5.7.82,4

is still visible in Chs 3 and 4 must hold a sufficient load of particles in a vertical column to block the passage of this radiation. The temperature of the dust particles mainly responsible for the IR images in the case illustrated is between those of the sea and the land. Since the particles' emission is detectable in the IR it must play a significant role in determining the temperature of the air containing them.

9.1.9 1140,21.8.80,CZ1

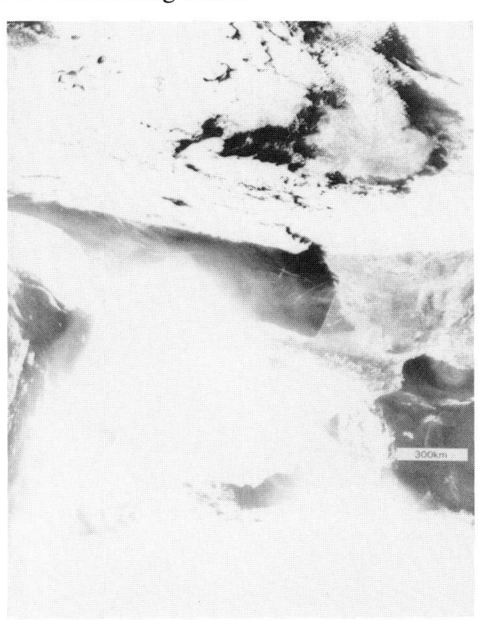

9.1.10 1140,21.8.80,CZ3

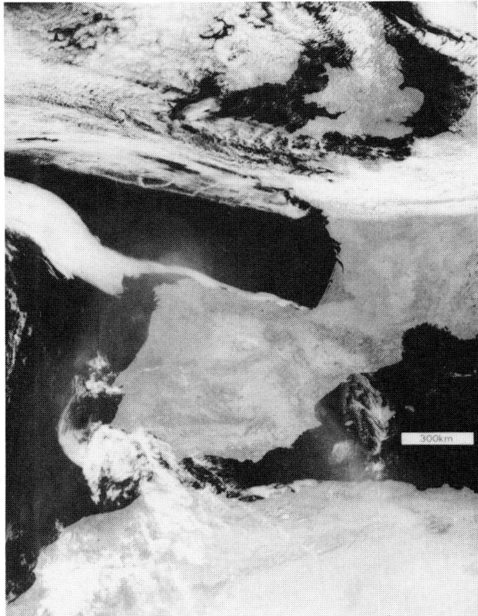

9.1.11 1140,21.8.80,CZ5

9.1.9, 9.1.10, 9.1.11
The CZCS channels in the visible are much more sensitive than Ch 1. CZ 1 is the most useful because the land is darker. CZ 3 is saturated in some areas even by the dust in the air and on the ground; and both channels show the contrail fragments over Biscay. CZ 5, which is very like Ch 2, but slightly nearer the visible, shows the dust area without interference from glint.

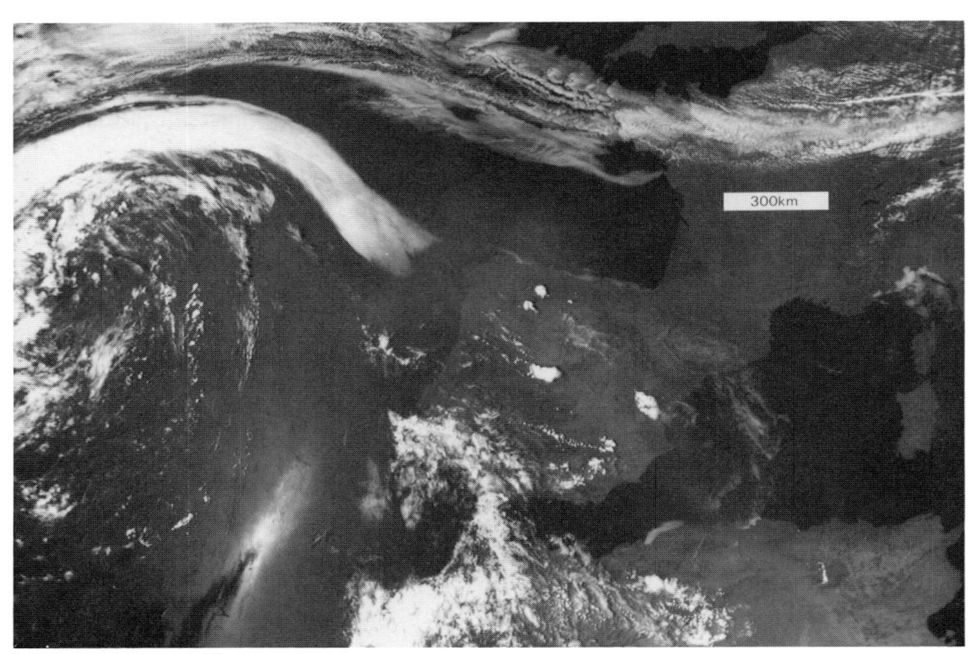

9.1.12 1538,21.8.80,1

9.1.13 1538,21.8.80,4

9.1.12, 9.1.13

The Ch 1 picture of the event of the previous picture but 4 h later shows how glint interferes with dust detection over the sea. The dust is not seen over Algeria because of the brightness of the land; but it is detectable because it is cooler in Ch 4.

9.1.14 1958,26.1.81,4

9.1.14
This initial surge of dusty air at the beginning of a Sahara dust episode (see [1], Chapter 23) shows that on a winter evening the ground becomes cooler than the dust, which is cooler than the sea.

9.1.15 0923,28.1.81,2

9.1.16 0923,28.1.81,3

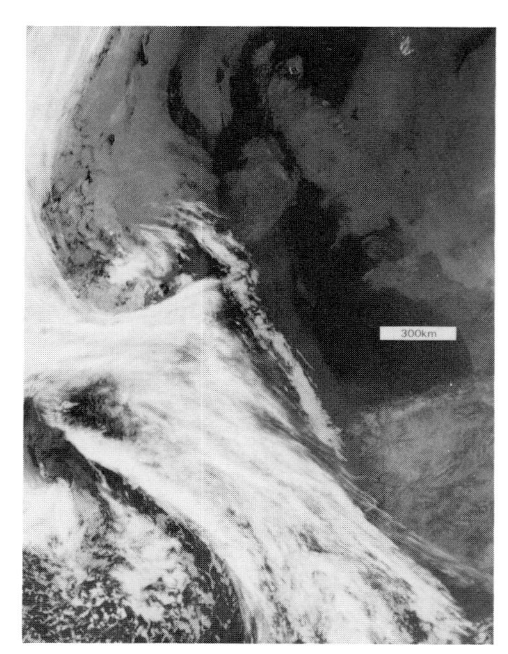

9.1.17 0923,28.1.81,4

9.1.15, 9.1.16, 9.1.17
Later in the episode referred to in **9.1.14** the dust could not be detected over land by Ch 4 because the air containing it had become cooler due to radiative exchange with the surface, which was cooler than the African desert and with outer space. But in the early morning the forward scatter in Ch 3 made it stand out clearly.

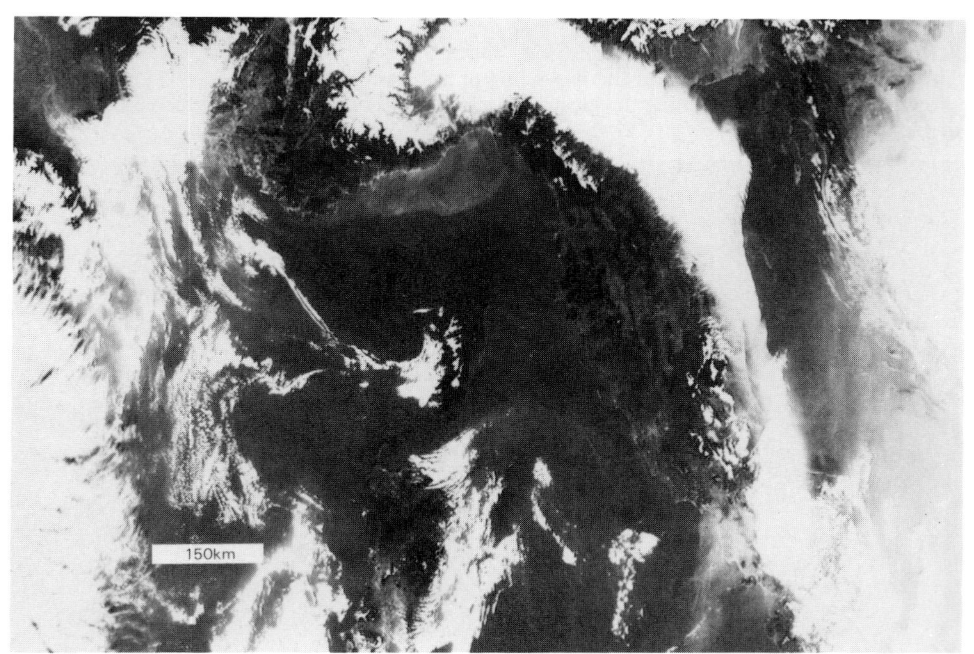

9.1.18 1052,10.2.82,CZ1

9.1.18, 9.1.19, 9.1.20
The CZ 5 picture shows some delicate rope clouds west of Corsica, but no suggestion of dust over the sea. CZ 1 and CZ 3 show much more sensitivity. Evidently the back scatter in CZ 5 is very slight. Because of the Coriolis force the cool polluted cloud from the Po valley hugs the Italian (right-hand) coast of the Adriatic. See also **12.5.5**.

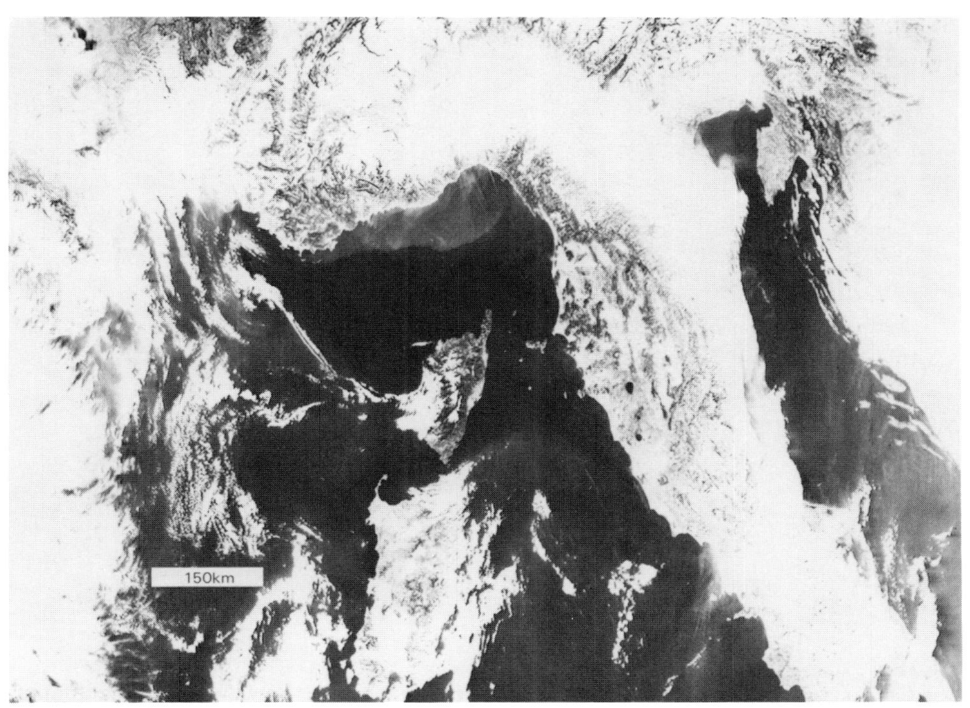

9.1.19 1052,10.2.82,CZ3

9.1.20 1052,10.2.82,CZ5

9.1.21 1309,10.2.82,2

9.1.21
The Ch 2 picture of the same area $2\frac{1}{4}$ h later, when the fog in the Po valley is getting thinner, shows that the change in the angle of sunshine make a great deal of difference to the scatter from dust in the very near IR and that the sea glow in CZ 1 and CZ 3 is not due to a strong blue sea colour.

9.2 DUST OR GLINT?

Although the CZCS pictures recorded the dust well, they were not always available. The edge of a glint area which arises because of a change in the state of the sea is usually along the wind direction, but so is the edge of a dusty area very often. In the Atlantic a single picture may not be sufficient to determine the cause of the glow. The different position of glint on afternoon and morning passes usually made it easy to confirm whether an area of glare was due to glint or dust. An even better comparison is between the two different views of the overlap area of two successive orbits. But with practice based on such observations the characteristic texture of the boundaries of glint and dust areas soon becomes recognizable. Glint is brightest in Ch 3 and absent from Ch 4 pictures.

9.3 THE ADVANTAGES OF CZCS CHANNELS

The CZCS satellite Nimbus 7 had a passage time between 0930 and 1230 GMT, which was between the early morning and early afternoon satellites and was therefore well placed to improve observational frequency. The camera was tilted about 20° towards the pole in order to eliminate glint on the sea; this had the advantage of making a clear distinction between glint and dust, the latter of which was well recorded as we shall see. It also has the effect of increasing the north–south dimension of the pixels and some pictures were made before this elongation had been programmed out of the printing process.

9.3.1 0846,20.8.83,1

9.3.2 1404,20.8.83,1

9.3.1, 9.3.2

The haze is much brighter in the morning picture where it is seen in forward scatter of the eastern sunshine. In the early afternoon it is seen in backward scatter and therefore with less intensity. The haze is scarcely altered, at least over the sea, while over land it may be added to by daytime emissions and spread vertically, but not much altered in its primary scatter properties. The wind direction, predominantly southeasterly is shown by the cloud streets. Note the sea breeze across the Swedish east coast, and weak high pressure over the Baltic with a weak low south of Ireland.

The main disadvantage of the visible channels was that they were too sensitive because they were designed to map out the variations in the colour of the sea, which is usually dark, the variations corresponding to the presence of biota and minerals in suspension. This meant that the clouds, some land, and even some haze was so bright that the radiometer was saturated and all detail lost in those areas. CZ 5, by contrast, was a bright channel, with a good range of intensities, and produced very good pictures. The swath was narrower than on the NOAA satellites, sometimes with no overlap of adjacent swaths, but this was counterbalanced to some extent by the advantage of a smaller pixel width of 0.8 km beneath the satellite.

The major uncertainty which made even CZ 5 a not very useful channel for routine use was that the satellite was under the control of oceanographers, and the message from overhead passes was only switched on for about one in three passes. It was, of course, not designed for meteorologists, who are grateful for such additional information as has become available.

Over land the relative darkness of land in CZ 1 compared with the others made it best for delineating haze over land, and an interesting case from the Mediterranean illustrates the use of CZ 1 to determine the haze top height by its position on mountains which protruded through the haze top. In the case illustrated the image obtained in Ch 1 was not very useful because of widespread and variable glint.

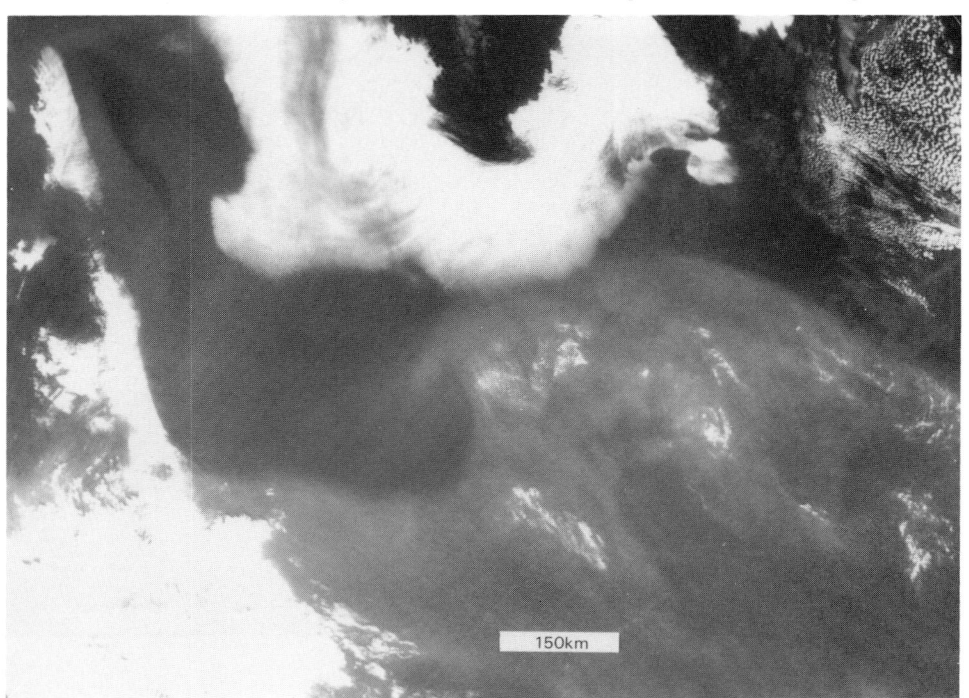

9.3.3 1013,20.8.83,CZ1

9.3.3, 9.3.4, 9.3.5, 9.3.6, 9.3.7
The CZCS pictures of this occasion show the greater usefulness of CZ 1, and the relative uselessness of CZ 4 in cases of this sort. From these and **9.3.1/2** the movement of the boundaries of the haze may be obtained. The usefulness of CZ 5 is primarily in locating what is seen in CZ 1 (where the coast is difficult to see) and in distinguishing cloud from thick haze.

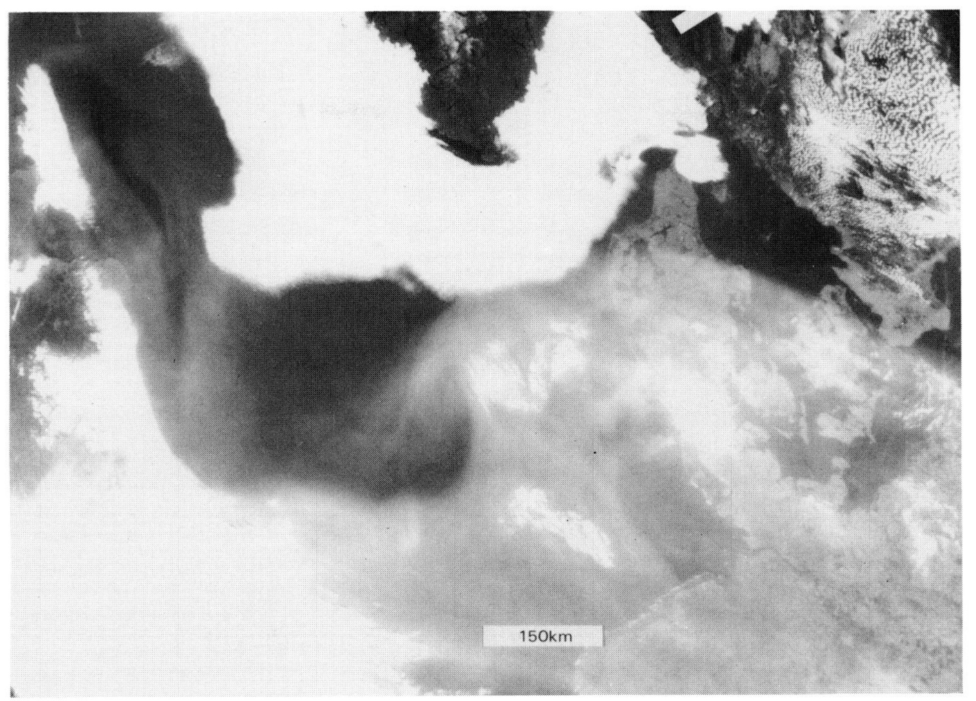

9.3.4 1013,20.8.83,CZ2

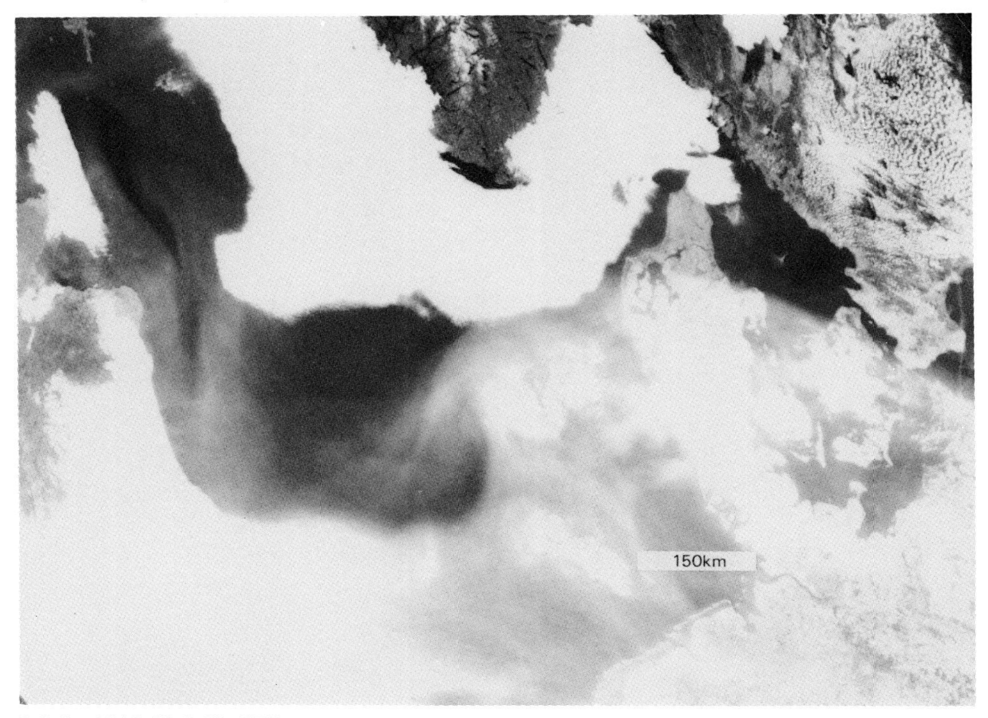

9.3.5 1013,20.8.83,CZ3

9.3.6 1013,20.8.83,CZ4

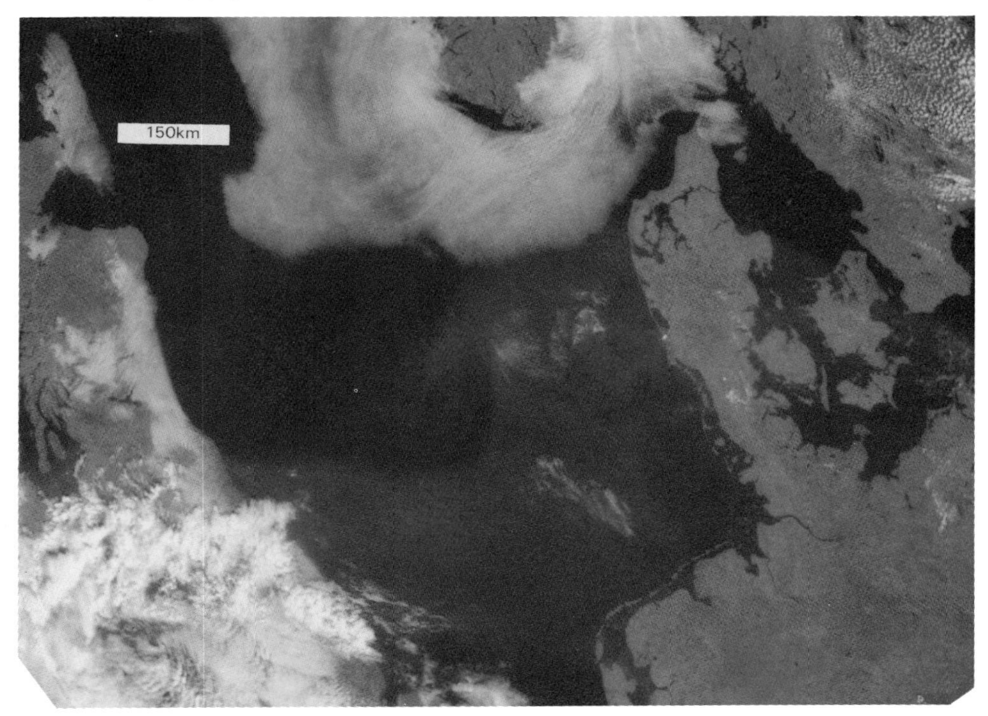

9.3.7 1013,20.8.83,CZ5

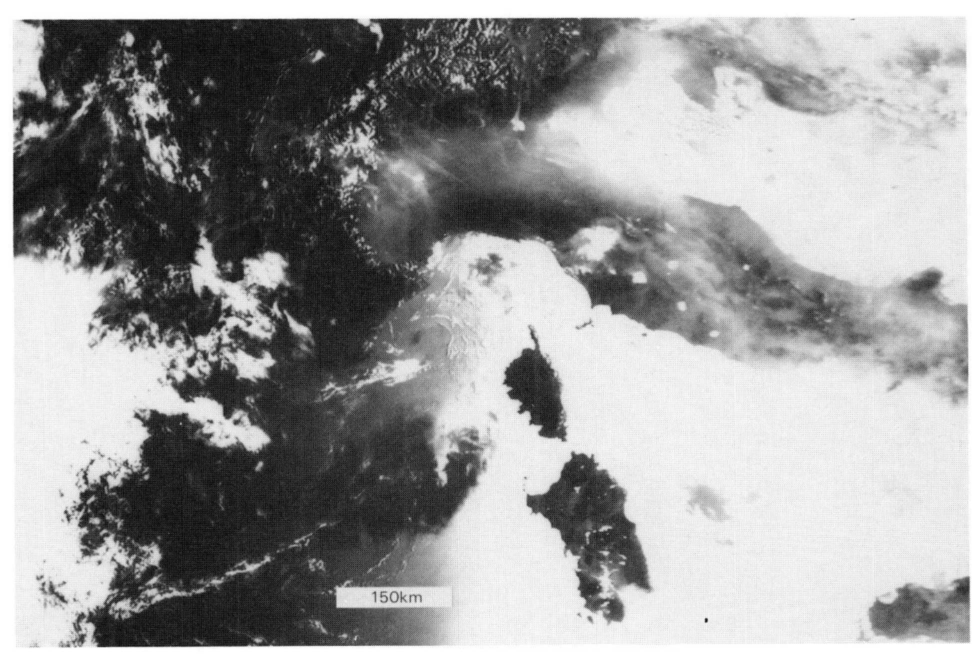

9.3.8 0804,22.8.83,1

9.3.9 1340,22.8.83,1

9.3.8, 9.3.9

During the day cumulus grows over the mountains and the glint moves from the east, where it is bright when the sun is low and the dark areas are clouds, to the west, where calm areas become black and the sea clouds are seen to have moved slowly in the light winds, and the haze in the Po valley is thinned out in the daytime warming.

9.3.10 1049,22.8.83,CZ1

9.3.11 1049,22.8.83,CZ3

9.3.12 1049,22.8.83,CZ5

9.3.10, 9.3.11, 9.3.12

In the visible CZ channels even the smallest, thinnest cloud is bright. In CZ 1 the only land areas appearing dark are those which protrude above the inversion at the top of the haze. The CZ 5 picture could, obviously, have been printed less brightly so that some detail would have appeared in the clouds, but that could have lost most of the thin cloud over the sea.

9.4 HAZE BOUNDARIES

Haze can be seen most consistently by the morning satellite, looking towards the sun, using Ch 1 to gather forward scatter. In the example **9.3.1/2** the same scene is seen in the afternoon, when the haze is much darker, not because it has been diluted, but because sideways scatter was much weaker. It is unlikely that the haze was much changed over the sea. Over the land the total amount of it was likely to have been augmented by daytime emissions. Assuming it to have been almost entirely primary scatter on the grounds that vision through it is not much spoiled, it is concluded that the particle size was in the vicinity of 0.1 μm, to fit in with section **2.9.1.**

In the late morning pictures by CZ 1–5 (**9.3.1–8**) the haze is very clear, but the intensities cannot be quantitatively compared to obtain total loads because of the many unknowns, although comparison with cases which happen to have been physically sampled by aircraft or in some other independent way might be developed for the routine meteorological pictures.

The positions of the boundaries are very well defined and their movements used to test model predictions based on meteorological measurements such as pressure field and assumptions concerning the departure of the wind from geostrophic. The movement of boundaries due to diffusion can be neglected by comparison with advection. These observations and many others of cellular convection both over land and sea seem to confirm that diffusion is certainly not influenced by the gradient of haze density. Plumes do not fan out with an angle of more than about 1 in 50, or two or three mixing depths in 100 km or so when cellular convection prevails (see also next section).

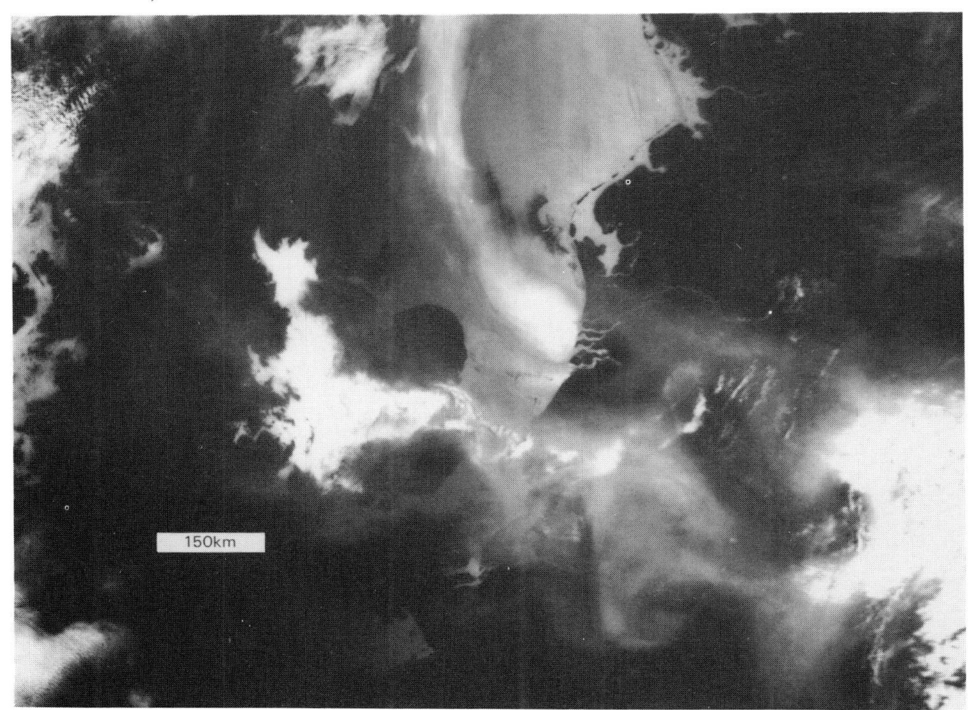

9.4.1 0852, 27.4.87,1

9.4.2 0852, 27.4.87,2

9.4.3 0852, 27.4.87,3

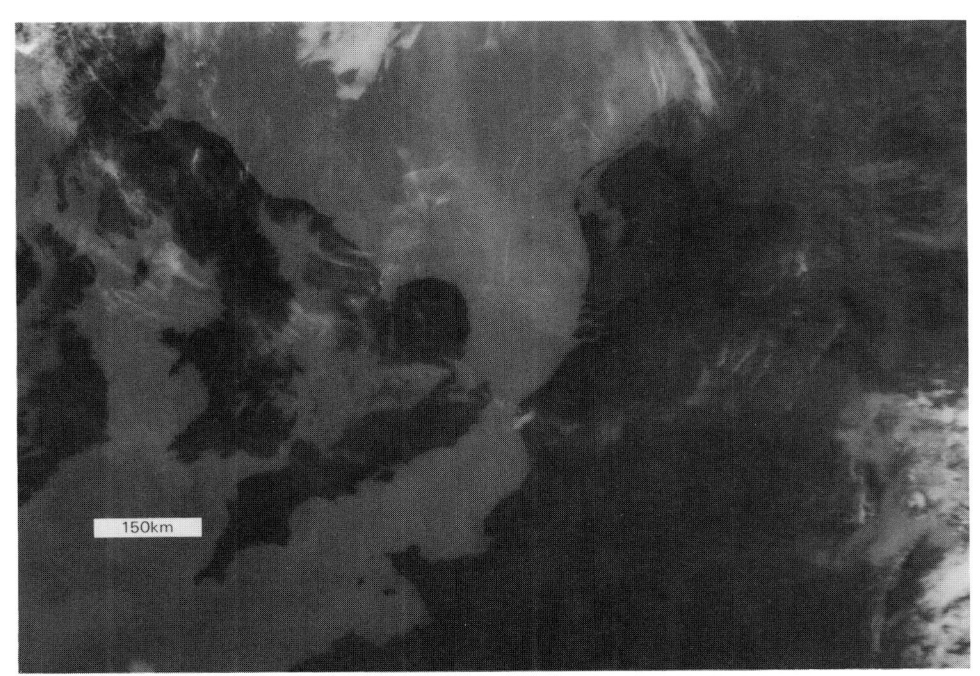

9.4.4 0852, 27.4.87,4

9.4.1, 9.4.2, 9.4.3, 9.4.4

The structure of a polluted area can be very complicated so that individual measurements on the ground cannot be taken as representative of a 'plume' presumed to have emerged from a source. In the case depicted an anticyclone centered over the north of East Germany was driving air from central Europe towards southeast Britain. The pressure pattern changed and moved considerably during the period of this episode and the haze probably had an origin more northerly than a backtracking of the existing pressure field would suggest. Poland, Czechoslovakia, and East and West Germany contributed with Holland and Belgium adding their quota as it passed overhead. The air arriving in England had a high ozone content and measurements were made upwind and downwind of London to assess its contribution. But such tests are made very difficult by the great spatial variations which this picture reveals.

The Ch 1 picture is printed dark to emphasize the contrasts due to haze. The strange behaviour at the coast of the Low Countries appears to convert the haze into cloud. A 'steaminess' is often seen off the east coast of England, but only in winds from the continent, and it may be due to the cooling of the air in cases when it is already very damp when over the continent.

To interpret these pictures the following is to be noted: Ch 3 shows the glint areas most brightly, the very dark patches of sea are calm areas with no glint, cloud is of intermediate shade, and haze is not seen; Ch 1 shows the particulate pollution (smoke, haze) with clouds mostly bright white, and glint is not as bright as cloud (in this case but not at the mirror point in near calm); Ch 2 shows the land more brightly, and the difference between Denmark and Germany reveals something about the agriculture; Ch 4 shows that the land is warmer than the sea, and displays as colder and higher those small bits of cloud which appear duplicated by their shadows on the glint in Ch 3.

The coastline of Schleswig on the German Bight is clearly delineated in Chs 1 and 2, but Ch 4 does not see through the high cloud of which the shadows are more prominent in 1 and 3 (and are displaced 26 km westwards relative to the clouds).

If simple answers to questions about this episode (of several days) are sought, they should not be put to someone who has studied the detail. Ask an enthusiast or campaigner for whom answers are simple and largely determined in advance!

9.5 PLUMES WITHIN BROAD AREAS OF POLLUTION: HOT SPOTS, WAVES

Plumes of water cloud within haze areas and plumes from isolated chimneys have been discussed in section 7.2. Here we note the streakiness of haze originating from large industrial areas such as the Ruhr and upper Silesia. The streaks show very slow widening and appear to indicate the wind direction as effectively as cloud streets when the mixed layer is confined by a strong inversion.

The significance of this type of observation is referred to in Chapter 10 and the search for the smoke produced by straw burning cannot claim to have been successful; this is possibly because suitable weather is similar to that which produces sea breezes, which prevent the smoke from drifting out to sea where it can best be detected by satellite. In the last few pictures of this chapter we take a closer look at one of these occasions.

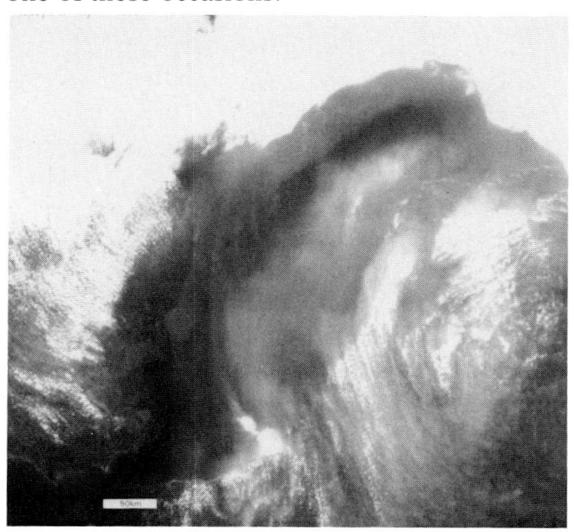

9.5.1 1127,8.7.82,CZ1

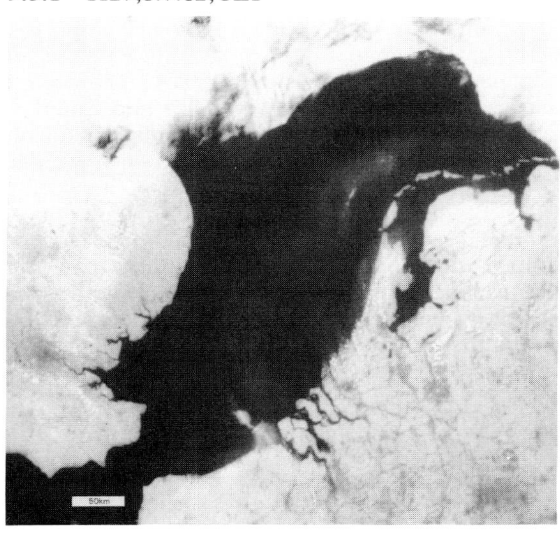

9.5.2 1127,8.7.82,CZ5

9.5.1, 9.5.2
The Frisian Islands and the east coast of England show the location of this picture. Air is streaming slowly from the Ruhr into the southern North Sea. There is a moderate sea breeze into England with cloud streets well inland.

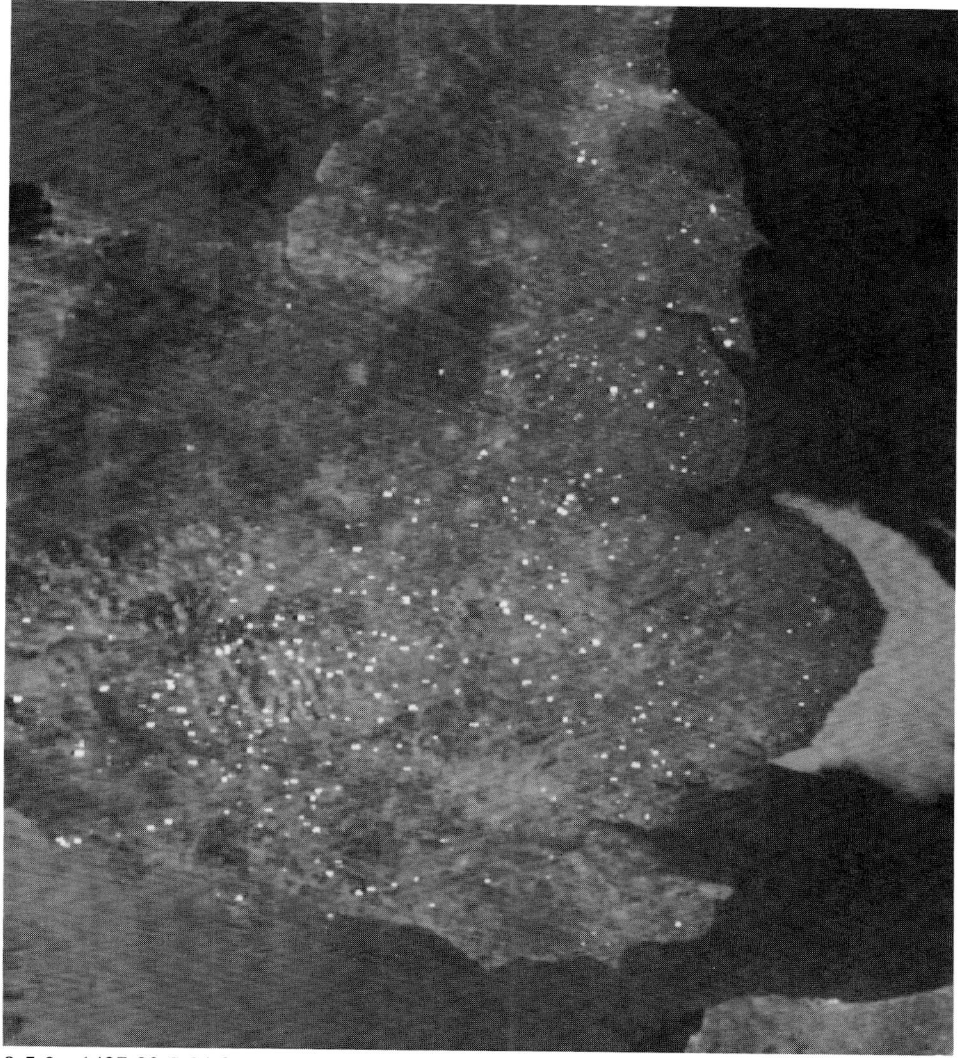

9.5.3 1437,20.8.84,3

9.5.3
In spite of persistent complaints about the pervasive smoke due to straw-burning at harvest time in England, it has not been possible to 'see' (using Ch 1) the plumes from individual fires on the farms, although they are easy to locate in Ch 3 as hot spots. Low-level smoke appears difficult to identify against the sun-illuminated ground, and it may be too well diluted by the time it reaches one or more pixel's distance. By comparison with the ease of seeing smoke over the sea the task is specially awkward, for when the sky is clear at this time of year there is frequently a sea breeze which keeps the smoke from the sea, and when there is a west wind it is usually fairly turbulent in the afternoon because of thermal convection and dilution is rapid. The individual plumes are not long-lasting like those from large industries, and even if the plumes have been visible by CZ 1 (e.g. **9.5.4/5**) such an observation is usually separated from the Ch 3 hot spot (**9.5.3**) observation by 2 or 3 hours. See also **6.4.5/6**.

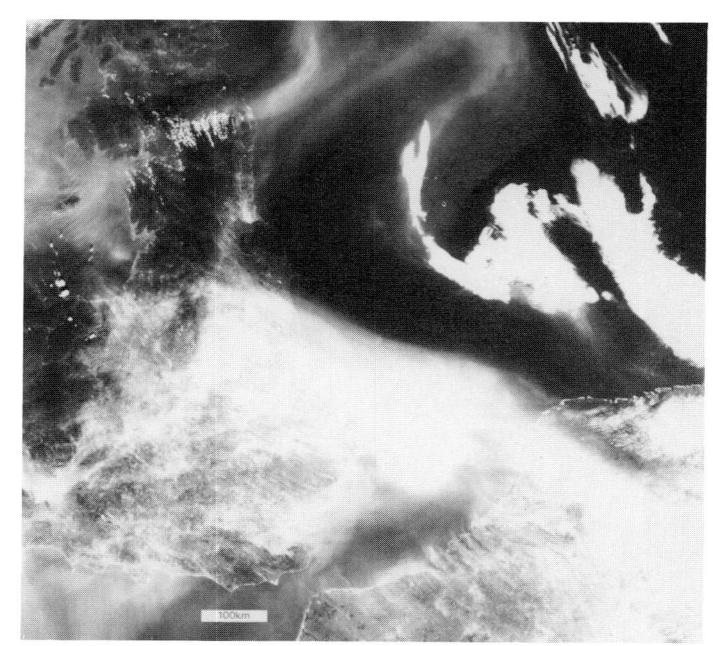

9.5.4 1110.20.8.84,CZ1

9.5.4, 9.5.5

CZ 1 makes the stream of industrial haze coming from the Ruhr area very clear on this occasion, and some plumes in England can be located, a few even dubiously associated with the sources revealed 207 min later (**9.5.3**).

An enlargement of part of this picture shows streets in the east leading to waves (wavelength 2.6 km) on the haze top in the west of England. With a strong inversion and cloud just forming in some wave crests, the crests below the condensation level become apparent in CZ 1 because of the enlargement of hygroscopic pollution particles as the humidity closely approaches 100%.

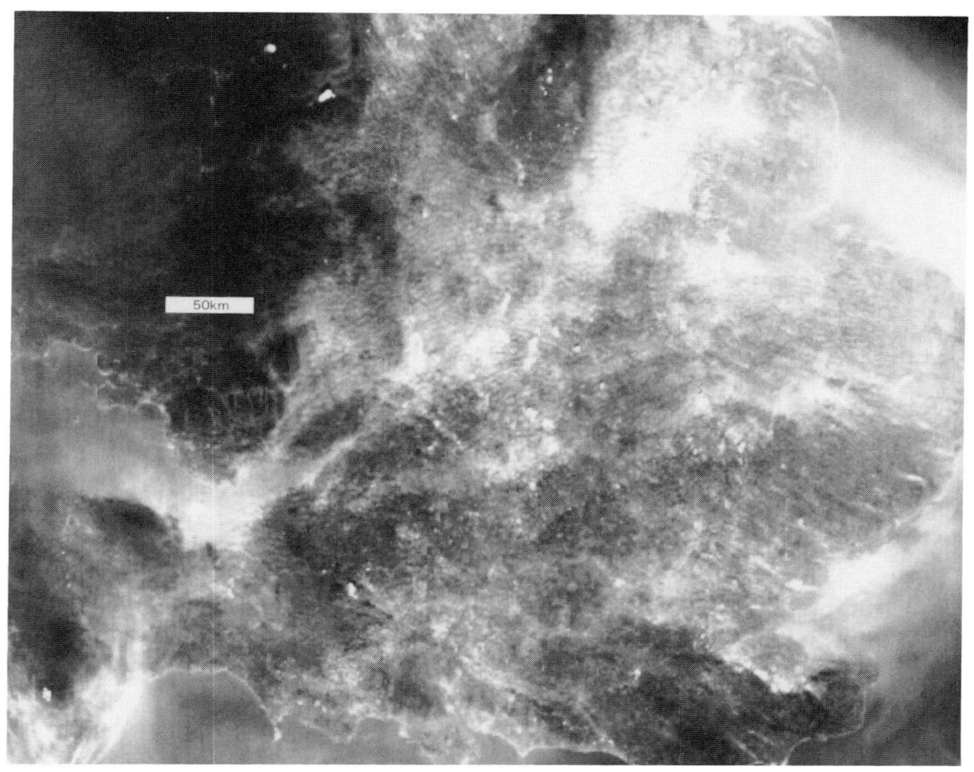

9.5.5 1110.20.8.84,CZ1

10
Deposition of pollution

10.1 RAINOUT

The most important effective mechanism for removing pollution particles and gases from the atmosphere is rainout. Pollution particles are, on the whole, either good condensation nuclei or fairly soluble gases. There are important exceptions, of which the chlorofluorocarbons are one of the best known, and they are removed only after being decomposed into more reactive substances by UV radiation in the stratosphere.

When condensation of cloud occurs most of the pollution is quickly incorporated in cloud droplets, and by agglomeration of these into raindrops is removed by gravity to the earth's surface. Cloud droplets, especially those larger than about 10 μm diameter, are very effectively collected and removed if they are in the path of falling rain or snow. In this way the air is continually cleaned by the formation of fallout.

By contrast, the same pollution, if it existed below cloud base, would be composed of such small particles that it would be carried aside by the airflow around the same falling raindrops. Thus 'washout', as the mechanism is optimistically called, is not a significant one for the removal of pollution, and the alternative meaning of its name does not belie it as a cleaner of the air.

In wet gas washing, which has been used at some power stations, about 96% of the SO_2 present in the flue gases is removed. This is achieved by making the gases rise through a downpour of water drops containing dissolved lime. This rain is very heavy compared with natural rain and the capture of small particles is not quite the same process as the dissolving of a gas which reacts readily with something already dissolved in the falling drops. However, the flue gases were cooled not only by evaporating water of the drops but also by heat conduction to the drops. This usually resulted in the gases emerging from the chimney, after a non-adiabatic cooling to below the dew point, containing a cloud of condensed water droplets. This had a serious effect on the buoyancy when these cloud droplets evaporated into the ambient air and a negative buoyancy was produced. Consequently the gases descended to the ground where the 4% of the sulphur oxides remaining were in higher concentration than would have occurred if no washing had taken place and the buoyancy had delayed descent to the ground until the dilution was negligible.

Even after removing pollution from the air rain is, in general, much less acid than the ground water, and so as such it cannot be a cause of increasing ground water acidity. Occasionally it can be very acid, as for instance when a large industrial plume is carried directly up into a rain cloud, in which case it would fall close to the source and would not be a contributor to the much more widespread 'acid rain' problem.

10.2 DRY DEPOSITION

The deposition of pollutant particles and gases directly on to the ground or buildings, trees, grass, etc., continually takes place. It only depletes the layers of air close to the ground and, although this may be important when concentrations are high close to the source, it is not a cause of increasing acidity of ground water. Close to the source the direct damage to vegetation and property has often been very costly but that is not part of the more general environmental deterioration.

In general the rate of deposition is roughly proportional to the air concentration, and the deposition velocity is defined as the deposition rate divided by the concentration. For typical rural surfaces this velocity is of the order of 1 cm s^{-1}, although it varies with the wetness and temperature of the surface, its stickiness, and in the case of vegetation its state of growth.

In order to reduce the deposition to acceptable levels as far as possible the policy in recent years has been to build as tall chimneys as possible and thereby reduce the concentration by prolonging the dilution into the ambient air. Even in cases where high air concentrations are achieved, as for example on the side of a deep valley where an inversion traps pollution when the movement of air is much reduced, it is not correct to describe the damage as due to 'acid rain'. This term is widely used to describe the damage which is assumed to be occurring, which is widespread and takes place often far from the source of the pollution, and which is brought down by rain, because it is thought that rainout is the most important mechanism for bringing the pollution to earth.

In some areas the evaporation of moisture from the vegetation is quite large and in this way the concentration of whatever is contained in the rain is increased in the ground water, but the acidity is generally neutralized by the erosion of the mineral base (rock or soil) in which the vegetation grows. Water long stored underground is not generally acid and the high acidity in lakes and streams is often the result of reactions in the soil with vegetable material.

Deposition having been defined as either wet or dry, it appears that no alternatives remain, which may be a reason why the other mechanism to be discussed next has been called 'occult deposition'.

10.3 CLOUD SCAVENGING

If pollution is emitted under low cloud it is likely soon to be mixed into the cloud and incorporated into the droplets. This produces polluted cloud, and we refer to it simply as 'dirty' cloud. If it is then carried by the drift of air on to a mountainside, trees and other objects with small dimensions, such as wires, long grass, hair, spider's webs, and so on, capture the cloud droplets and collect the pollution much more effectively than if they had been exposed to the same concentration of pollution but without the droplets. We call this cloud scavenging, which is a much more useful name than occult deposition. It certainly is not occult, for it has been known and understood for many decades. In some forests it causes a greater deposition of water than ordinary rainfall: raingauges sited under trees have often collected more rain than those in correctly exposed positions.

Low cloud is often as dirty as wet smog (smoke and fog), which the experience of cyclists and others showed to be extremely filthy in the days before urban smoke was reduced following legislation to control its emission. Because they are very effective

10.3.1

10.3.2

10.3.3

10.3.1, 10.3.2, 10.3.3
The Jisersky Hory forest destroyed by dirty cloud and bark beetle. Small branches are damaged on the northwest side facing the pollution source, and protected by snow at the lower levels.

collectors of cloud water, conifers with thin needles are very efficient collectors of air pollution when cloud is dirty.

Scavenging is a very important factor in the 'acid rain' problem. It has increased the total deposition on a forest and correspondingly increased the extraction of minerals, particularly those containing calcium, from the soil to neutralize the acidity. Almost all soils contain enough minerals which would perform this task given time. In conifer forests there is a large accumulation of dead material, and if the rock is not rapidly weathered the ground becomes progressively more acid. This happens under natural conditions even in the absence of artificial or other acidic fallout or dirty cloud, and the end product of a pine forest, after a few thousand years in which the debris of all old trees accumulates, is a peat bog.

This degeneration takes place much more slowly if the underlying rock is easily weathered to become soil and contains calcium in plenty. If its acidity is increased, either by natural forest mechanisms or by the addition of pollution, aluminium is dissolved from many kinds of rocks or the soil, and is harmful to trout, which have considerable economic importance to mankind. The build-up of acidity in ground water occurs if the soils are not helpful to its prevention.

As for the trees themselves, they may be subject to damage by frost and/or drought from time to time. Because the weather varies considerably over the years most forests have undergone the selective elimination of those trees which flower too early and lose a large number by late spring frosts, and those which flower so late that they cannot take advantage of a long growing season. Some similar mechanism operates to favour those trees which make the best compromise between slow growing, which protects well against drought, and fast growing, which kills in periods of drought. Thus frost and drought will always take a toll and will leave some trees in a state of stress. If the stresses are aggravated by pollution, the way may be prepared for infestation by bark beetle, from which the trees may not recover. The addition of pollution to the list of likely stresses in recent decades has greatly increased the chance of the destruction of a forest by insect pest. The pictures show a forest in Czechoslovakia which has suffered this fate.

Frost is a stress which forests can to some extent survive without any damage. In valleys, fairly high concentrations of gaseous and particulate pollution can be suffered without harm. As we proceed up a mountain we may notice an increase in damage to the trees, even though the measured average air concentration of pollution (both gaseous and particulate) significantly decreases; the deposition of pollution by scavenging, however, significantly increases. Although valley fog is common, it is only rarely supercooled, and the wind is usually very light or calm so that the displacement of air over the trees is small. Further up the hillsides the dropsize in wet fogs is greater and the average air displacement in cloud or fog greater. On the highest mountains the dropsize and wind strength on clouds are greatest and the simultaneous occurrence of freezing conditions is also much more frequent.

The importance of freezing needs special emphasis because it means that the scavenged cloud droplets are captured as supercooled liquid and freeze solid on impact. The pollution, which is in the form of solution in the droplets, is expelled as a highly concentrated fluid onto the surface of the trees, while the water becomes almost pure ice with the particulates entrapped within it. The trees suffer in the same

10.3.4
Dirty layers in rime collected on wires at the Milesovka Observatory, and derived from local industries. *Photo. M. Koldovsky.*

way as the rivers at the time of the spring snow-melt, when the first water to run into the rivers contains a much higher concentration of polluting salts than did the snow when it was deposited during the winter.

The rime or glazed ice deposited on the trees may often partially melt in winter sunshine because of the absorption of sunshine by the foliage within the ice. This would cause the worst combination of pollution and frost. In the area of the forest shown in the pictures there are plenty of young, healthy-looking trees growing among those destroyed or in areas where the dead trees have been cleared. The smallest of these are completely prevented from scavenging dirty supercooled cloud by being buried under snow, which is like most rain in being much cleaner than polluted low cloud. Trees of earlier years in the same area which were too big to be completely buried under the snow show many symptoms of frost damage on their upper branches only and thereby provide a rough measure of the winter snow depth.

Scavenging is a mechanism which seriously affects only the wind-facing side of the collecting objects. This is made obvious where the scavenging of sea salt by trees at the coast prevents the growth only on the sea-facing side of the trees and they grow leaning over inland. In the case shown in the pictures, at the tops of the worst-damaged smaller trees the branches facing the northwest were the damaged ones. In this direction lie the four large power stations in the DDR whose plumes are shown in **7.2.1–4.**

10.4

10.4(i) Long range transport

The transport of pollution over long distances is not in itself a harmful event, even if the recipient region is sensitive, provided that sufficient dilution occurs on the way and the natural turnover of material by the vegetation and drainage is not overwhelmed by the amount of it. Dry deposition is greatly reduced by reduction of the concentration in the air. Dilution is increased by vertical mixing and shear flow, particularly shear across the wind, for this widens the plume and vertical mixing takes place over a greater area.

Polish and other East European pollution is often transported to Norway. In many anticyclonic situations there is a strong inversion at the top of the mixed layer so that its depth and the consequent dilution is not increased much by daytime convection. Yet it is much more clearly seen by the forward scatter of low morning sunshine than in the middle of the day. A typical example of this is shown in the accompanying pictures. The pollution was generated in Poland and the surrounding countries and was transported to Scandinavia. But it was not until the air first arrived over the sea that cloud was formed at the top of the mixed layer. This was almost immediately followed by impingement on the mountains of southeast Norway, which is the part of Norway most afflicted by acid deposition.

In situations like this it does not make much difference to Norway what is the height of the chimneys in Poland, and so they might as well be as high as possible in order to decrease the pollution concentrations at ground level in Poland close to the sources.

Pollution carried towards Norway from Britain, Belgium or West Germany is usually diluted to a greater degree because in the cyclonic situations which bring surface air from these directions to Norway there is a great deal of shear and more

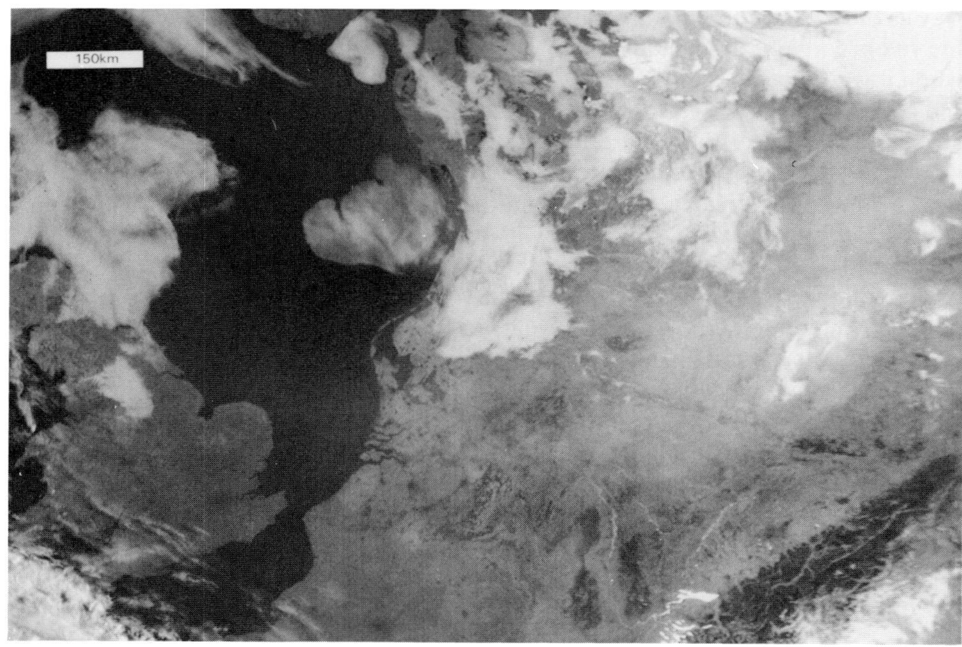

10.4.1 0806,21.8.84,1

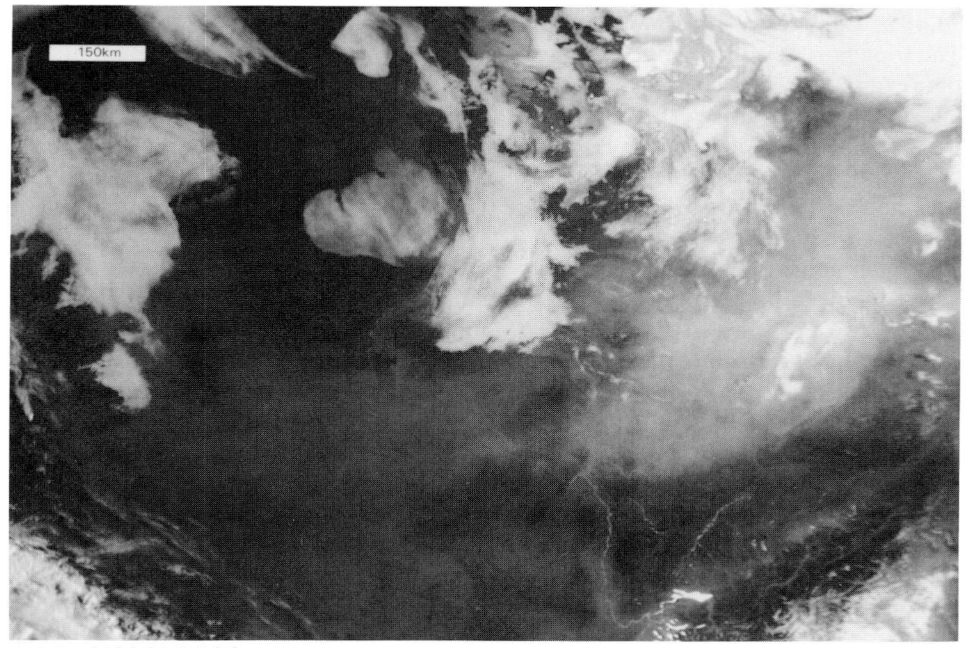

10.4.2 0806,21,8,84,2

10.4.1, 10.4.2

Pollution revealed by the low morning sunshine generated mainly in East Germany, Poland and Czechoslovakia, where large quantities of solid fuel are burned. These pictures are similar except that in Ch 2 the land is brighter.

frequent strong winds with upper winds coming from the northwest. Also the cloud base is usually higher and the low cloud not confined by a strong inversion. The rainfall which occurs when the low-level air is arriving in Norway from between west and south is mostly generated in air at a higher level, which could have passed across Scotland or further north. It might even have come from the southwest, having left the surface to ascend a warm frontal surface somewhere over the Atlantic Ocean. Furthermore these are not circumstances when pollution would be accumulated in a bottom layer thin enough to avoid dilution by shear. Consequently it is unlikely that pollution from the west or southwest travelling towards Norway and Sweden will be visible from satellites.

Another means by which pollution confined by an inversion may be deposited in high concentrations at a large distance from the source was illustrated by the dispersal of radioactive material all over Europe after the explosion at the Chernobyl nuclear power station in the Ukraine about 100 km north of Kiev during April–May 1986. After being trapped below an inversion for a few days and gradually depleted by dry deposition and some lateral spread, the upper air became cooler over Central and Western Europe as a result of a high level cold front moving across from the north on 1 May. During the next few days convection clouds penetrated the inversion and grew into cumulonimbus. This generated rain showers which also rained out pollution. As a result of the horizontal convergence of polluted air from below the inversion into the base of the showers the pollution rained out had been gathered from a large surrounding area.

This convergence mechanism has been known for a long time by meteorologists and was highlighted by the discovery that it is used by desert locusts. By their random flight they remain airborne in converging air, become concentrated in swarms under the showers, and are thereby enabled to lay their egg fields in ground that has recently been watered. The large number of the emerging hoppers concentrated in a semi-desert area means that any predators make a quite negligible impact on their numbers. Their trick is to be concentrated by the horizontally converging air and at the same time to avoid being carried up to great heights by the ascending air. Pollution performs the same trick by being rained out, so that the shower acts as a filter removing the pollution from the upcurrent. At first the large numbers of locusts found under the showers made it seem that they possessed some navigational expertise, and even ability to forecast showers, but now that we have become used to their behaviour being explained by their simple automatic flying behaviour, it comes as no surprise to find that radioactive pollution can be concentrated, against all the laws of dilution, which is an irreversible process normally. Similar observations cannot be made with more common forms of pollution because there are too many sources emitting the same substances. In the case of the Chernobyl disaster we knew where all the radioactive pollution had come from. The emissions lasted about 10 days, so that the trajectories of successive emissions crossed over each other in a complicated manner, and the incident provided information of unparalled accuracy and detail from which we can learn much.

In Britain storms generated just outside the polluted area were not many miles from others which rained out dangerous doses; this showed that the boundary of the 'cloud' of pollution was still quite sharp in some places even several days after emission.

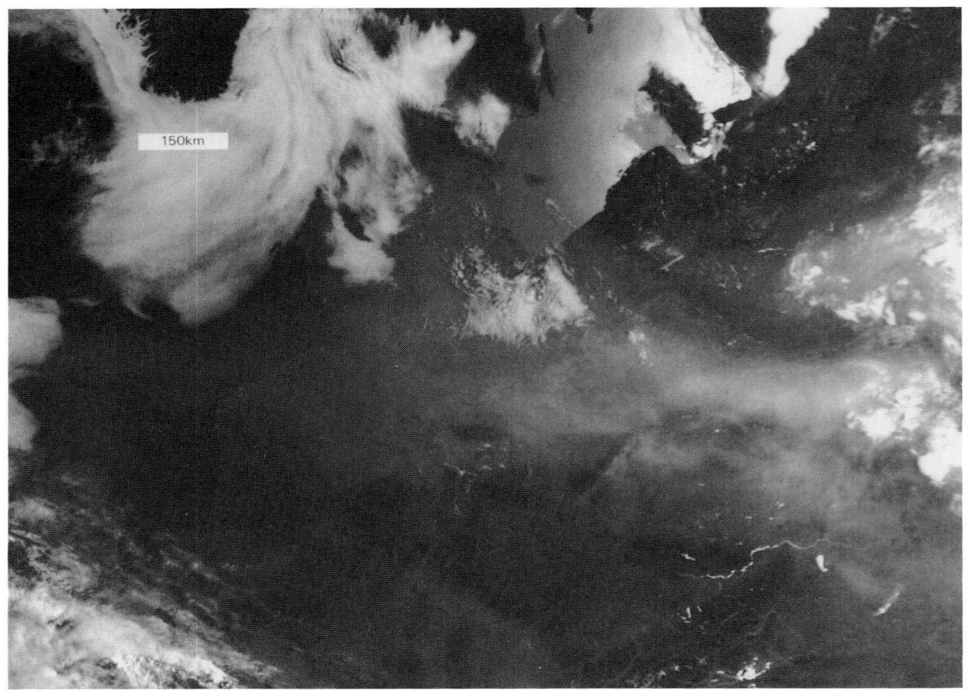

10.4.3 0742,22.8.84,1

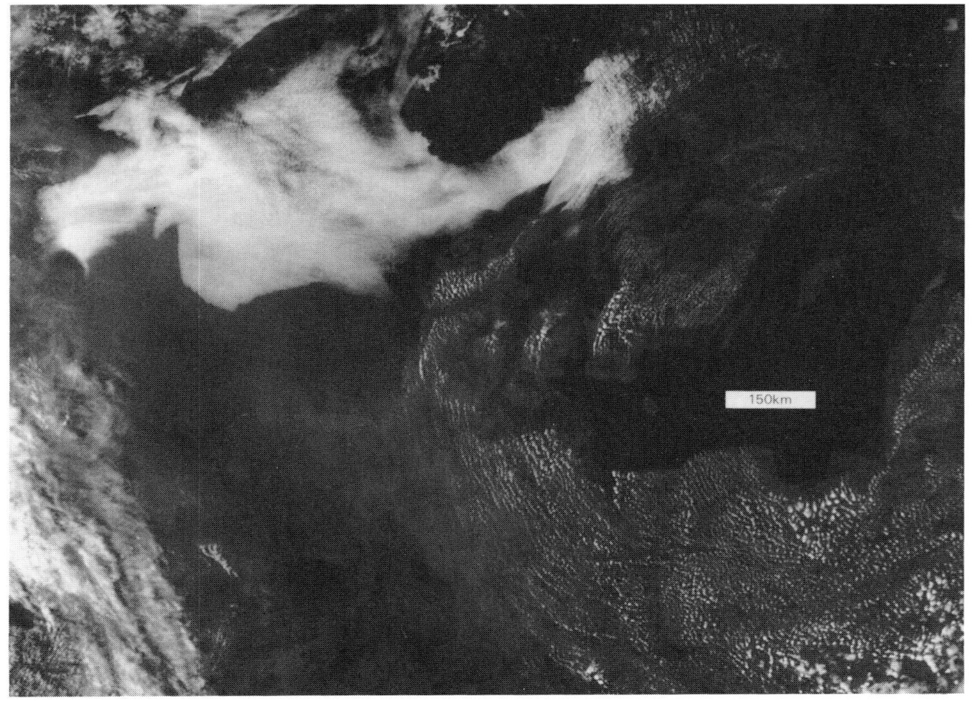

10.4.4 1413,22.8.84,1

10.4(ii) Long term changes in the environment

Most of the time most pollution becomes well-dispersed. It is important to appreciate that no steady state is ever reached, but changes on the geological scale all the way down to daily changes in the weather are gradually transforming the face of the earth. A pine forest may have its demise as a peat bog advanced by one or two thousand years and accelerated, as a result of one or two hundred years of increased acid pollution. To us this appears as a major disruption of the normal state, but even normal states are slowly changing. Some peat bogs are older than our civilization (say, modestly, 5000 years) and yet there was no pine forest before the end of the last ice age (say 10 000 years ago).

The cultivation of forests and the removal of timber cause ecological changes in the water which drains through into lakes and rivers. These changes are sometimes harmful to fish; their causes may be as much due to forest mis-management as to the deposition of pollution and it may not be correct to assign the consequences to a single cause or mechanism. For example the mobilization of aluminium into solution in ground water by the reduction of the pH below about 5.0 appears to be harmful to trout. It was, for a time, blamed also for harm done to some trees, but this now appears not to be the mechanism responsible.

In so far as mankind is interfering with the long-term evolution of species and their ecological environment, the most serious consequence is when a species is made extinct. It may take many centuries for its genes again to become part of the store available to contribute to an ecology in which several species fit together, each in its own niche in which its activity contributes to the mutual welfare of the community of life. Rapid changes wrought by the powerful machinery which today spreads human population into almost every corner also create vast tonnages of waste, of which nuclear radioactive waste is only a tiny fraction. The advance of the frontiers of human habitation may destroy not only forests cut down to make room, but also the soil on which the forest grew. When the removal of trees causes erosion in every rainstorm, the accumulation achieved by the forest in thousands of years may be swept away in a single season.

Much of the damage arises from people having simplistic ideas about how the ecology is held together, and it may fall apart if one link or thread is carelessly broken.

By contrast it can also be said that many ecologies have survival power because they are complicated, and it is only by the power of our mechanical devices and control of fire that we destroy forest ecology so easily. Also waste products may be consumed by vultures, rats, or flies, and one of these may take over from another

10.4.3, 10.4.4

Other major sources of pollution exist in the Ruhr area of West Germany, Belgium and Holland (see **6.4.13–16**), and in this case south of the Danube (glint) in Hungary and beyond. More important in the greater part of these western countries is the pollution from road traffic, which is scarcely detectable by the satellites under study here and which is subject to photochemical changes in sunshine. The particulate products of the eastern countries is to some extent due to the use of lignite, which has a much higher proportion of sulphur and ash than black coal. But it has been argued that it continues because the industries have not been as widely modernized as in the West. The case illustrated here is characteristic of summer anticyclonic weather, and shows the haze drifting towards southeast England.

according to circumstances caused by mankind as much as by climate or geological changes.

All over the world we find evidence that windborne pollution has been a major influence in forming the landscape. Sandstorms are the result of selection by wind mechanisms of particles of a narrow range of sizes. Stratified rocks show that the other major fluid system — the sea — spread layer upon layer of particulate material to build masses of rock during centuries. Gravel and sand moraines remind us of the shovelling power of glaciers. Barcan dunes, which migrate for hundreds of miles without loss of sand, show that movement without dispersal is possible and without the interference of life forms which have their own purposes.

Today we see vast clouds of dust travelling across and outwards from the desert areas of Africa, Asia, and the less extensive deserts of North and South America. The dust from volcanoes is carried around the world at a much higher altitude. Even Iceland has its windborne dust storms.

We know that the radioactive strontium from nuclear explosions of the early 1960s was deposited in rain all over the world in the following few years. Only feigned outrage was justified when DDT was discovered in Antarctic penguins, for if the natural dispersion mechanisms had not made that inevitable there would have been many more disasters from the accumulation of air pollution

There are, certainly, special problems due to the dispersal of only very slowly destructible substances, but we should not be upset simply because of a disturbance from what we thought was a norm, which many generations of people have naively thought to have been a created, rather than an evolved, norm.

There is irony in the fact that the changes we have brought about which threaten our safety are entirely due to our success, which is hurried excess, which means achieving aims much more quickly than the evolutionary mechanisms can react to it. That seems to be exactly what the designers want!

We need to act with a better understanding of evolution, so as to work with it, and avoid initiating changes which threaten aspects of our culture which we have made precarious. The potential for disaster is increased by exploiting possibilities to the maximum; any failure in the complex systems which support each other can halt the operation of many components.

Before we became almost completely dependent on the once-for-all consumption of mineral fuel, the various life forms fitted together in a slowly evolving pattern of exchange, with sunshine providing the opportunity to break the rules of a closed system and enabling us to reduce entropy on earth and engage in creative and purposeful living. The waste of one species was made the raw material for the niche of another. But we create too much waste nowadays; the dispersal power of the air and the sea do not act fast enough to prevent obnoxious accumulations of waste.

10.5 THE SPREAD OF NUCLEAR POLLUTION FROM CHERNOBYL

We begin with some general considerations which set the scene and then give a detailed description of the developments as seen by meteorological satellite.

The emission of radioactive pollution began in the early morning of 26 April 1986 (2223 GMT on 25 April) at Chernobyl (51°18′N 30°12′E) and continued significantly for 10 days, the emissions on successive days being roughly in the proportions as follows:

Day	Apr	26	27	28	29	30	1	2	3	4	5	May
Proportion		12.0	4.0	3.4	2.6	2.0	2.0	4.0	5.8	7.8	8.8	

with relatively very small amounts on days after 5 May (10th day). There have been many authoritative accounts of the measurements of radioactivity in the ambient air, of deposits on the ground, and of models used to amplify and illuminate discussion of events. The purpose of this account is to describe some of the essential features as seen by AVHRR. The pollution was not directly detected by AVHRR, but certain features of the weather were very important in determining where and when deposition occurred.

The US Secretary of State, Mr Schultz, is reported as saying that no one in Western Europe had need to worry because the prevailing winds would blow the debris eastwards into the Soviet Union. Such comment proved to be very ill-informed, especially as the first report of the emissions came from Stockholm where it was detected in the air; but it revealed a widely propagated concept of dispersion of pollution as a progressive enlargement of the single identifiable plume. Thus it was commonly supposed that the cause of the disappearance of plumes was their gradual dilution by eddies smaller than the plume itself, or, if the eddies were not smaller than the plume, they would cause it to wander about so that it covered any particular place only part of the time.

This is an erroneous concept, based on the vague idea that the 'turbulence' which causes the dilution consists of random stirring motions which are like the eddies which cause the gradual disappearance of smoke from a fire. This leads to the idea that the most effective way to 'disperse' any effluent in an environmental fluid, such as a river, the atmosphere, or the ocean, is to ensure that there already exist (or if not to create) random, chaotic motions which mix the pollution into the surrounding unpolluted medium. These motions, which are not usually described in any detail, are presumed, in the formulation of most models designed for the computation of the consequences, to reduce gradients of concentration of the pollution in close analogy with molecular dispersion or any conduction mechanism which tends to create a uniform distribution of the pollution or whatever the intrusive quality may be, such as heat, acidity, or dust.

The events following the disaster at Chernobyl well illustrated that these concepts are too simple for understanding the events and are in error concerning the mechanism. The basic mechanism is shear flow, which has the effect of stretching any compact block of fluid into a long streak or sheet. This increases (see Ref [1]) the gradients of the intrusive quality, if only because the dimensions of the polluted region are reduced in at least one direction as it is stretched in others. Thus a compact mass becomes a thin sheet or elongated tube or column, where the greatest concentrations in the middle of the polluted region are placed closer to parcels of the ambient unpolluted medium. This juxtaposition makes the motions of smaller eddies, and of molecular mixing in particular, more effective in producing a more uniform distribution of the pollutant.

In the case of the atmosphere the most important shearing motion is the variation, with height, of the horizontal wind. The most important smaller-scale mixing motion is that due to thermal convection, in which warmer parcels are carried vertically upwards and cooler ones downwards, at the same time mixing with the surroundings through which they are passing. (For an extended description see [1], chapters 7–11). There is also mechanical turbulence, which is the wakes of obstacles such as buildings and trees which are present in the path of the wind and eddies caused by instability in the shear layer due to the drag of the ground. Mechanical eddies have the property that they degenerate into smaller eddies as a consequence of larger-scale shearing motion. To model the track of a pollution particle the most satisfactory procedure is to assume that every particle is carried along with the wind and to impose some sort of spreading on that; this is usually referred to as a Lagrangian method, tracing puffs, which are assumed to be diluted according to some formula designed to simulate reality. In so far as theories contain any way of predicting the value of the peak concentrations to be found in the wind, they are based on observations of many cases and not on any basic theory of the mechanism.

Thus, for example, in following a single particle (according to a Lagrangian procedure in which 'the wind' is deduced from the pressure field, which is the best-measured meteorological parameter) it is readily recognized that several approximations have to be made. Every method using finite time or distance intervals is subject to errors because of the averaging that is necessary in any calculations of this kind. If departures equal in magnitude to these errors are artificially introduced, variations in the track of a particle being carried along by the air are discovered, and the imposed numerical errors are seen to produce a kind of dispersion which is sometimes exploited by modellers to represent the real dispersion mechanism. It must be appreciated that random variation procedure is quite arbitrary, has no connection with the real dispersion mechanisms, and is usually justified on the grounds that it sometimes appears to produce a result not entirely dissimilar to observed events. This justification is often muddled up with a theory of randomness in which arbitrarily introduced variations are assumed to produce results similar to actual fluctuations, which are called random solely for reasons of complexity (and our ignorance of their precise causes), which does not necessarily have anything to do with artificial randomness.

In the case of the atmosphere it is an error to assume that such a spreading of a Lagrangian track is a good method of generating a widening plume, the likeness being very superficial but taking no account of atmospheric stratification which inhibits vertical spreading. However, such procedures are widely used because of the absence of any alternative which is as easy to program, and it is not difficult for the operator to fabricate a plausible justification, usually in terms of randomness concepts.

Undoubtedly useful, and often convincing, models can be synthesized and are more helpful than none. But from what has been said earlier in this chapter it can be appreciated that atmospheric precipitation mechanisms are the opposite of dispersive. Rainfall is a very effective concentration mechanism, and, with it, so is the deposition of pollution by rain. In the aftermath of the Chernobyl disaster there was, of course, a very serious situation due to the deposition by gravity of the larger particles (greater than 50 μm) of radioactive pollutant within a short distance of the source; a short distance in this case being 30 km, which appears to have been the distance beyond which the drastic measures of decontamination taken within this distance had to be reconsidered.

At greater distances it was possible for rain showers to collect and deposit airborne pollution present in a very large area, such as 100 km^2, into about 1 km^2. This might happen at a front, but from the satellite viewpoint would be most serious in regions where large cloud particles, which act as the collecting filter, are airborne pollution present in a very large area, such as 100 km^2, into about 1 km^2. This might happen at a front, but from the satellite viewpoint would be most serious in regions where large cloud particles, which act as the collecting filter, are predominant. These are areas where, in Ch 3, clouds become very rapidly very black. In a shower initiated by ground level heating, the low level air which has been warmed by proximity to the ground and converges into the shower is the most important source of pollution. Then, the deposited pollution would be similar in terms of the isotopes present to pollution measured by ground level instruments in the air. At fronts, by contrast, surface air may well be converging towards a rain-producing area, but only the air from one side (the warm side) of the front would be subject to lifting while the colder air mass would usually be subject to subsidence and divergence. In this latter case pollution similar to that deposited by the rain may be completely absent from the surface air which may have been sampled.

The tracks of some parts of the emissions from Chernobyl were so bizarre as to confound confident prediction; indeed they have been credibly documented only because very small quantities of radioactive substances can be identified. On several occasions black snow has been deposited in eastern Scotland, and this has been traced back by reversed trajectories to regions in eastern Europe where black coal is widely used as an industrial fuel (e.g. upper Silesia). But the backtracking procedure is so unreliable that it has been easy to question conclusions of this kind. On the present (Chernobyl) occasion, however, there is no doubt that radioactive particles deposited in Scotland came from Chernobyl, which is much further from Scotland than Silesia.

The events of 26 April–6 May 1986 demonstrate the wide range of possibilities which may well give greater credibility to backtracking analyses in the future.

We next describe the meteorological events during this period. There was no occasion when a hot spot was indubitably recorded by Ch 3 at Chernobyl, although some pictures, as we shall see, do seem to allow this possibility. The initial explosion probably propelled some radioactive particles to greater heights than other subsequent emissions, and placed them at levels above the well-mixed layer which existed from the ground up to whatever height was determined by the surface temperatures achieved in sunshine the next day. Such higher level particles tracked by a Lagrangian method may legitimately be assumed to have followed isentropic (i.e. constant potential temperature) trajectories (see, for example the backward and forward trajectories in the paper from Berlin referred to at the end of this chapter). Particles within the well-mixed layer can realistically be assumed to be fairly uniformly distributed within it by the thermal convection due to sunshine on the ground. A backward track is obtained by simply reversing the displacement in each time step.

Rain-producing clouds can be assumed to be deep black in daytime Ch 3 images, either in fronts or showery areas. The direction of the wind is very reliably revealed by cloud streets, but these must not be confused with lee waves, which usually lie across the wind direction. The thermal wind (which is the shear direction above the well-mixed layer) is reliably indicated by the streaks of middle- or high-level cloud.

Thus the satellite view often shows much more than a weather chart because it shows the wind at several levels.

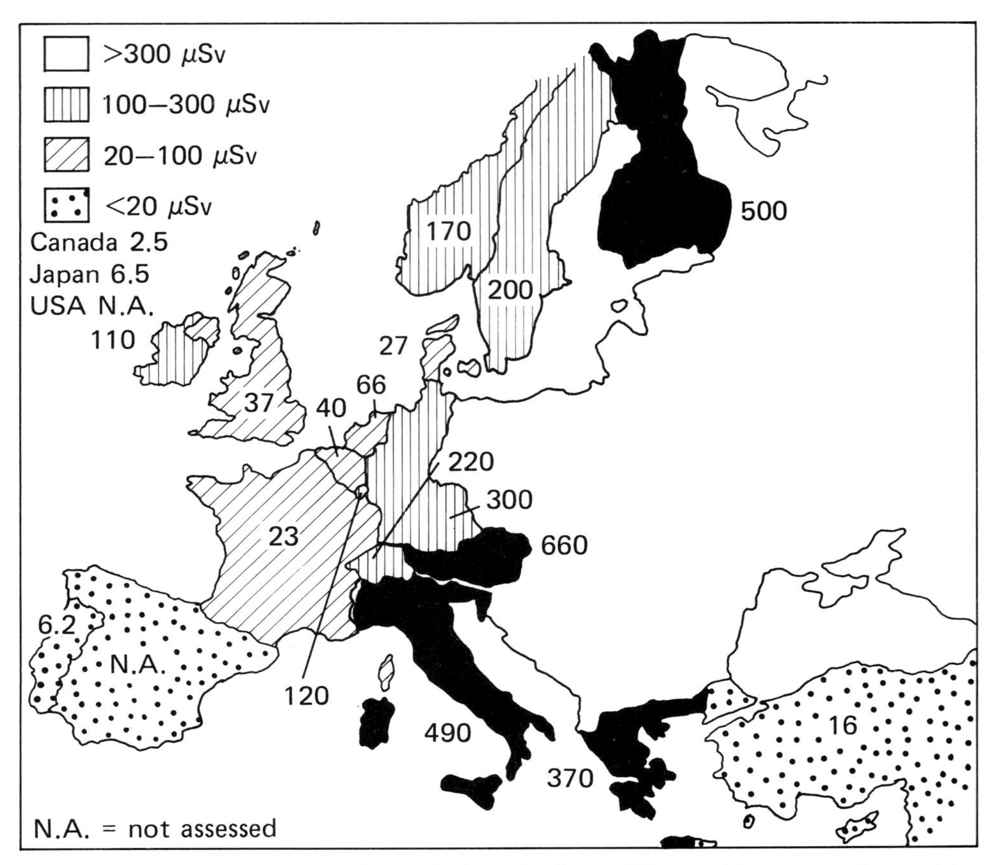

The radiological impact of the whole episode on Western Europe is pictured in the accompanying map (from Atom **377**, March 1988). As would be expected according to the behaviour of a progressively diluted plume meandering across the territory downwind of the source, the deposited activity decreases with distance from the source, with the notable exception of Ireland. But it is important to note that such politically defined maps obscure the great variations which occurred within each country and were not related to the concentration measured in the air when the plume was over the country. The rain, which was responsible for most of the deposition, was very patchy and intermittent, so that some districts received doses which were many times as large as the average for their country. The importance of this relates to conclusions drawn by official bodies. Thus the average for Britain is quoted as being only a small fraction of the normal annual dose of radiation from natural sources, or equivalent for someone in northern England to a three-week holiday in Cornwall (where the natural dose from radon gas is the greatest in the country). The small total dose figure is produced by averaging the dose in the mountain areas, which experienced rain showers, over the whole population, including those in London, which suffered negligible deposit.

Similar arguments are used concerning the incidence of cancers induced by a dose, which would probably have been taken as a result of having eaten polluted vegetables or meat of animals which had grazed on the mountains. It is correctly said that a few thousand cancers which are likely to have been caused by the dose cannot possibly be identified as being due to the Chernobyl emissions when as many millions will occur in the same population due to natural causes during the half century following the event. At the same time it must be appreciated that the method of calculating the number of cancers probably caused by the event is open to very serious doubt because there is good evidence that small doses reduce the chance of getting a lethal cancer (e.g. [14]).

Arguments based on averages and extrapolations to small doses to obtain information about effects which cannot be investigated individually are very dangerous if they are to be the justification for important administrative decisions. It is clear from this incident that those most seriously affected were within a short distance at the time, but at greater distances the effects may be very variable. For example animals grazing in Finland, where rain fell early on, were at much greater risk than those very much closer where it did not rain and who only suffered a much smaller dose by dry deposition. In many parts of Europe the greatest deposits, due to rain, were close to places which suffered almost no deposit: this is specially true in Ireland, where there was no significant concentration in ground level air, but serious deposits from showers drawing pollution from air at higher levels.

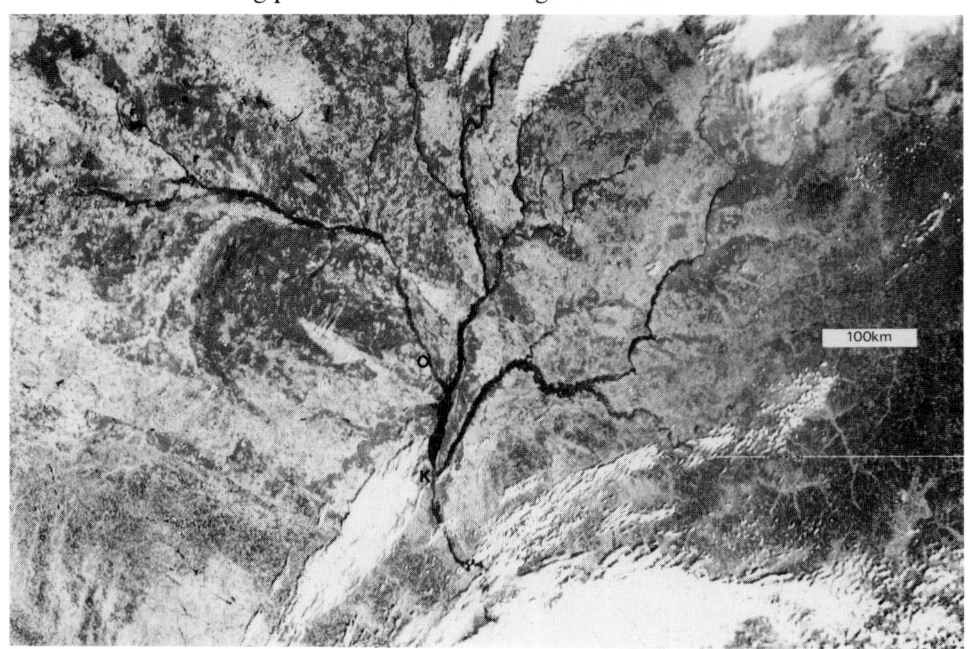

10.5.1 0921,25.4.86,CZ5

10.5.1

This picture of the previous day shows Chernobyl (O) to be just outside the edge of the cloud system of a cyclone centered to the west of the Crimea. The cloud streets at 0921 indicated a convergent inflow, the wind over Kiev (K) being from NNE. There appear to be three plumes about 280° and 75 km from Chernobyl, indicating a wind from the north.

10.5.2 1218,25.4.86,2

10.5.2

Three hours later a system of cloud streets has developed over Russia, indicating that particles becoming airborne at Chernobyl would be carried along the streets towards Lithuania and the Baltic Sea. There was an active front over Sweden extending into Poland, and cloud over southern Finland. The low to the south was obviously filling up because the cloud streets now indicate subsidence and outflow.

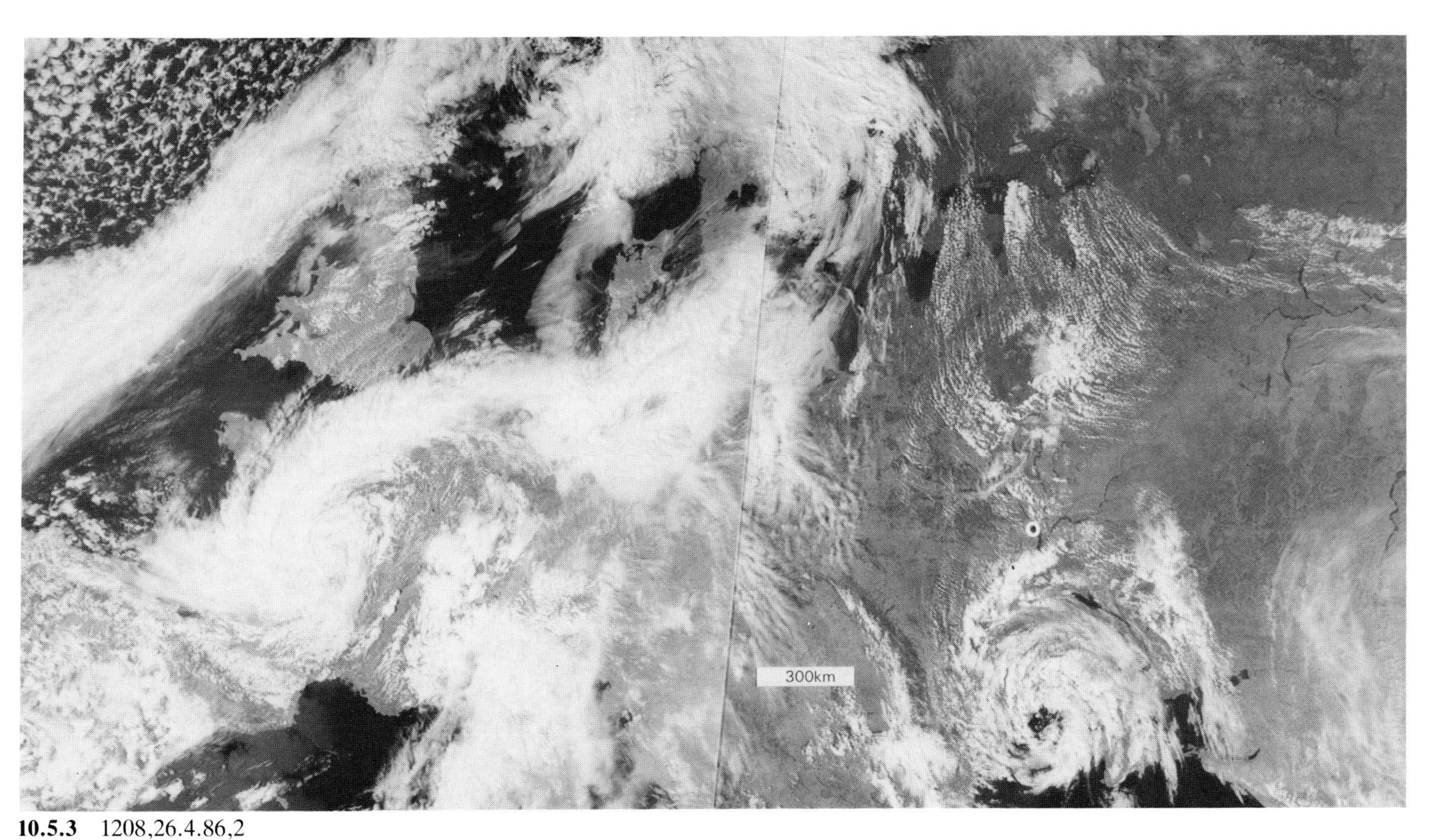

10.5.3 1208,26.4.86,2

10.5.3

The scene remained very much the same 24 h later, even in respect to some minor details. Clearly some rainout was likely on both sides of the Baltic Sea. Although the front over Ireland indicated the likely dominance of southwesterlies over Britain, England had come under a northeasterly wind associated with the low over western France. The well-developed streets indicated a fairly uniform well-defined mixed layer in Russia.

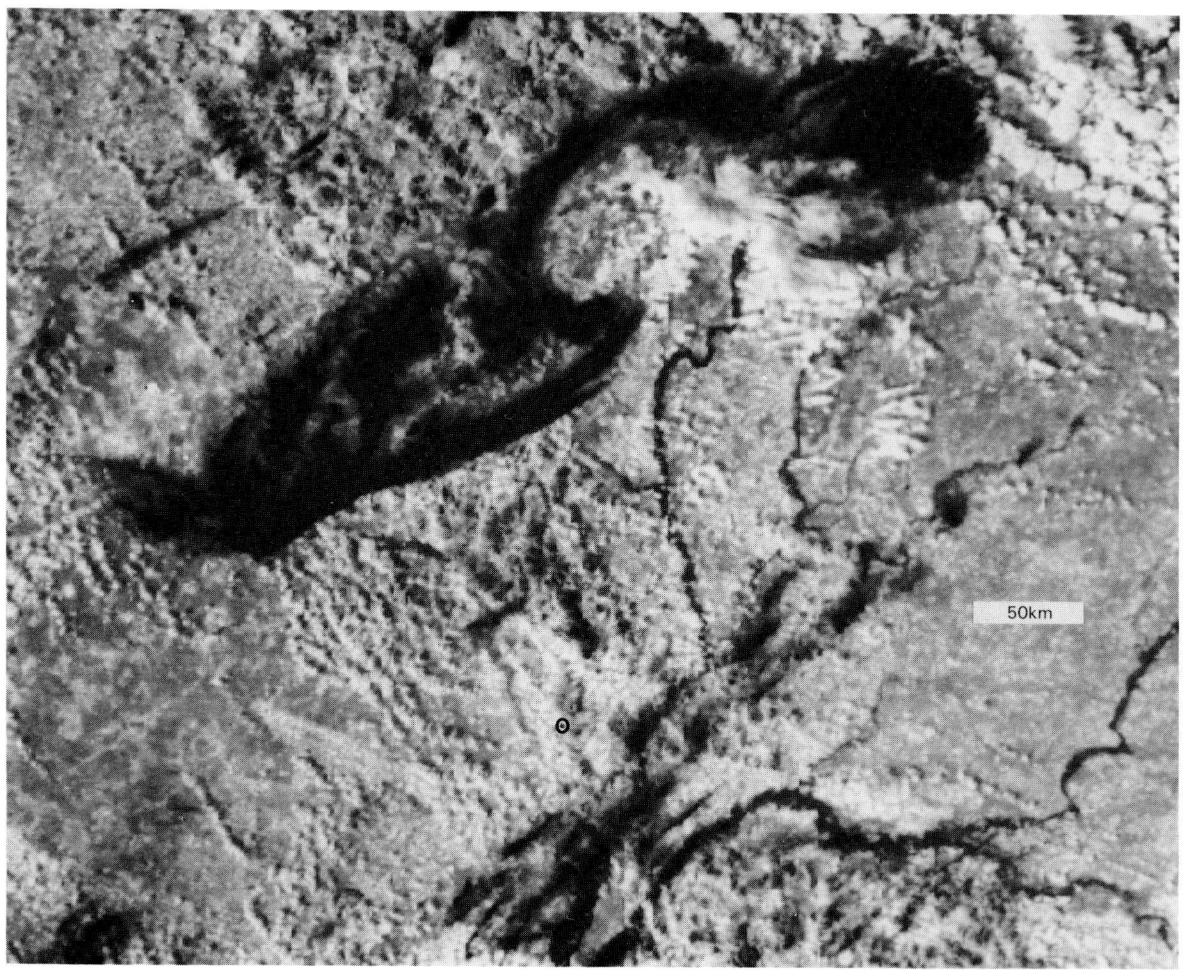

10.5.4 1208,26.4.86,3

10.5.4
This Ch 3 picture shows, enlarged and in black, the patch of cirrus seen in the previous picture, which was thought at one time possibly to have been generated by the explosion about 12 h earlier and about 100 km further south; but there is no corroborative evidence that there was any connection.

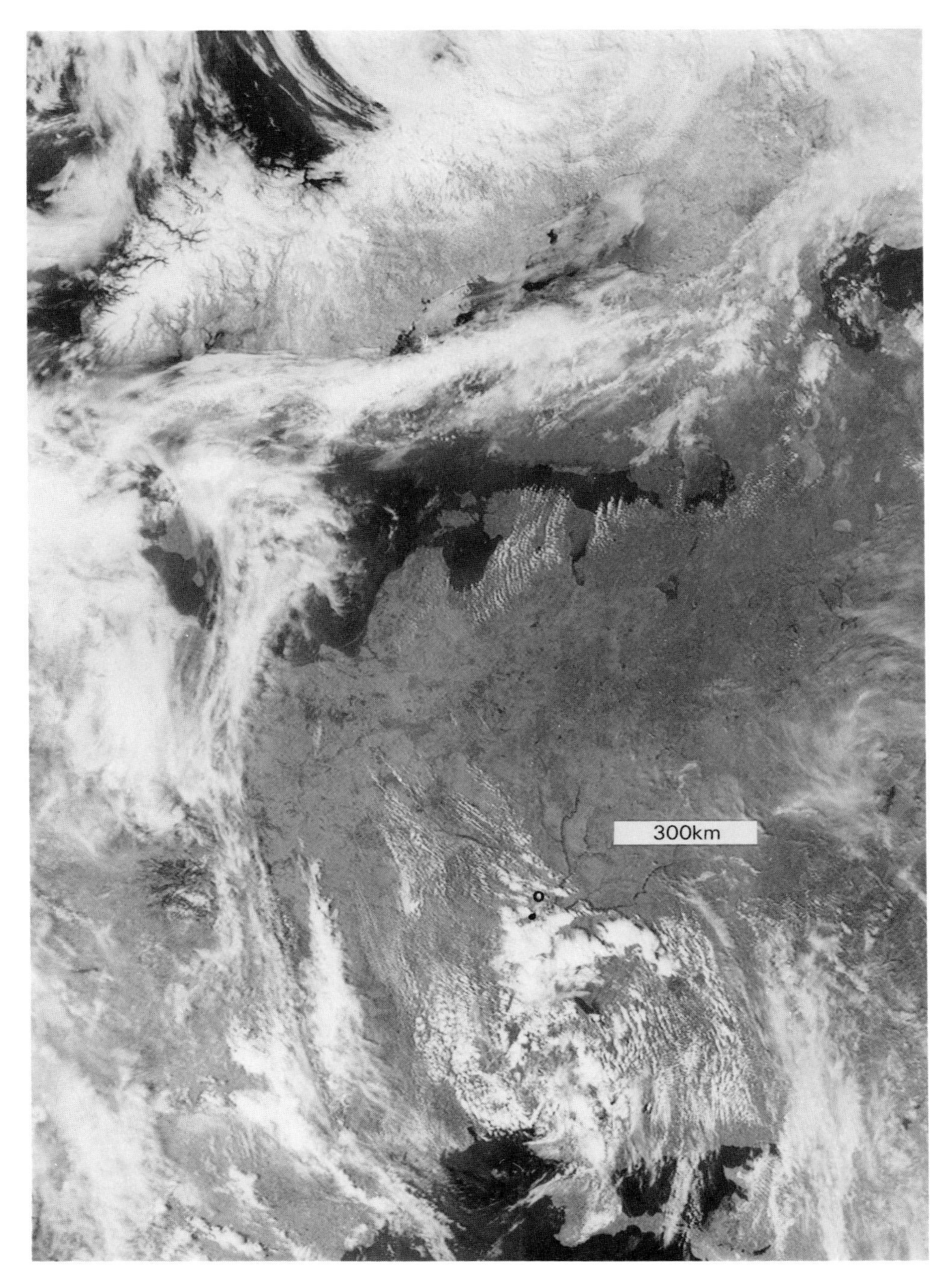

10.5.5 1158,27.4.86,2

10.5.5

For the next day the same situation prevailed, with almost all the pollution being carried in the mixed layer travelling towards the Gulf of Bothnia. By this time the deposit in Sweden and Finland by rain could have been considerable.

10.5.6 1147,28.4.86,2

10.5.6

The weather in the Ukraine is now appearing more stormy, and the effluent is being carried around the new cyclone centre east of Chernobyl. The skies are clearing over the Baltic states with a southeastwards drift being established across Central Europe. The streets show a southwesterly wind over England and a flow from between west and WNW over France.

10.5.7 1137,29.4.86,2

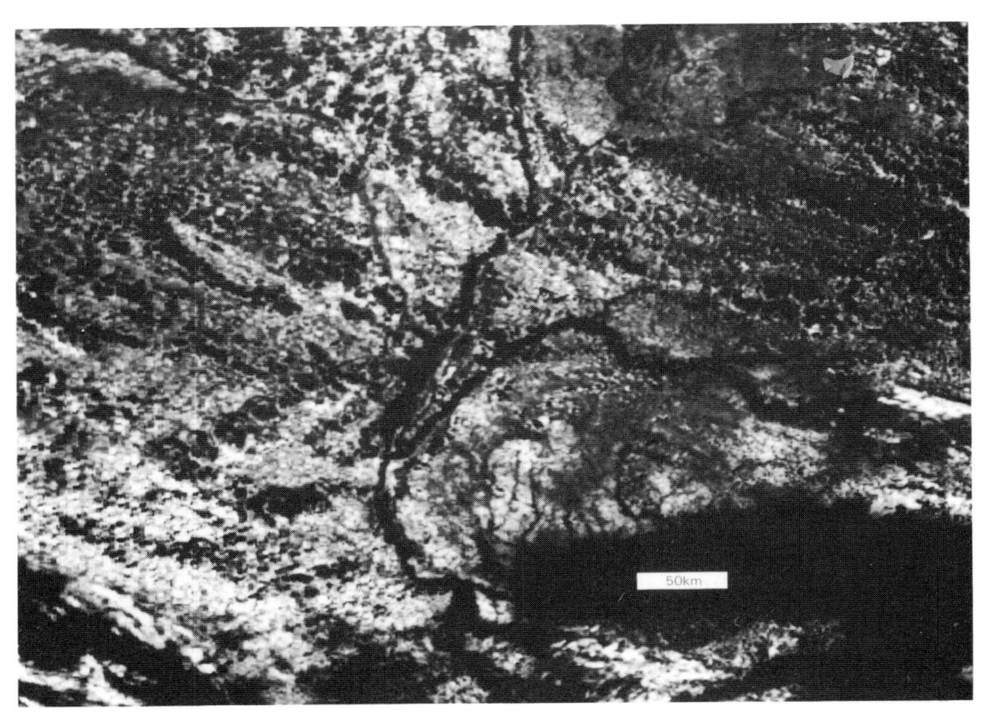

10.5.8 1137,29.4.86,3

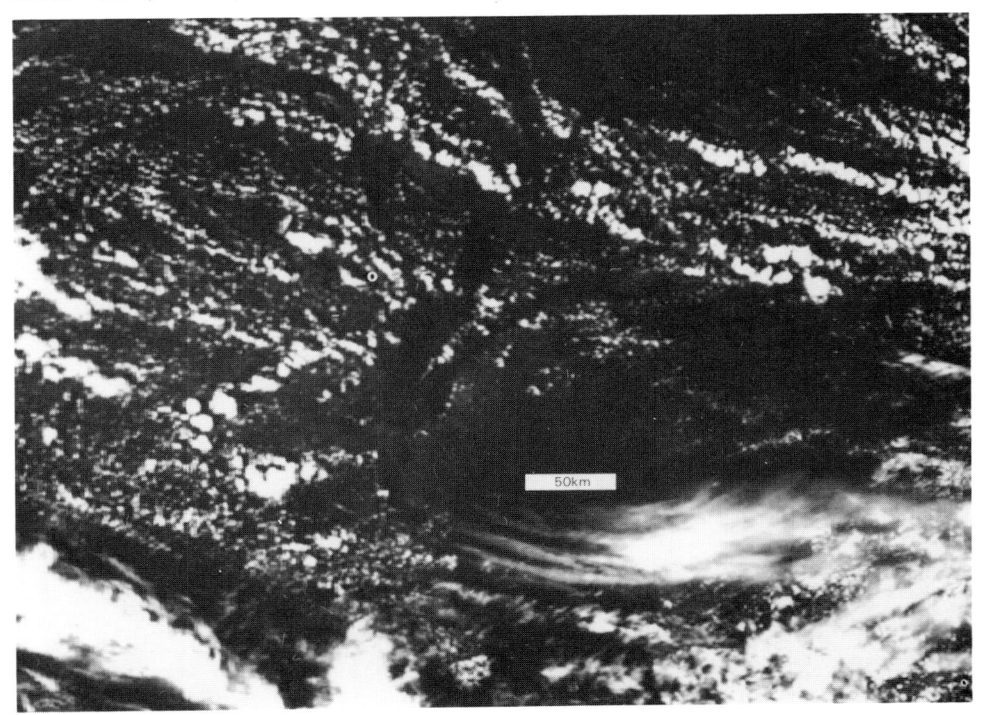

10.5.9 1137,29,4,86,1

10.5.10 1305,30.4.86,2

10.5.10
No possible hot spot was seen at this time. A flow from the east has now been well established, especially in Italy and Yugoslavia. Weak glint covers the North Sea and Western Mediterranean. A westerly or southwesterly appears to be established over the British Isles with cirrus across the North Sea.

10.5.7, 10.5.8, 10.5.9 (**10.5.8** is an enlarged part of **10.5.7**, and **10.5.9** is the same area)
High pressure is now being firmly established over the southern Baltic with easterlies over Southern Europe and westerlies over England and Wales. A few large cumulus are being created in Southern Germany and Austria, the drift becoming very variable over Chernobyl itself, and the low moving eastwards further into the Soviet Union. A bright spot was detected over Chernobyl in Ch 3 at this time, but this could possibly have been a very reflective cumulus cloud: in Ch 1 at the same time there were many bright cumulus but most of them are dark in Ch 3, which suggests that a hot spot was detected at Chernobyl. The air over the southern Baltic was becoming hazy with air pollution, probably from Poland and East Germany. The plumes to the west of Chernobyl (see **10.5.1**) are no longer in evidence.

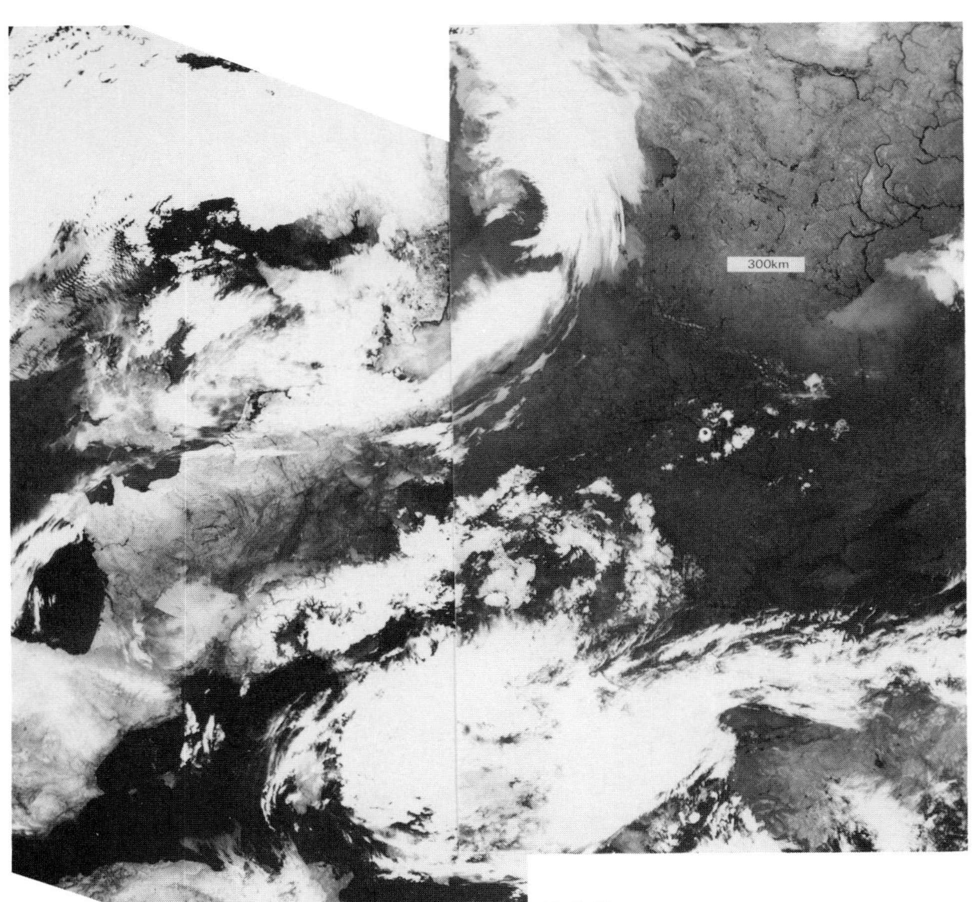

10.5.11 0119,1.5.86,4

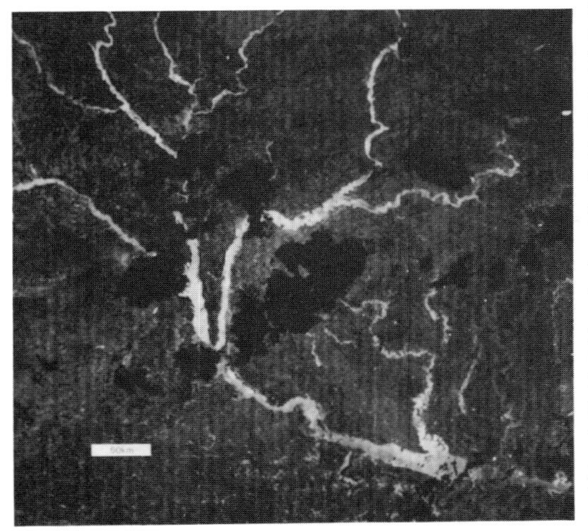

10.5.11 0119,1.5.86,3

10.5.11

This picture shows convective activity across Central Europe which persisted after midnight. It shows clearly also the high cloud of the front lying across Northern Europe from Cape Finisterre, Britanny and the Netherlands to the Baltic States and Finland. The Ch 3 enlarged version of the Chernobyl area shows that several hot spots were detected east of the River Dnieper, and the glow at the edge of the large cloud over Chernobyl is suggestive of a hot spot obscured by the cloud. At this time it was not unreasonable to believe that areas north of this front would not receive any more deposition. But the streets and waves show a northwesterly wind over Sweden and Lithuania, a northerly over Poland, northeasterly over Slovakia, and an easterly wind over Southern Germany, and even a southeasterly in southwest France.

10.5.12 1254,1.5.86,2
Visible air pollution mainly from East Germany, Poland, Czechoslovakia and neighbouring areas, can be seen spreading into the Bay of Biscay and English Channel, and across Corsica and Sardinia into the Western Mediterranean. Rain continued from time to time across Italy and the Balkans. Glint dominates the sea west of Gibraltar.

10.5.13 1244,2.5.86,2
During this day there was strong development of convective rain in Switzerland, south Germany, Italy, and parts of the Balkans. While there was some escape of dust haze from Tunisia, there was continued pollution around Corsica, and the escape from France is replaced by strong flow of industrial haze from Europe into the North Sea and a strong flow across the Channel from France, as the front over Ireland begins to develop a cyclonic circulation off southwest Britain.

10.5.14 1233,3.5.86,2

10.5.14 1233,3.5.86,3

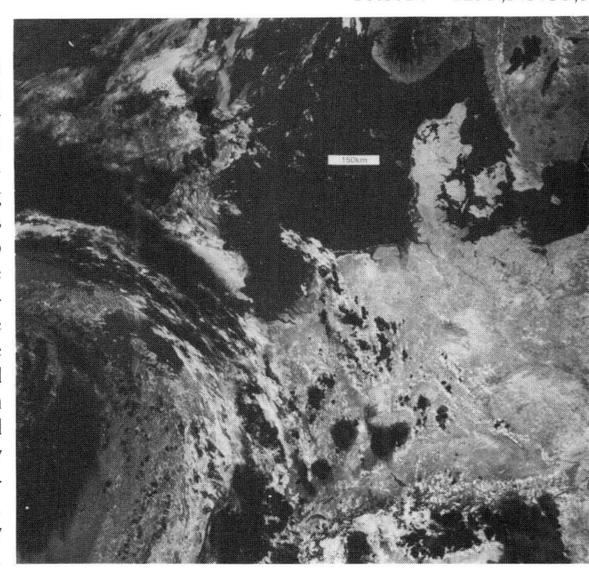

10.5.14
Cumulus crossed Holland into the North Sea and showers developed over northeast Britain and later over Wales, Ireland and Western Scotland. A cyclone over the eastern part of the Black Sea is carrying Chernobyl effluent southwestwards into the Balkans and later into Greece and Turkey, although the absence of a dominant street structure indicates weak winds in the Ukraine. The enlarged part of the Ch 3 scene shows the wet cloud carrying pollution in the direction from the Alps towards Britain and the North Sea, which is now covered with pollution from Eastern Europe and probably beyond. The emission at Chernobyl is now building up to its second crescendo.

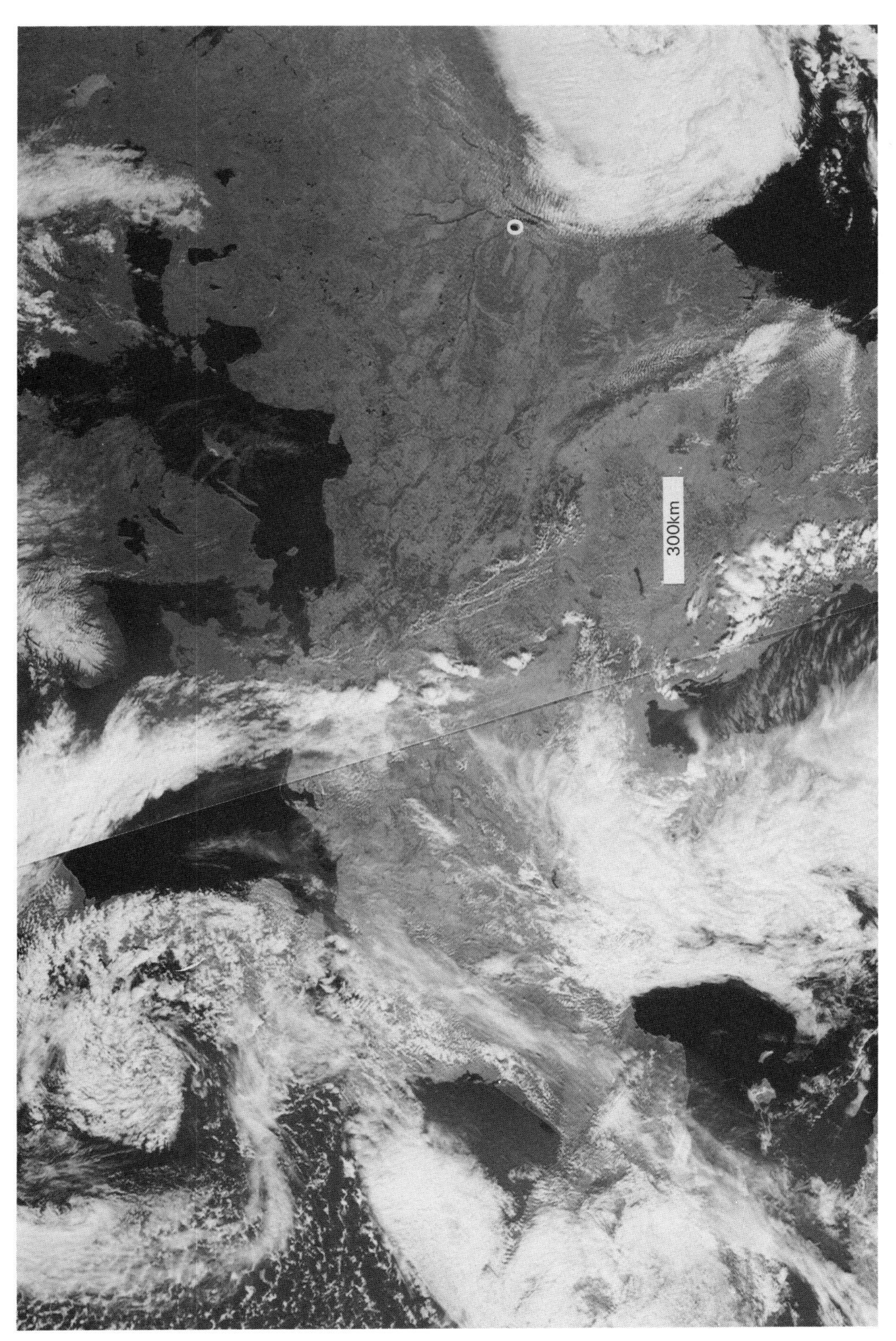

10.5.15 1225,4.5.86,2

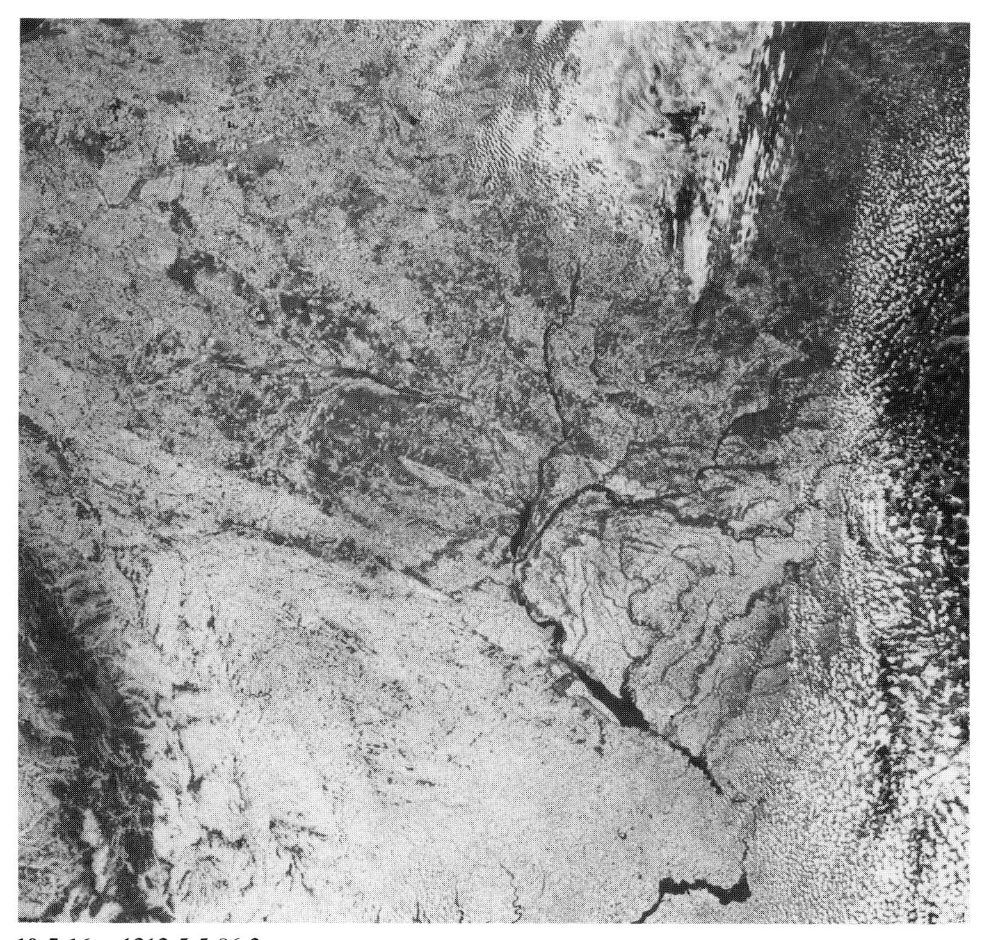

10.5.16 1212,5.5.86,3

10.5.16
At this time there is a slight suggestion of a hot spot at Chernobyl, where the cloud has now been evaporated by the subsidence in the anticyclone now centered southwest of Leningrad. This is the day of the last large emission from the damaged reactor. Almost the same scene was repeated on 7.5.86 at this time of day, but there was no emission of importance after 5.5.86. The feeble white spot at Chernobyl did not appear to be due to a local cumulus because it cast no shadow, as did the small cumulus elsewhere in the Ch 2 version of the same scene, and in the east of this picture.

10.5.15
On this day there was a scattering of rain all over the region west of a line from Greece to Denmark. Waves and streets show a drift of air from Chernobyl towards Bulgaria where the wind is northwesterly.

Reference

Trajektorienbestimmung zum reaktorunfall in Tschernobyl. E. Reimer, U. Langematz, E. Hollan; *Beilage zur Berliner Wetterkarte*. Freien Universitat Berlin, Jan. 1987.

11

More-natural air pollution

In Chapter 6 we described how air could be made very clean over the polar ice cap or in a Pacific Ocean desert. In Chapter 9 we investigated desert dust, which Ch 4 can detect more easily than industrial pollution. Here we shall be concerned with volcanoes, forest fires and agricultural sources of particulate air pollution, not because we have natural events of great importance to describe but because recent occasions in Western Europe revealed some interesting properties of the various channels in detecting some of these emissions.

11.1 VOLCANOES

The rapidly changing scenes produced by volcanic eruptions are best observed by geostationary satellite, and from the recordings it is possible to make time-lapse cine films of the events. Here we shall be concerned with the detectability of the products of an eruption by routine meteorological satellite.

In three eruptions of Etna in recent years we have pictures of the hot spots in Ch 3 and the plumes of dust and cloud in other channels. The pictures 'speak' for themselves, and we seek only to draw attention to the most important features which the pictures record.

ETNA 4.8.79; 15.5.83; and 25.12.85

The eruption of 4.8.79 was feeble and the pictures are of no special interest here. There is a single nocturnal picture in [9] p. 207.

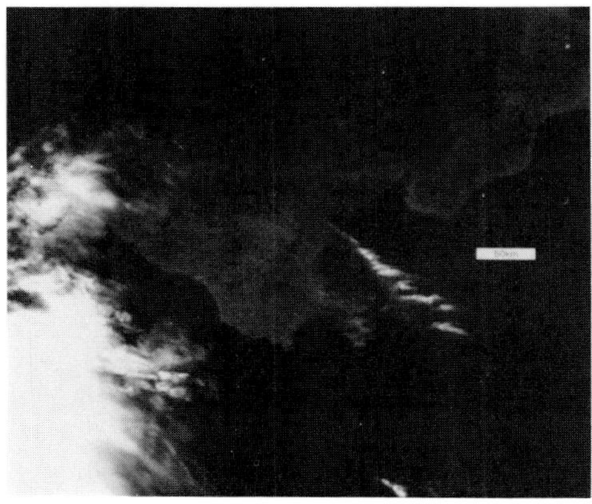

11.1.1 1340,15.5.83,1

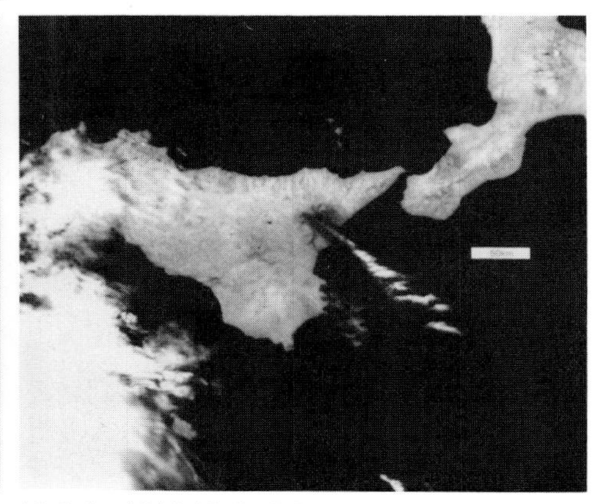

11.1.2 1340,15.5.83,2

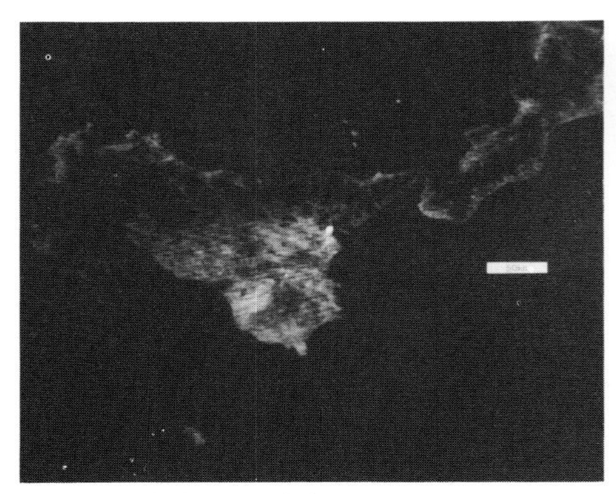

11.1.3 1340,15.5.83,3

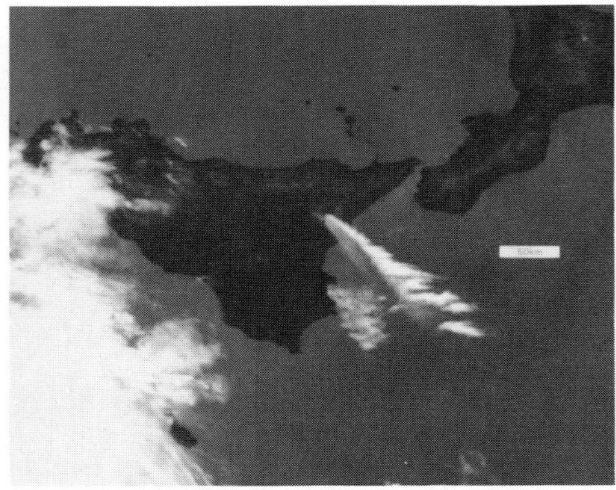

11.1.4 1340,15.5.83,4

11.1.1, 11.1.2, 11.1.3, 11.1.4

Only on the pass at 1340 was any record obtained of this brief eruption. In Ch 1 a dark plume emerges, reaching the cloud base at a distance of about 20 km where water condensation begins to make the plume white. In Ch 2 the black ground close to the volcano is evident, and this is enhanced on the north side by the plume's shadow. The only significant feature in Ch 2 is that all the land is brighter than in Ch 1.

In Ch 3 there is a large hot spot (about 10 pixels and the nearest part of the plume (the first 30 km or so) is brighter than the rest, which implies that it is significantly hotter. Stromboli, the smallest of the islands visible north of Sicily, is the brightest of them because of its hot spot, which is usually present (see later pictures), although it is much less than Etna in this case. The white cloud part of the plume is presumably high enough to have condensed large droplets, and no small droplets are present, and so it is dark.

The Ch 4 picture is the most dramatic, the dusty part of the plume evidently being diluted and cooled on the outside. The white (water) part of the plume appears to begin at the volcano, and this appears to contradict Ch 1, where it is dark, and Ch 3 where it seems to be hotter in the first 20 km or more.

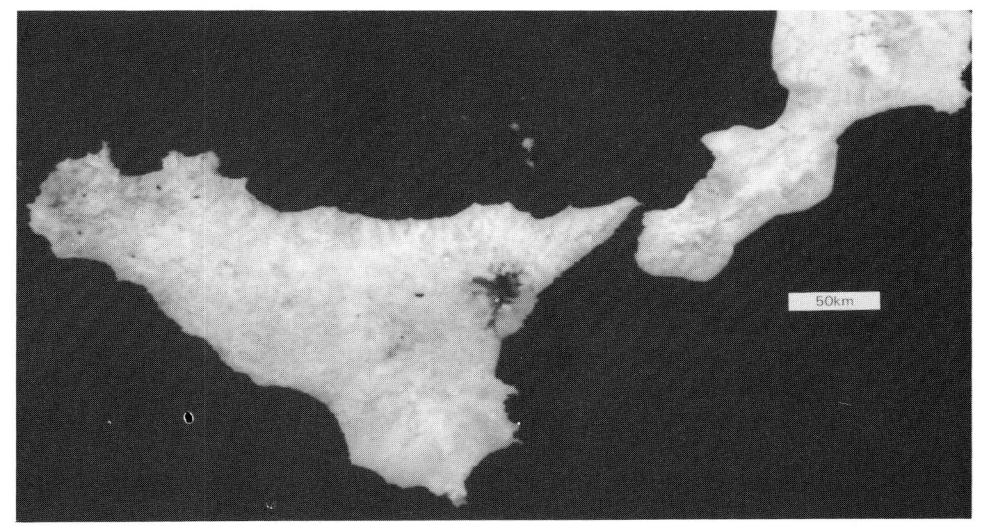

11.1.5 1342,14.5.83,2

11.1.5

On the day before the eruption the mountain is black due to the lava of previous eruptions, and there appears, possibly, to be a steam cloud on the south side of the peak.

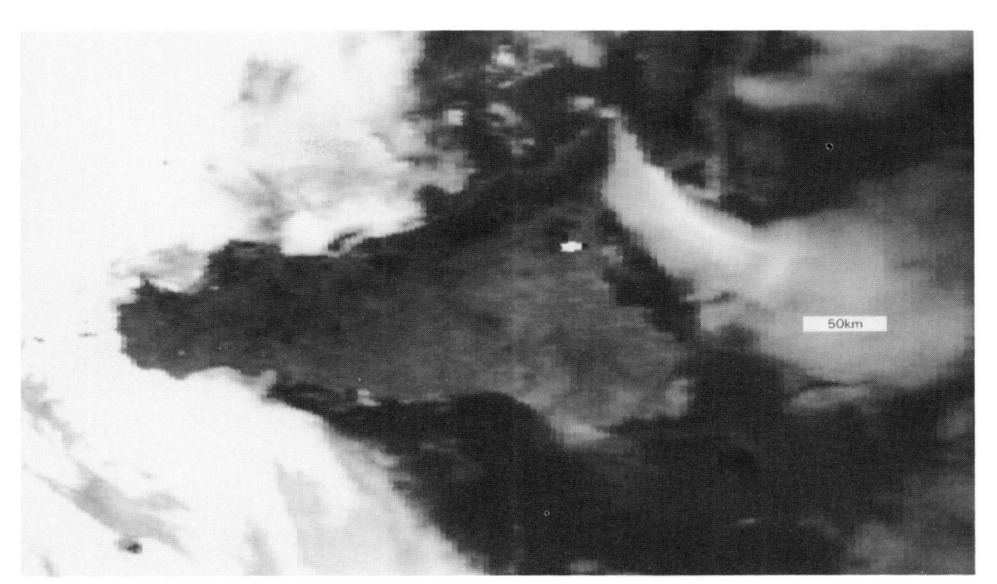

11.1.6 0751,17.5.83,3

11.1.6
Two days after the eruption the hot spot appears to occupy 21 pixels, each elongated to about 4 km east–west by nearness to the edge of the swath. The black spot on the east side is probably a reproduction effect and is certainly not a shadow of a cumulus cloud over the hot spot because the sun is in the east at this hour.

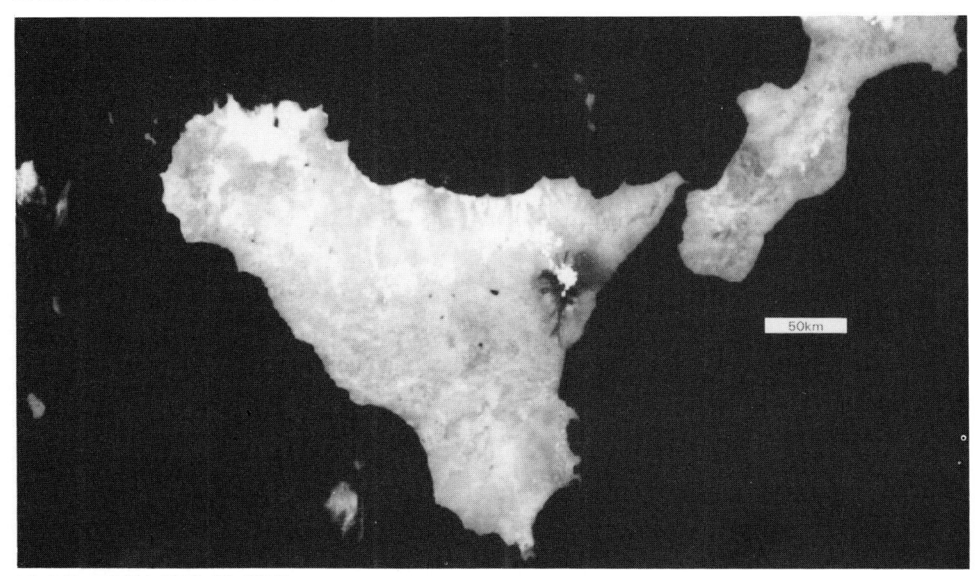

11.1.7 1034,17.5.83,CZ5

11.1.7
A different (CZ 5) look at the mountain after the eruption shows a big cumulus cloud over the peak, occupying as much as 60 pixels (each CZCS pixel being about 0.53 of the area of an AVHRR pixel). The picture is stretched in the north–south direction because of the CZCS tilt of the camera.

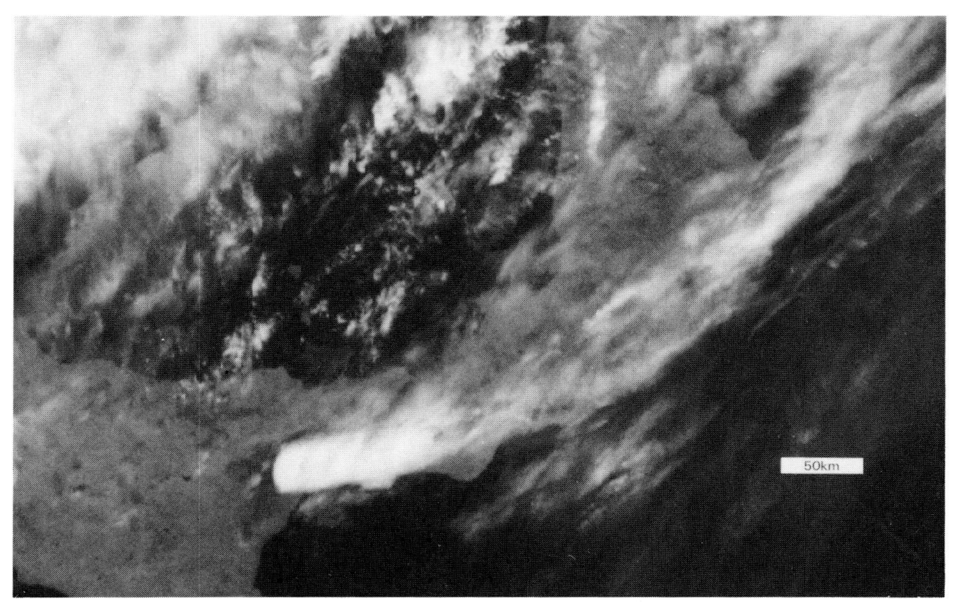

11.1.8 1013,12.4.83,CZ5

11.1.8

This more familiar picture of Etna shows it with a banner cloud in a strong west wind, with a more southerly component in the wind at higher levels. No eruption is occurring.

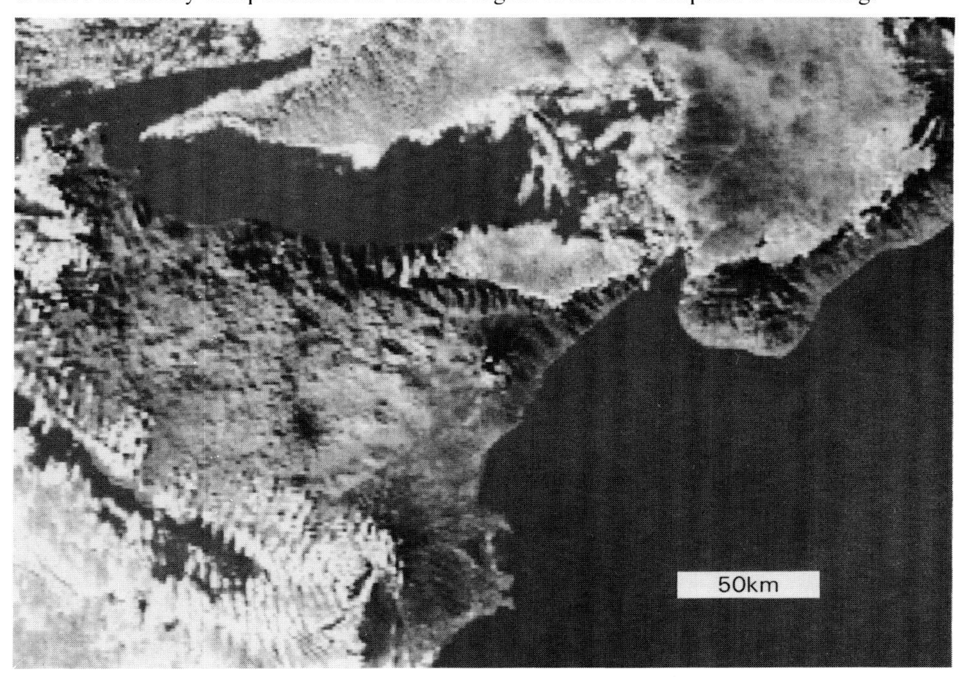

11.1.9 1318,18.12.85,3

11.1.9

The hot spot on the top of Etna was already detected a week before the next eruption.

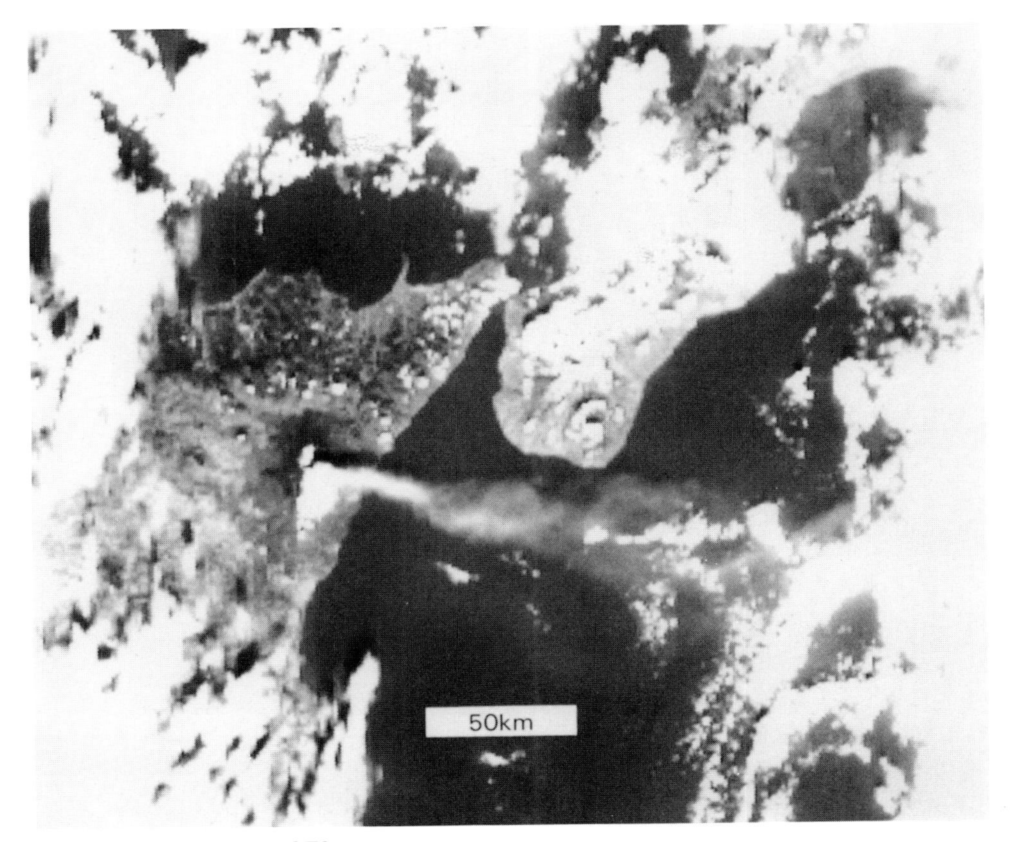

11.1.10 1033,25.12.85,CZ2

11.1.10

Quite a large area of cumulus, most of which was colder and lower than the plume, existed, with the plume emerging from its northern edge. But the higher level wind was westerly, and the dust plume and its shadow are easily visible in CZ 2. None of the other CZ channels showed any added detail which had not been seen by CZ 2.

11.1.11, 11.1.12

The Ch 1 picture shows two vortices in the plume, and these may have been shed by the mountain. They had presumably occurred since the previous picture, which requires a wind speed of about 35 km at the cloud altitude to reach the position of the eastern eddy, just over 190 km from the mountain, in the three-hour interval.

Only the cloud (not the smoke) of the plume is visible in Ch 3, while the hot spots on Etna and Stromboli are prominent, with cumulus on their south sides; and there are many dark cumulus showing large droplets elsewhere in the picture. The plume smoke was detected by Ch 4 at this time, which shows that the smoke's thermal emission in Ch 3 differed from that of the sea (its background) less than it did in Ch 4, compared with the general illumination level of the picture, which in the Ch 4 case was due to emissions and in the Ch 3 case to scatter of sunshine. The particles were too small to produce significant scatter in Ch 3, whereas in Ch 4 there was a significant temperature difference from the sea, and the difference was in the emission in Ch 4.

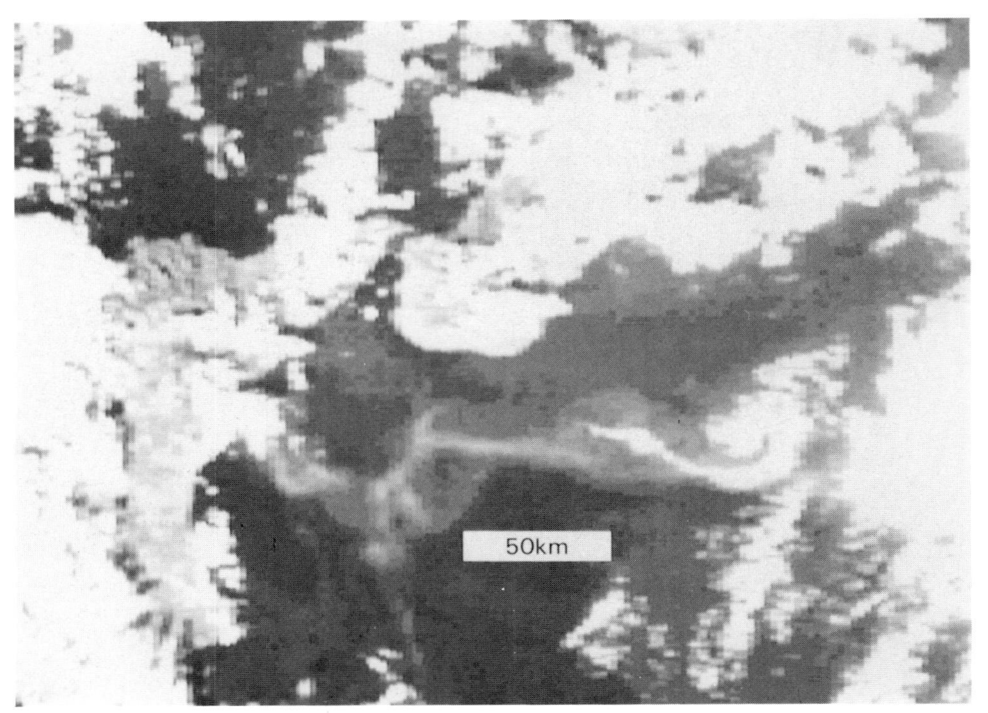

11.1.11 1343,25.12.85,1

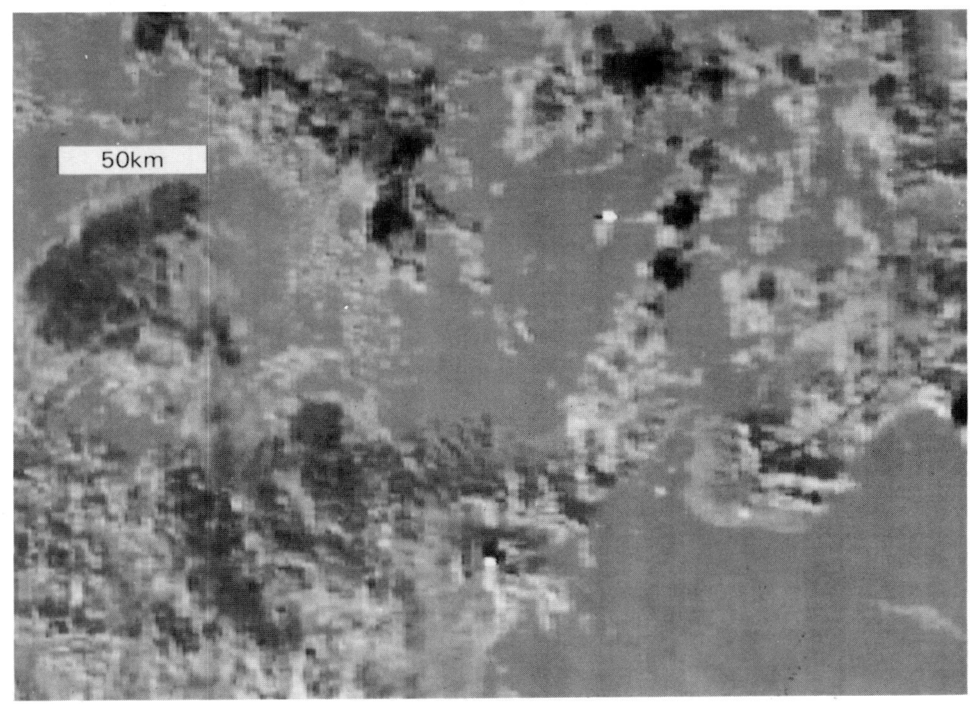

11.1.12 1343,25.12.85,3

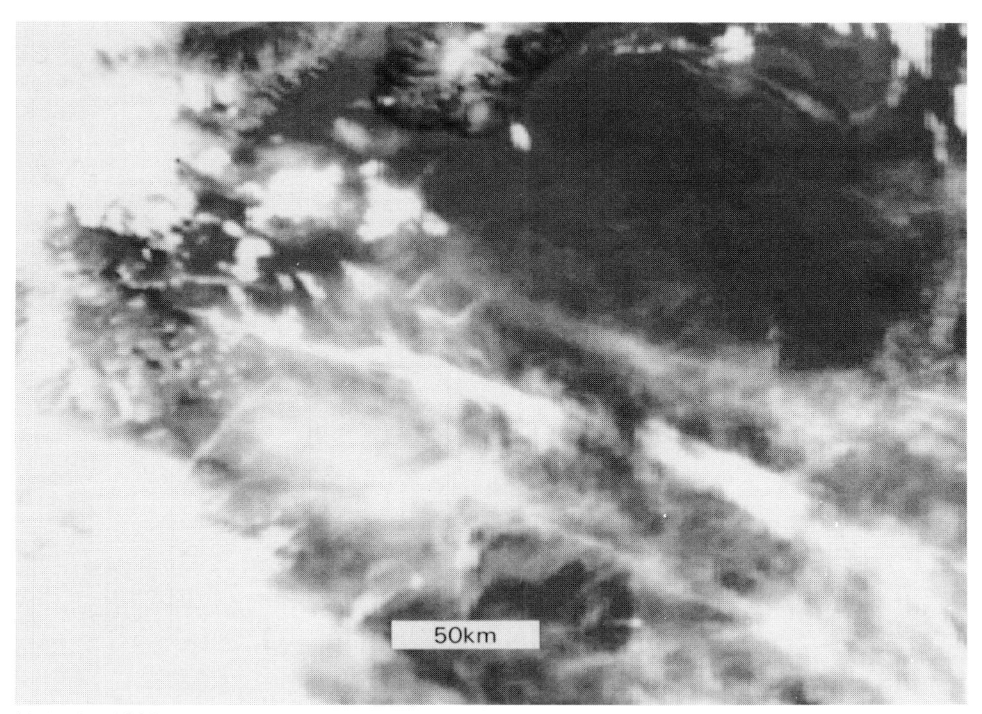

11.1.13 1333,26.12.85,1

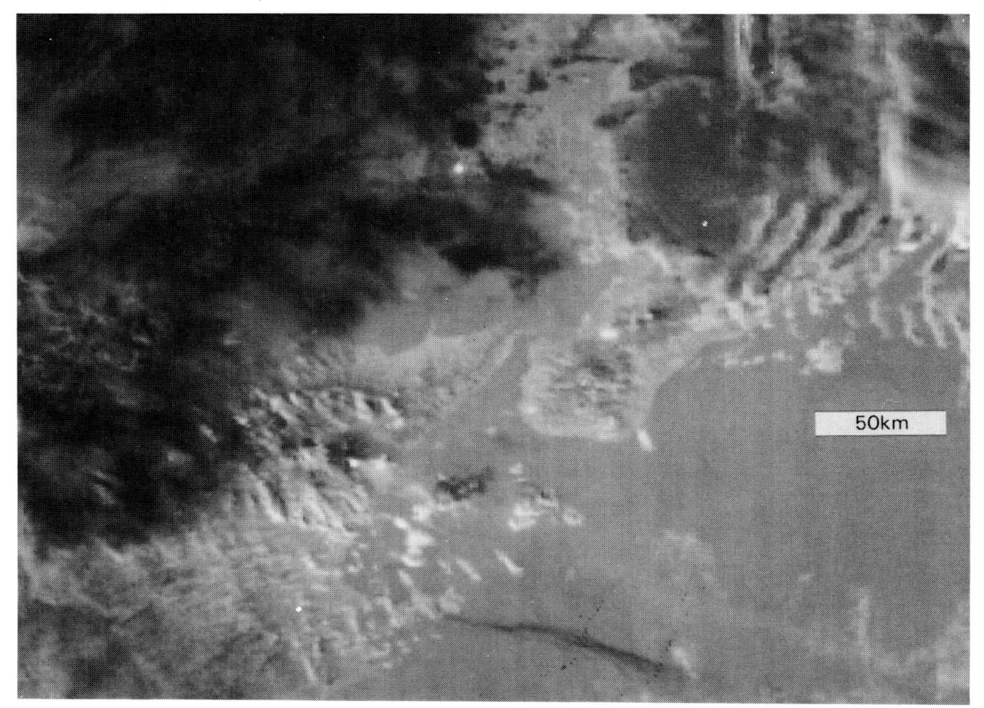

11.1.14 1333,26.12.85,3

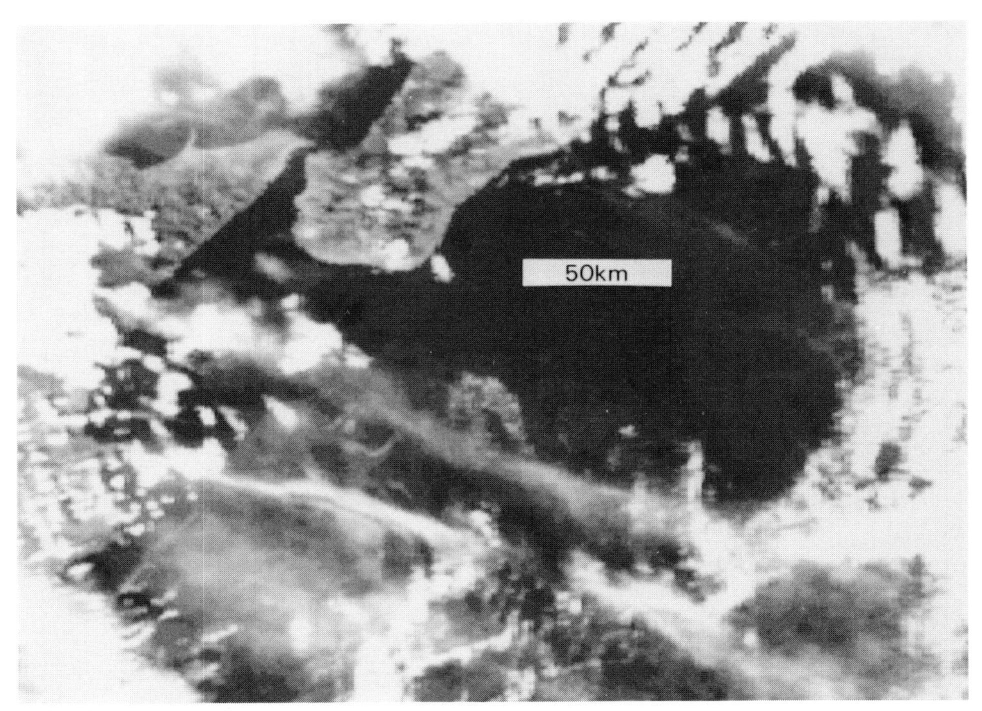

11.1.15 1333,26.12.85,4

11.1.13, 11.1.14, 11.1.15

On the following afternoon the smoky plume was obvious in Ch 1 (and in Ch 2, not shown here), and is parallel with a line of cirrus to the south of it. In Ch 3 the plume is not seen, while the cirrus is black. In Ch 4 the plume is just visible and the cirrus is a very cold white.

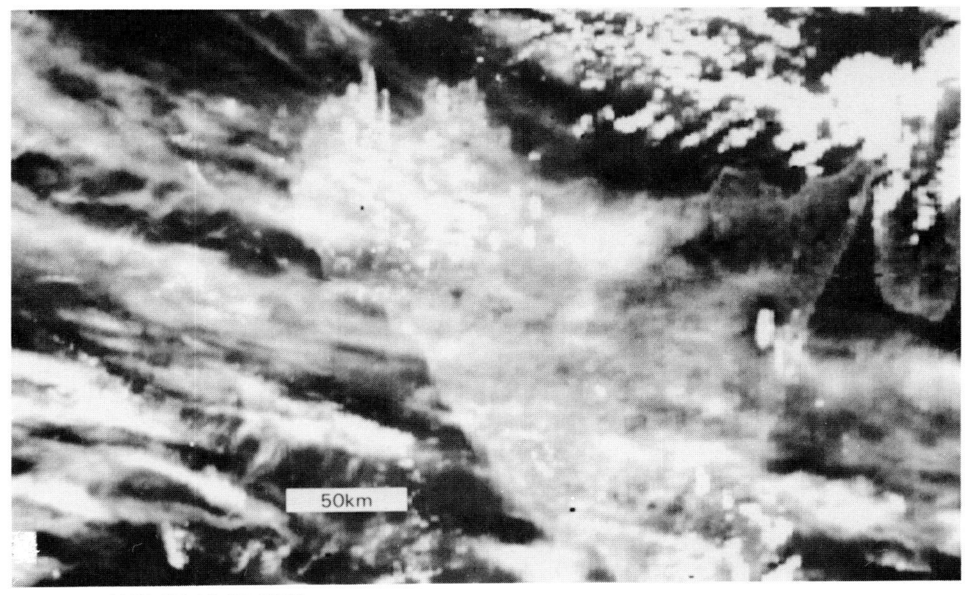

11.1.16 1109,27.12.85,CZ2

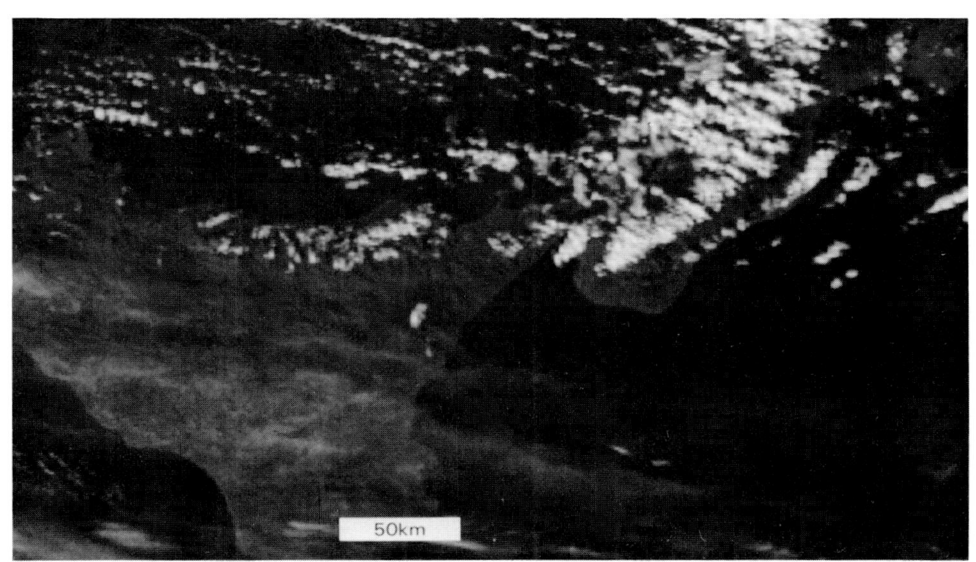

11.1.17 1322,27.12.85,1

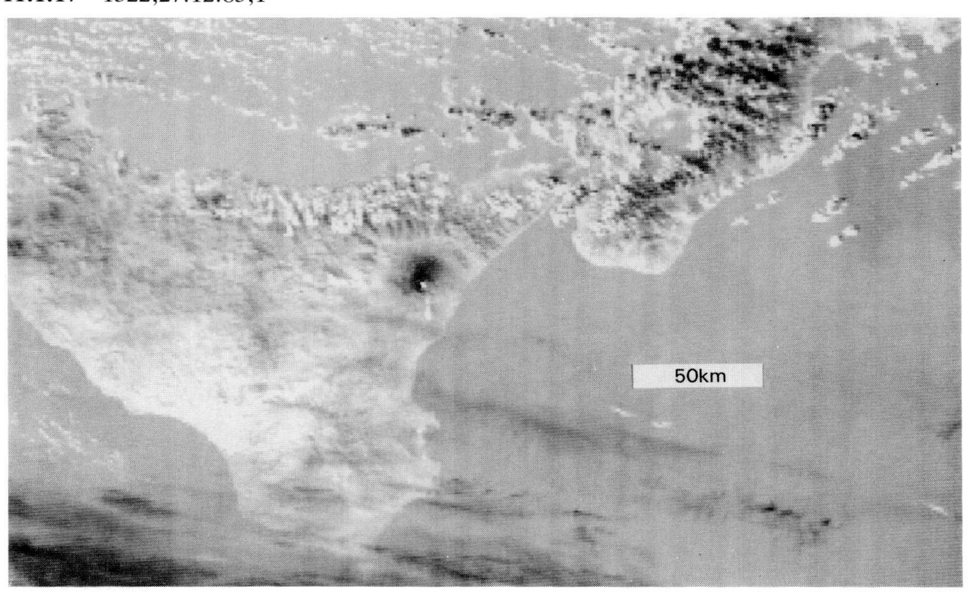

11.1.18 1322,27.12.85,3

11.1.16, 11.1.17, 11.1.18

The CZ 2 picture shows the plume carried away in a wind from WNW, while the lines of cirrus come from almost due west. There is a large cumulus patch on the mountain, with a smaller north-south white streak on the southeast side.

In the Ch 1 picture 143 min later, the same white areas are present against a much darker image of the ground. The lee waves of the toe of Italy are parallel with the mountain range rather than perpendicular to the wind. The plume is not visible in Ch 3; the white streak on the southeast side of the peak is in the same place, and the hot spot appears very small because cumulus obscures the western part of it. Being over a hot source that cloud has risen high enough to be black in Ch 3.

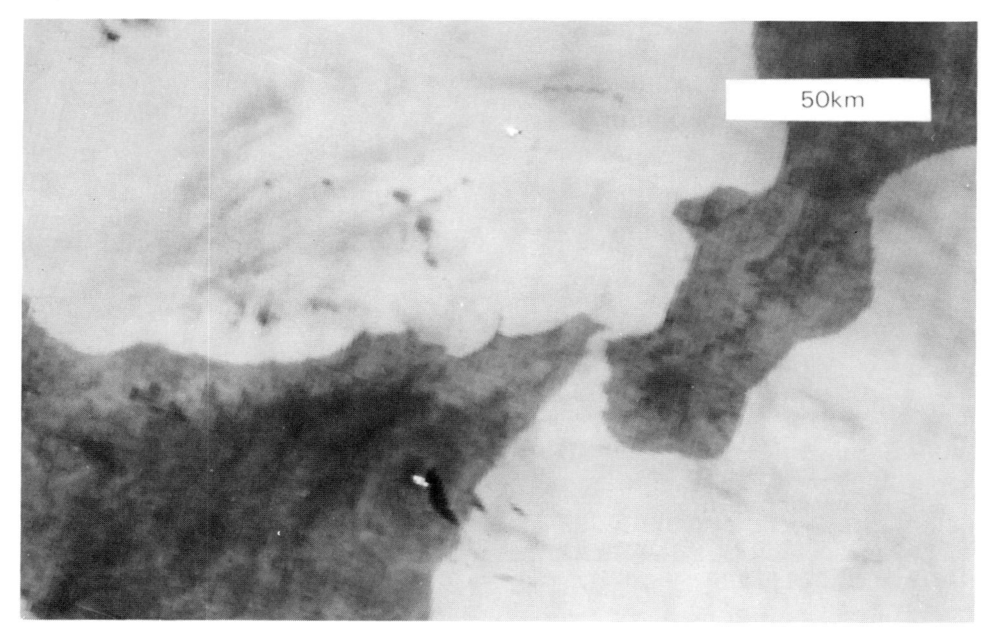

11.1.19 0137,28.12.85,3

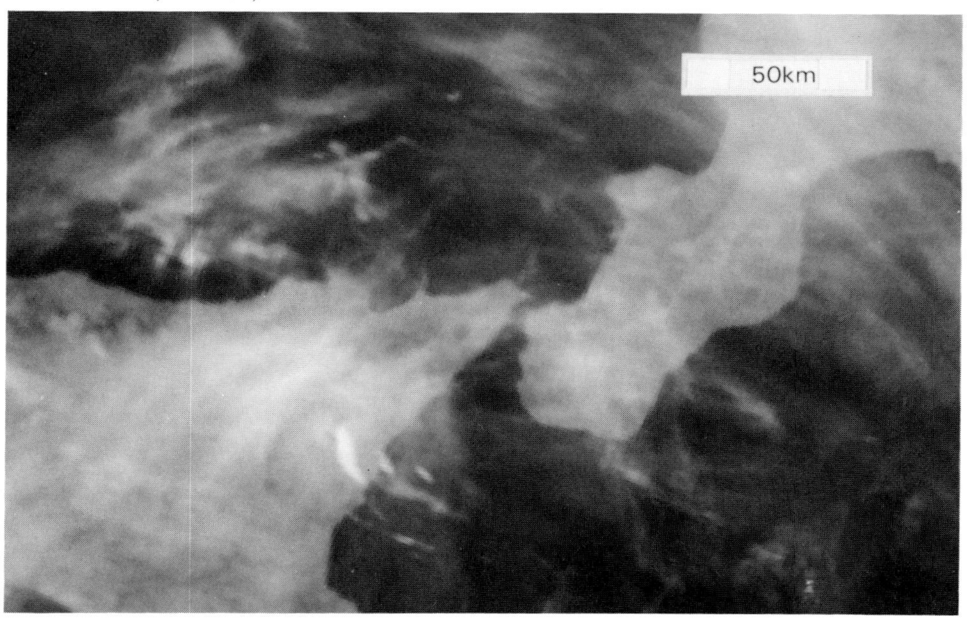

11.1.20 0137,28.12.85,4

11.1.19, 11.1.20
The following night when the cumulus at the peak was no longer present, the hot spot appears wider. Wave clouds, including three lee waves, are dark in Ch 3, but in Ch 4 they are cold and white, and, apart from the Etna and Stromboli hot spots, Ch 3 is almost a negative of Ch 4. Three nights later (not shown here) the Etna hot spot was less than a quarter of the size of the one on Stromboli.

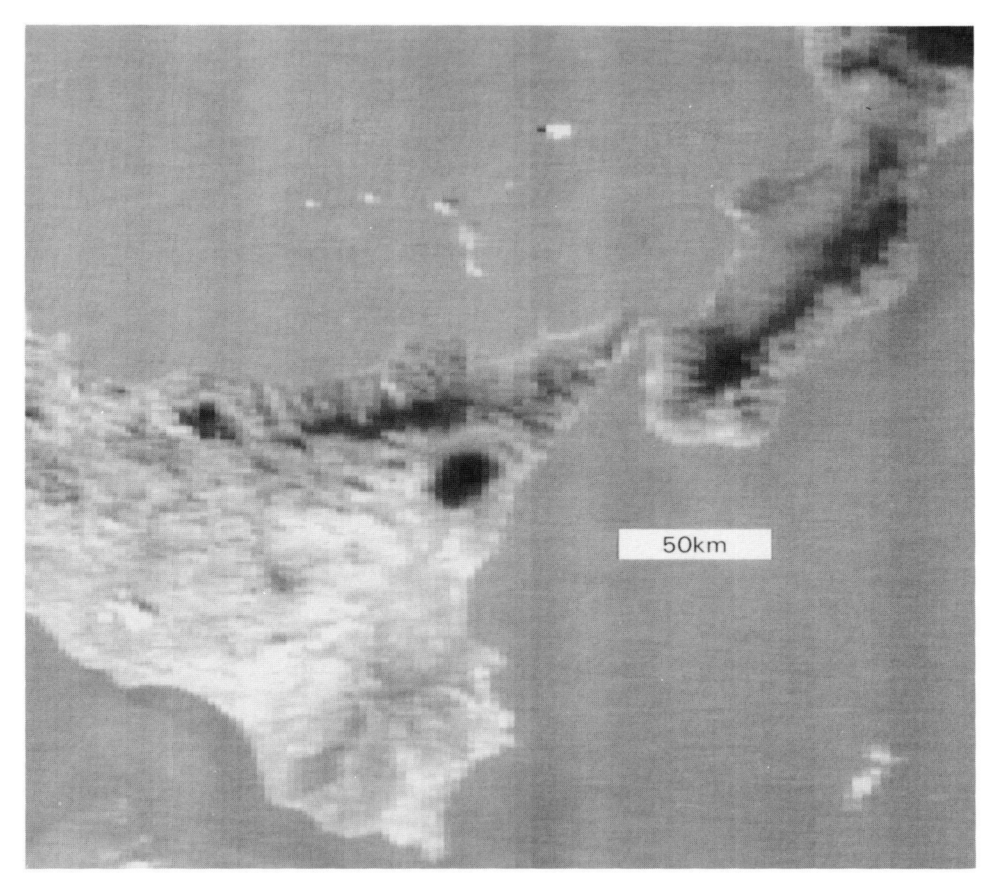

11.1.21 1403,11.1.86,3

11.1.21
This is the more normal scene: a large hot spot, probably with some small cumulus, on Stromboli, small cumulus on the other islands, but black-topped cumulus over the higher ground, including one hiding Etna's hot spot.

11.2 FOREST FIRES

Forest fires are part of the global ecology which reduce the proportion of oxygen in the atmosphere when it becomes too large. Not only do they convert some oxygen into carbon dioxide, they also reduce the population of oxygen producers — the trees. If the proportion of oxygen is reduced, fires become less likely to be started by lightning or any other cause, including human carelessness. They are more likely to be started in dry areas, and the two cases illustrated occurred after a long period of warm,dry weather, typical of the Mediterranean climate after midsummer.

Algeria

In this first case several fires were detected by satellite in Northern Algeria in a deep southerly current of air from the Sahara, which was particularly warm and dry after crossing the Atlas mountains.

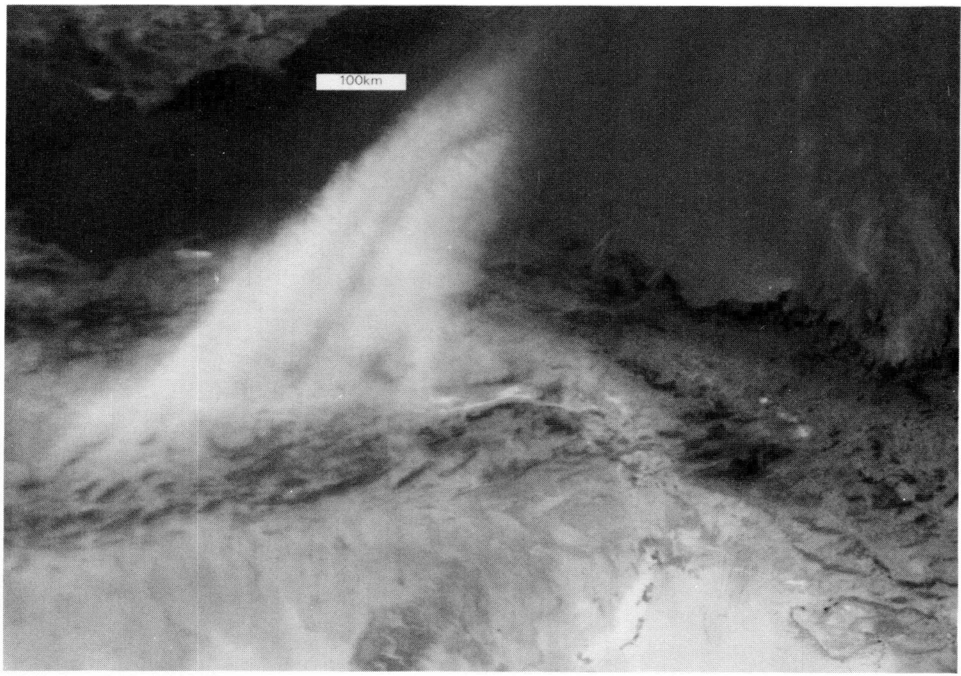

11.2.1 1355,27.7.83,1

11.2.1

A few small fires can be seen between Bougie (Bejaïa) and Algiers (Alger) and very extensive ones closer to the border of Tunisia. There is a large patch of orographic cirrus in the lee of the Saharan Atlas in the wind which is slightly west of south. The white patches in the mountains and the southern desert are dry salt lakes and the dry desert is whiter than where there is vegetation. The smoke is drifting out to sea.

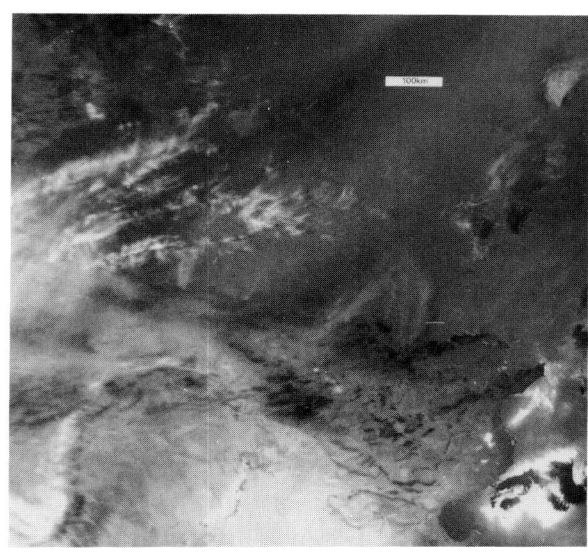

11.2.2 1343,28.7.83,1

11.2.2

On the following day the whole Western Mediterranean is covered by fine dust or smoke haze with bright glint out to sea from the gulf of Gabes (the whole picture is taken from the western edge of the swath), and patches of calm or cloud shadow produce black areas within this glint. Corsica and Sardinia, largely obscured by haze, seem to have some orographic cirrus also.

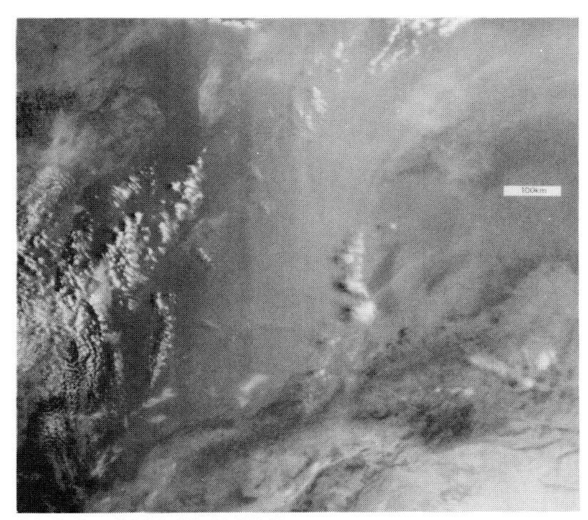

11.2.3 0821,29.7.83,2

11.2.4 0821,29.7.83,3

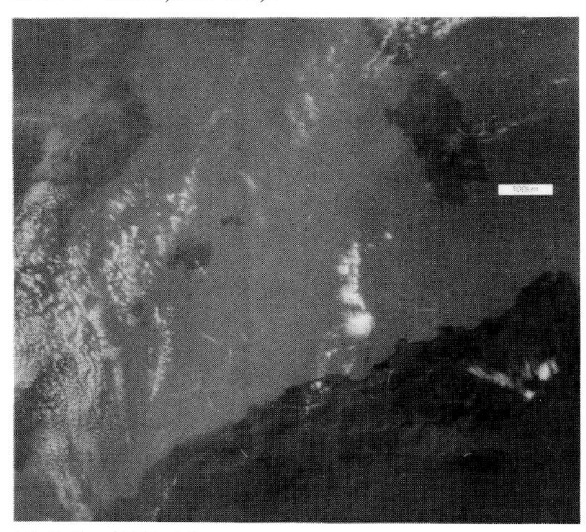

11.2.5 0821,29.7.83,4

11.2.3, 11.2.4, 11.2.5

Chs 2 and 3 show great differences the following morning, the glint now being centered north of Bougie. In Ch 2 the white altocumulus contrasts with its very dark shadows in the west, but north of Bougie the shadow is blurred by being as much on the smoke haze as on the glint, whereas Ch 3 does not show the haze (nor the shadows on it), so that Sardinia's coastline is very sharp. The salt pans, bright in Ch 2, are darker than other patches of dry desert in Ch 3. Hot spots abound among the smoke sources along the coast, and there are some in Sardinia also. Ch 3 shows the high cloud (white in Ch 4) as black, like its shadow.

Ch 4 simplifies the scene; the dry desert is not yet warm early in the day, but the temperature indicated may be affected by the moisture in the 400 m of air at a lower level than the high ground but over the low-lying desert.

11.2.6 1025,29.7.83,CZ1

11.2.6, 11.2.7
Even after two days the fire smoke, which is mostly from small area sources, appears streaky over the sea; while the more pervasive desert haze is more uniformly diffused, presumably because it is from a more widespread source area which is also further from the sea than the fires are.

In CZ 5 the cirrus southwest of Sardinia is bright only in its very densest parts, but it casts a good shadow on the smoke. Compare the appearance of this cirrus with it in the other pictures.

The mysterious body north of the island of Pantellaria (**11.2.2**) has now disappeared.

The cloud bands in the Gulf of Genoa seen in the CZ 5 picture are billows with strands along the shear direction (at right angles to the waves). A further pass at 1511,29.7.83 by NOAA-7 showed nothing of additional interest except that the fires near the coast continued fiercely.

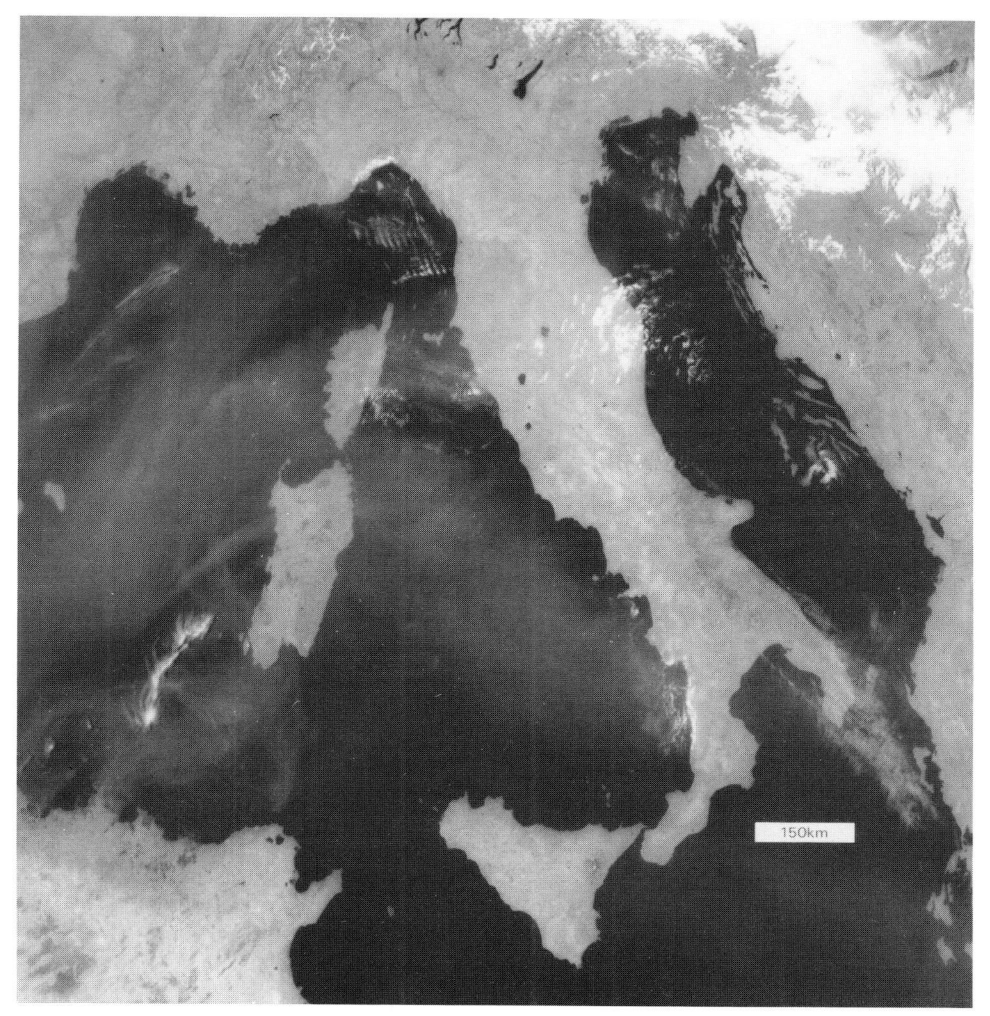

11.2.7 1025,29.7.83,CZ5

Cote d'Azur, Provence

These fires caused much more of a stir in the news media than those in Algeria because they occurred in a very fashionable area with expensive social habits.

11.2.8, 11.29

In this Ch 4 picture the temperature contrasts on land have been emphasized with all the cool areas of sea or cloud appearing white. The simultaneous Ch 3 picture shows the area near Cannes to be the hottest patch, most of the others not being a saturated white. There appears to be very little wind.

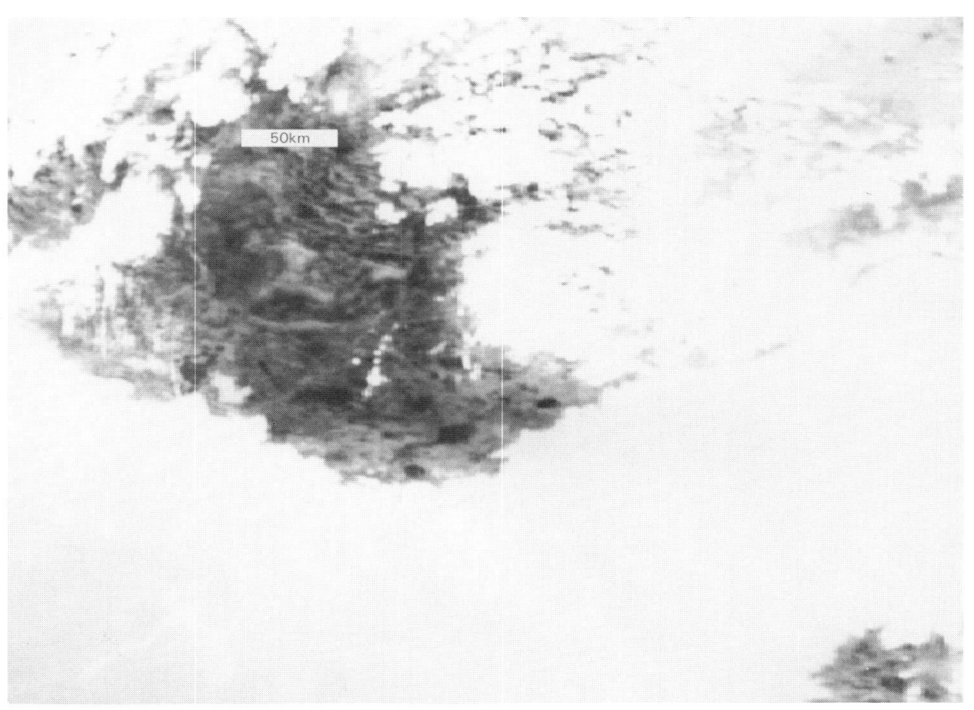

11.2.8 1431,22.8.86,4

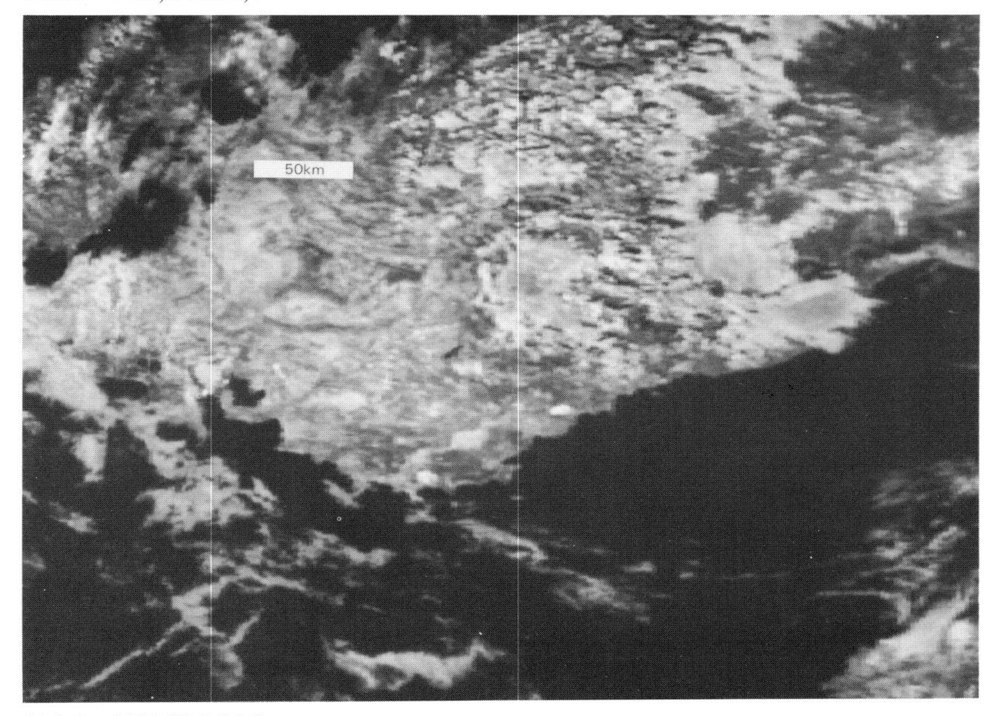

11.2.9 1431,22.8.86,3

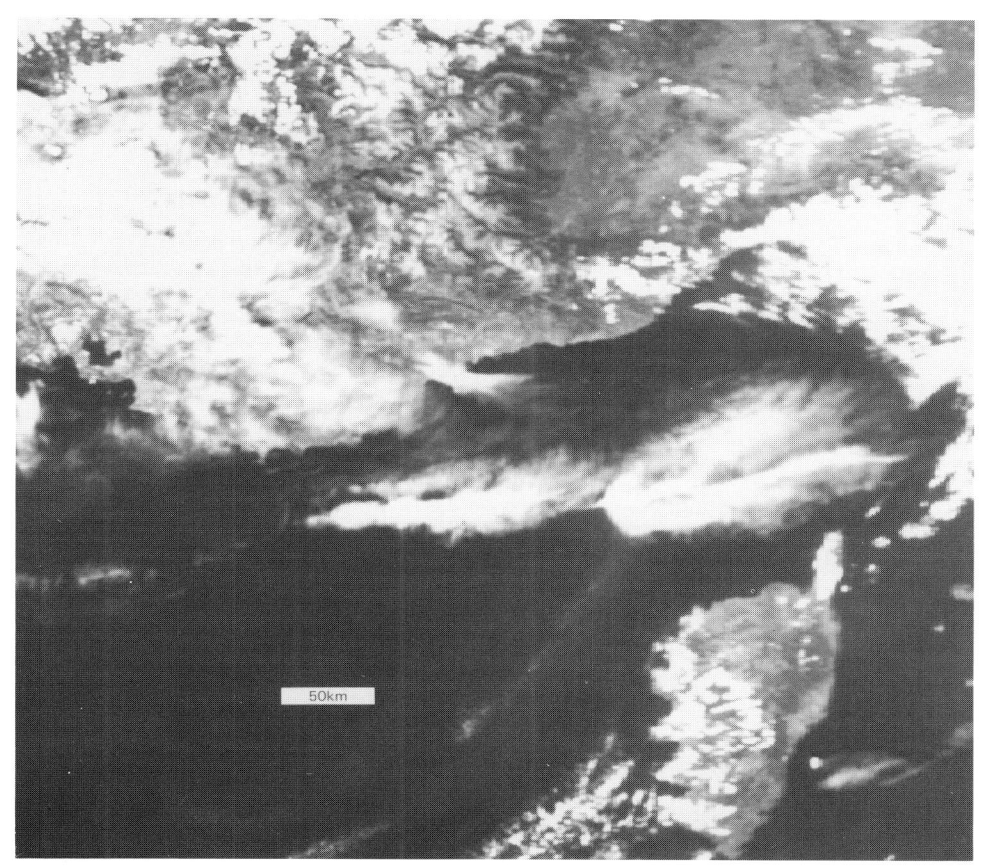

11.2.10 1420,23.8.86,1

11.2.10
On the following day a dense smoke plume extends over 60 km out to sea from Cannes, and cirrus covers much of the locality. The plume and the cloud northeast of Corsica indicate a wind from the west or southwest.

11.2.11, 11.2.12
To the west of Cannes the cloud streets over Provence indicate a wind from the northwest, and the smoke, well-detected by Ch 1, is being carried by this wind out to sea from Cannes. But the 'finger' mountain at the north end of Corsica has lee waves which, although they are parallel with the mountain, lie in a direction which implies a wind from the WSW. The wave pattern is influenced by the flow over the main mountain mass of the island, so that the fourth wave is not parallel to the mountain.

The smoke plume from Cannes is having fragments blown off its top in the same direction as the wind over Corsica.

In Ch 3 we do not see any smoke, but notice that the other very hot spot near Monaco is also producing smoke seen by Ch 1. There are other hot spots in France and Italy, but they are very abundant in Corsica, but these last were unnoticed by the media; I was informed that a certain amount of forest management by unorthodox methods is carried out there. The air to the east of the island seems full of smoke, which in this case is dense enough to be seen by Ch 3.

The lee waves have bright edges, indicating smaller droplet sizes there than in the crests of the waves.

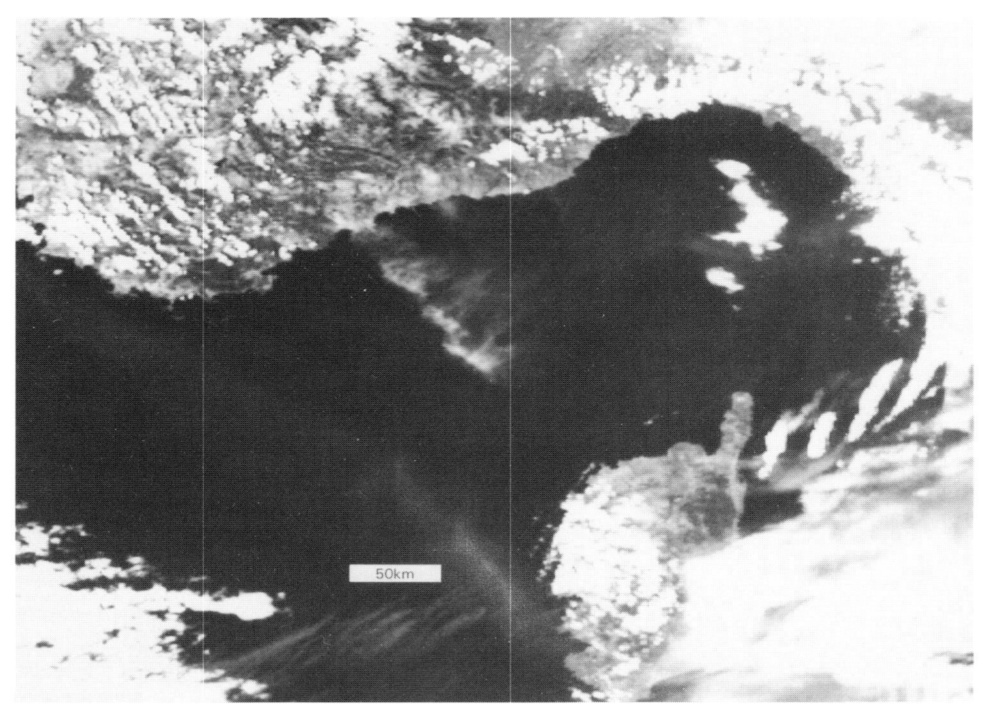

11.2.11 1409,24.8.86,1

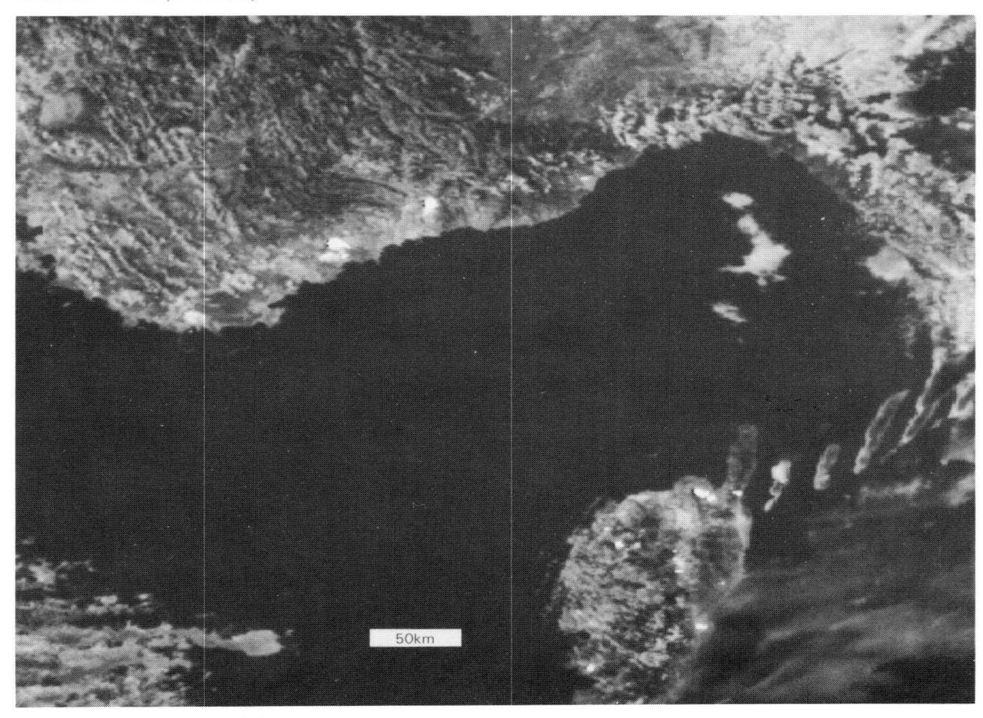

11.2.12 1409,24.8.86,3

11.3 DUST FROM CULTIVATED LAND

Dust is most easily seen by satellite over the sea, and Biscay provides some good examples. It is likely that some of the smoke is of industrial origin from the neighbourhoods of Oviedo and Bilbao, but there is also a source of comparable magnitude in the more rural areas of Galicia and perhaps the northern part of Leon where the rivers drain southwards on the south side of the Cantabrian mountains in northwest Spain.

Southern Biscay and Northern Spain

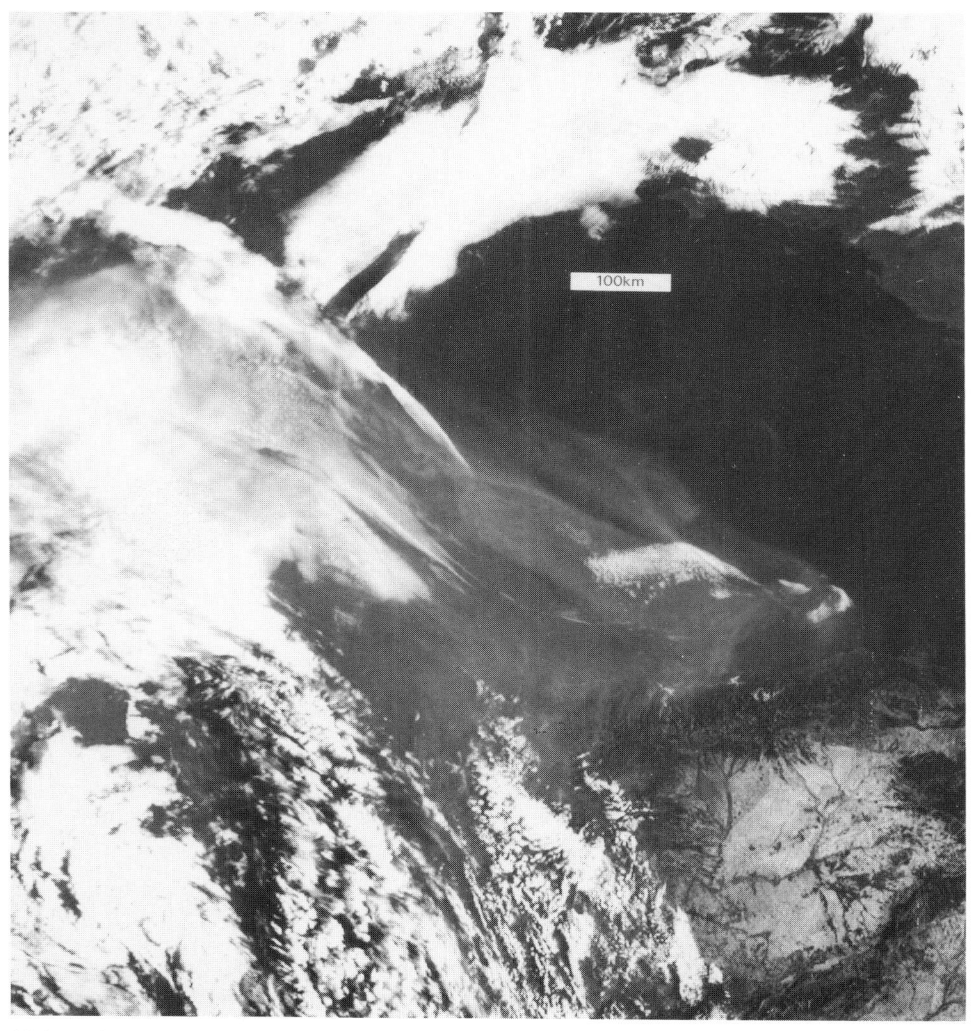

11.3.1 0731,18.9.85,1

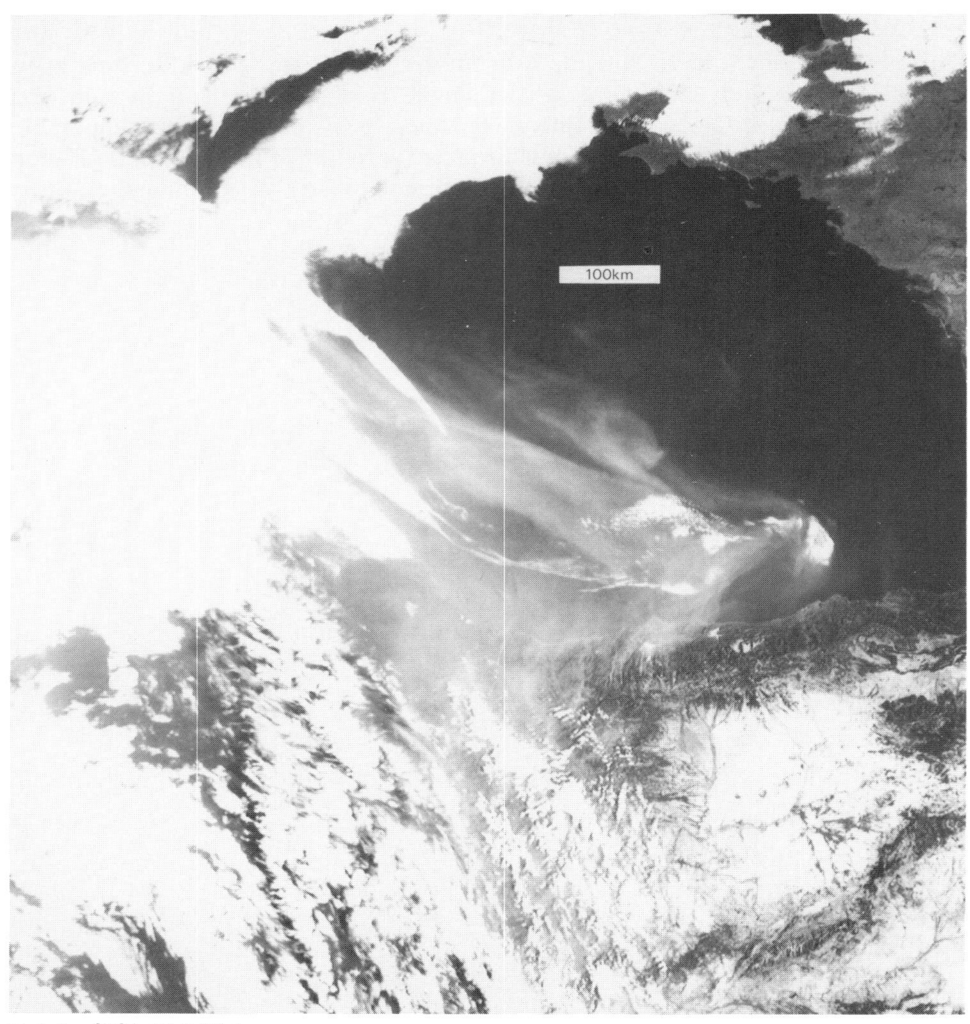

11.3.2 0821,18.9.85,1

11.3.1, 11.3.2

Early in the brief operational life of NOAA-8 it doubled for NOAA-6 by passing half an orbit later. Thus these two pictures represent the same scene under increasing solar illumination at a time of day when it is increasing rapidly. The smoke appears to have originated in northwest Spain from agricultural fires and perhaps others whose hot spots are seen in the accompanying Ch 3 pictures (below). The changes are slow and the smoke has drifted with a variable wind.

The dryness of the area is indicated by the pale colour of the lowland, south of the mountains, where the rivers are lined with dark trees.

11.3.3, 11.3.4

The hot spots, probably agricultural, stand out most clearly when the sun is low, but can mostly be identified in the later picture. The haze is not detected over the sea, but there is a dark strip of sea close to the coast, which is upwelling cool water. The distribution is the same in the two pictures and has not altered much three days later (see **11.3.9**) which indicates that the dark area is not an atmospheric phenomenon.

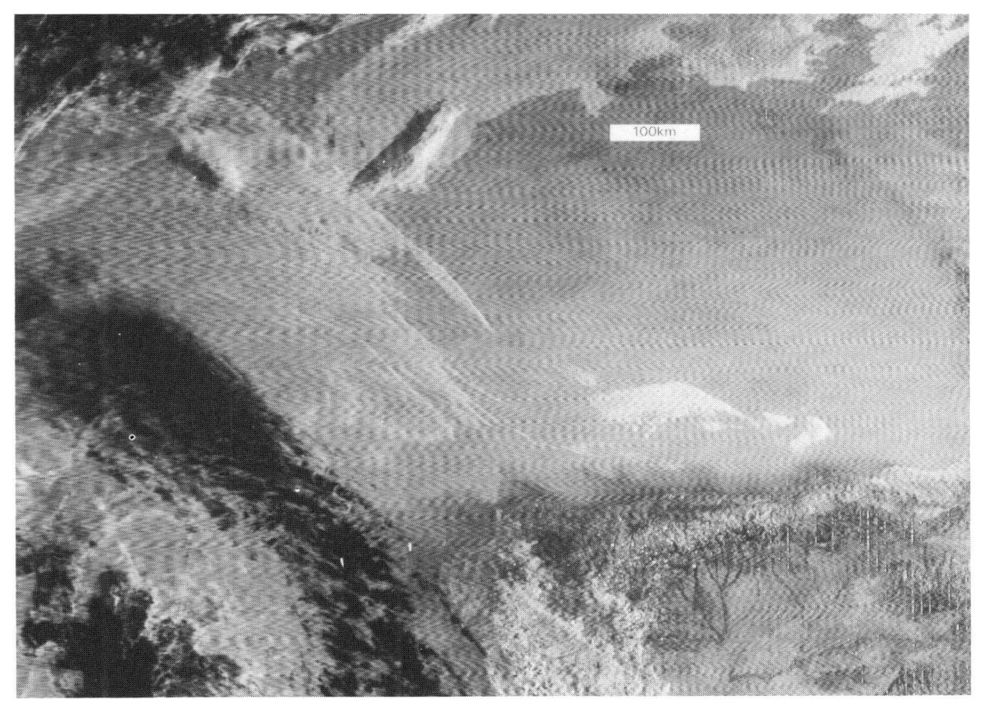

11.3.3 0731,18.9.85,3

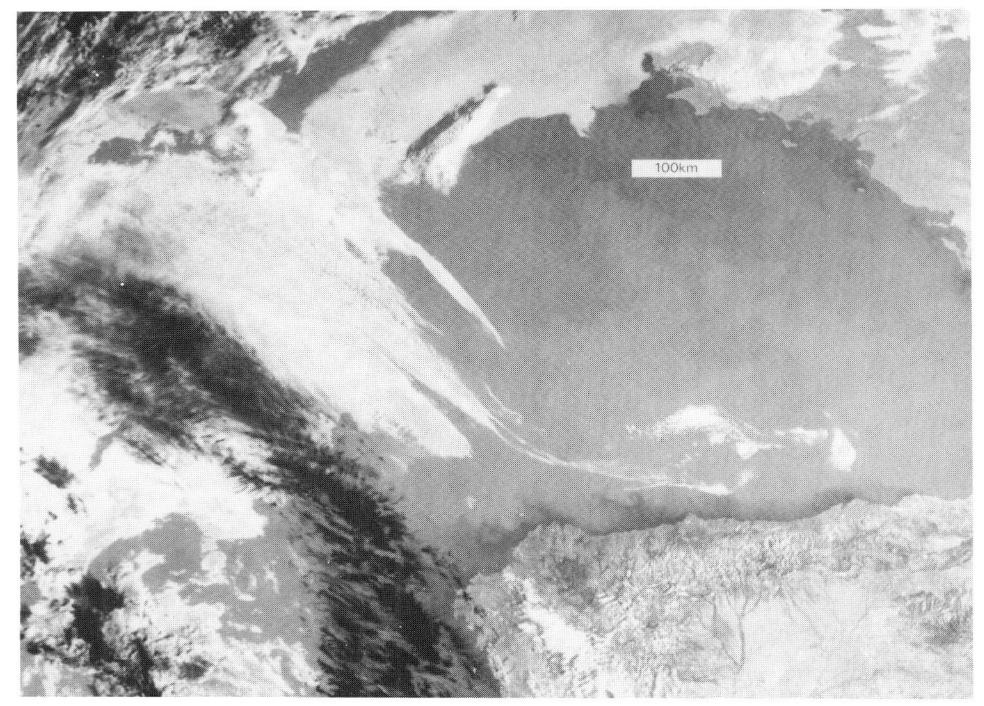

11.3.4 0821,18.9.85,3

11.3.5 1150,18.9.85,CZ3

11.3.5

In CZ 3 the haze is very evident. The rivers Gironde and Charente seem to be spilling coloured water into the sea, and it is likely that it contains agricultural debris. A solitary plume from a combustion source close to Cape Machichaco, just east of Bilbao, is being blown out northwards. The north–south dimension in these CZCS pictures is stretched.

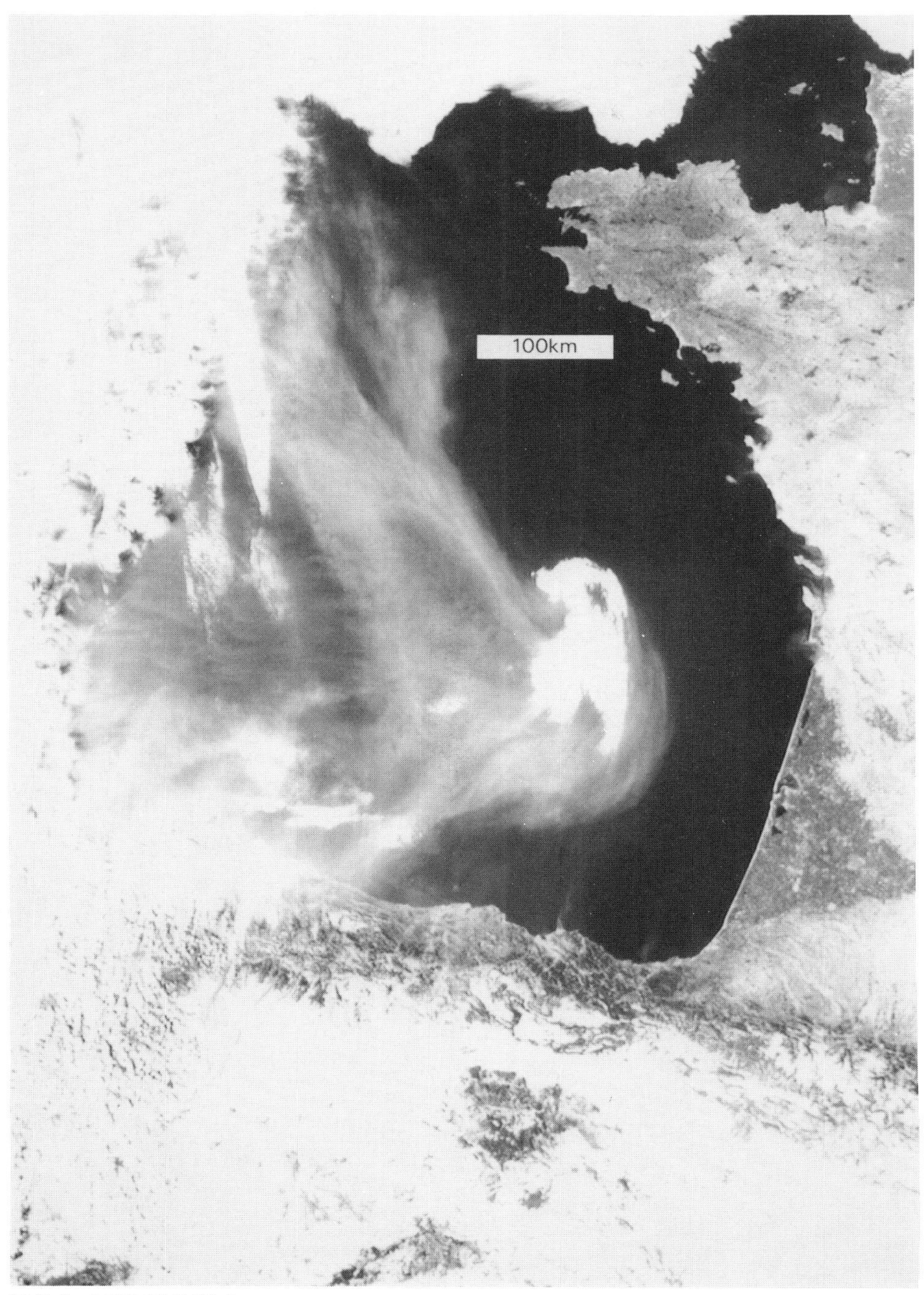

11.3.6 1423,18.9.85,1

11.3.6
Comparing this with the previous picture we see that Ch 1 has many features in common with CZ 3 in detecting haze.

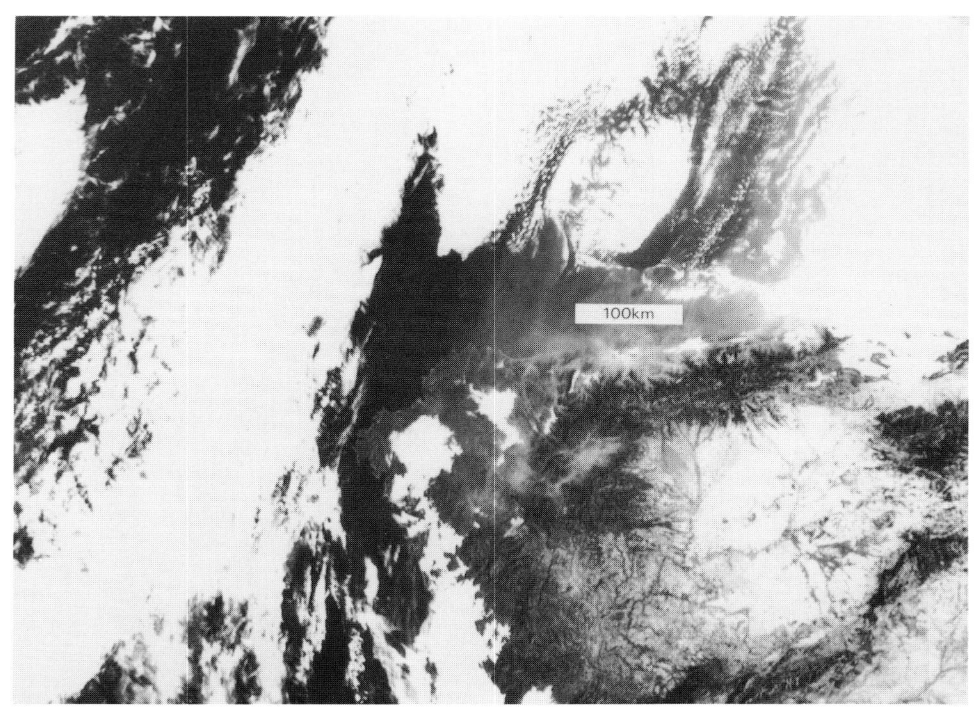

11.3.7 0835,22.9.85,1

11.3.8 0835,22.9.85,3

11.3.9 1341,22.9.85,3

11.3.9
The land is now much brighter, but a few hot spots still stand out and the cool water is still evident.

11.3.7, 11.3.8
The wind direction is now from the SSW, and the sources of haze in northwest Spain show higher concentrations over the land, which are favourably detected in Ch 1.

Some of the hot spots are still visible in Ch 3, which also delineates the cool coastal water moving slowly westwards, but having scarcely changed from the position shown in **11.3.4**, although the water in that position may have been replaced at least once.

11.3.10 1120,22.9.85,CZ2

11.3.10

As expected, this visible channel shows the smoke well. Such white spots as appear are small cumulus in this channel.

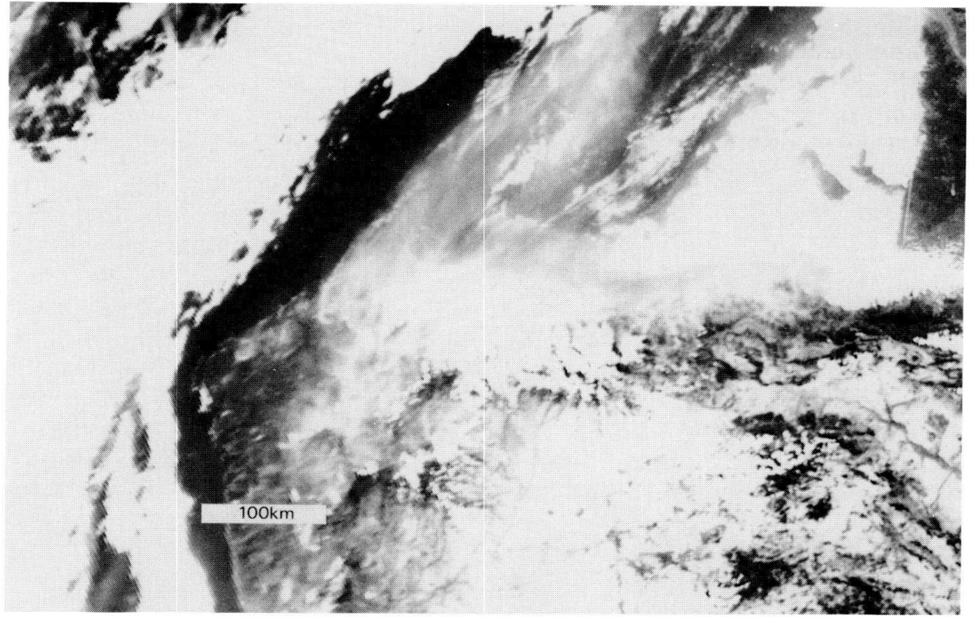

11.3.11 1341,22.9.85,1

11.3.11

There is more cloud at this time than in midmorning, but the sources of smoke seem to have been intensified.

12

Satellite as mesoscope

A satellite picture is more than a record of the state of the atmosphere at one moment. The shapes and forms of clouds imply a history and a contemporary motion pattern, and this in turn, for the meteorologist in particular, implies a frontal analysis and a pressure field.

On a smaller scale than the kind of analysis which is made when the meteorological observations on a weather chart are robed in isobars and fronts, the satellite picture contains cloud patterns which are not perceptible either on the chart or in the reports of the individual reporting stations. Even when the observer is aware of a broader structure in the clouds the reporting codes do not usually permit the transmission of this information. The kind of information which the forecaster imbibes as he looks at the part of the sky visible to him is often contained in the satellite pictures.

The details strike the observer's consciousness, and in many cases the three-dimensional picture and the motion pattern which he inscribes mentally on the picture are consolidated by inspection of the other channel (i.e. whichever of Chs 2 or 4 was not looked at first). After that he can get down to the surprises and secrets of Ch 3. It is a very stimulating exercise to inspect Ch 3 only at first to see whether one could do without the other two channels.

The geography beneath the clouds is all-important, and an observer must have a good knowledge of it and be ready to look at a good map from time to time to confirm details. The wave clouds and orographic cirrus tell whether it is a north or south fohn blowing across the Alps, and the stratus which appears to be glued to the north coast of Spain does not deny that there is a strong southwesterly wind, for it has been seen there many times before. The cirrus of a warm front crossing the southern tip of Greenland does not imply the same relative direction of the surface wind as would be found in mid-Atlantic because the walls of the great snow-plateau prevent it being invaded by air from the sea. Convection or mountains can transform the appearance of an air mass, and this often makes it easy to 'see' the topography beneath the clouds.

In drawing attention to some clues of this kind which make up the face of the troposphere, whose mood we wish to assess as quickly as if it were a human face, it must be emphasized that in such interpretations lies a great store of experience. As these themes are developed it will emerge that most situations are complex and contain several simultaneously functioning mechanisms.

12.1 STREETS

Any cloud formation which does not cover the whole sky demands an explanation; why has not the cloud forming mechanism covered the whole sky and is it merely a matter of time before it does ? A scene in which new individual clouds are continually being formed is also one in which continual evaporation is taking place. This usually implies that moisture is being carried upwards and that warming, either by the mixture of rising warm air parcels or by adiabatic subsidence, is increasing the saturation mixing ratio as fast as the actual ratio is being increased.

Motion and cloud shapes are affected by shear, and convection in streets aligned along the shear is usually also along the wind direction because the shear is caused by

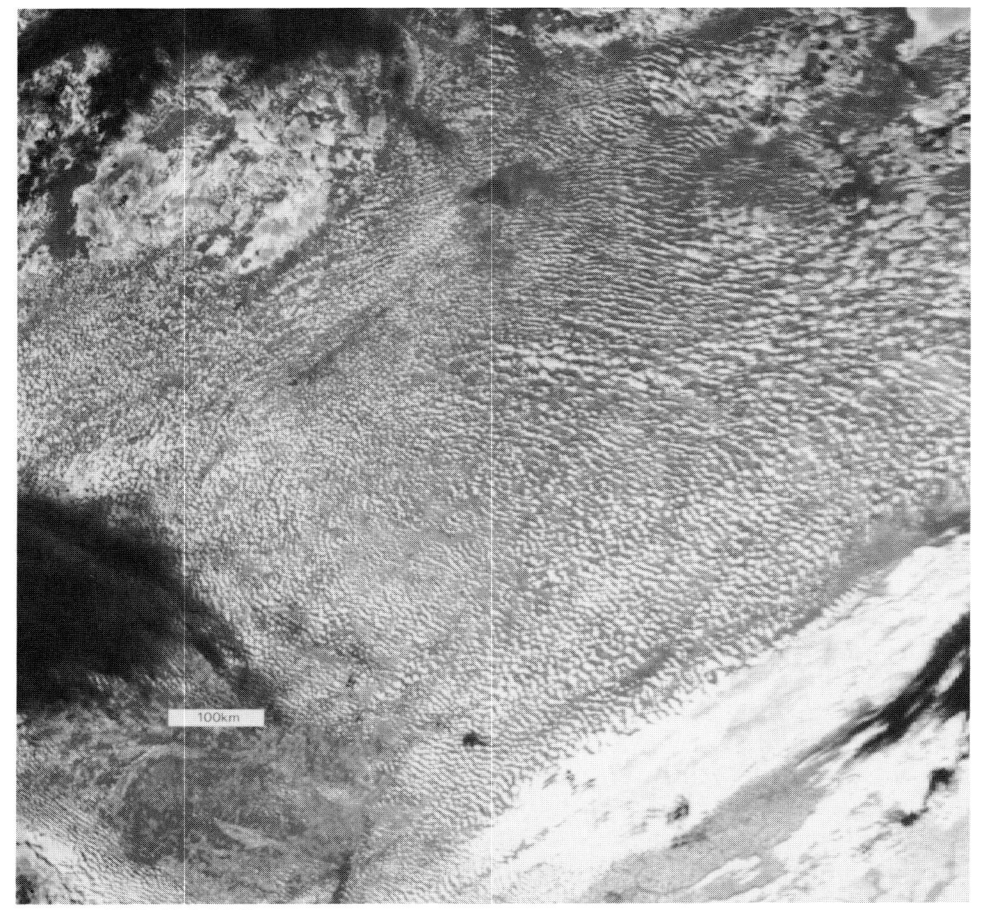

12.1.1 1248,10.9.86,3

12.1.1
This East European scene is not dominated by coastlines, mountains, or fronts, and so the street structure can reveal an anticyclone. The very stable air mass indicates that the rotation is not cyclonic. The clouds are lined along the wind direction, which, in the well-mixed layer, is the direction of shear due to surface drag. The cloud, which is cumuliform, must be evaporating because it occupies less than half the sky and it is well past midday, so it can be assumed that subsidence is probably the mechanism causing the evaporation.

surface drag. At higher levels the thermal wind dominates, with the streaks usually lying with cold air on the left in the northern hemisphere. Ageostrophic flow, with streets or streaks significantly out of line with the isobars, is caused by surface drag, but is also important at high altitude when development (conversion of gravitational potential energy into wind systems) is taking place.

12.1.2 1352,11.5.82,2

12.1.2
The cumulus streets begin well inland but extend beyond the coast where the air moves over the sea from the British Isles. A warm front approaches from the southwest and the low level air is becoming more stable, the streets turning to waves over and in the lee of southern Ireland, and disappearing in France. The striations in the advancing cirrus, over France particularly, imply a thermal wind parallel with the edge of the frontal cloud, with colder air over Britain.

12.2 LEE WAVES

Waves Lee waves tend to lie across the wind direction, but are also often caused to lie parallel to a mountain ridge which the wind carries the air across. A given airstream usually has characteristic wavelengths in which the waves remain stationary relative to the mountains. Most airstreams have a single such wavelength, some have none, but some have more than one which have their maximum amplitudes at different altitudes. That maximum tends to be at the top of a cloud layer because radiation from the cloud top establishes an inversion there and this is a layer of much greater stability than at other levels.

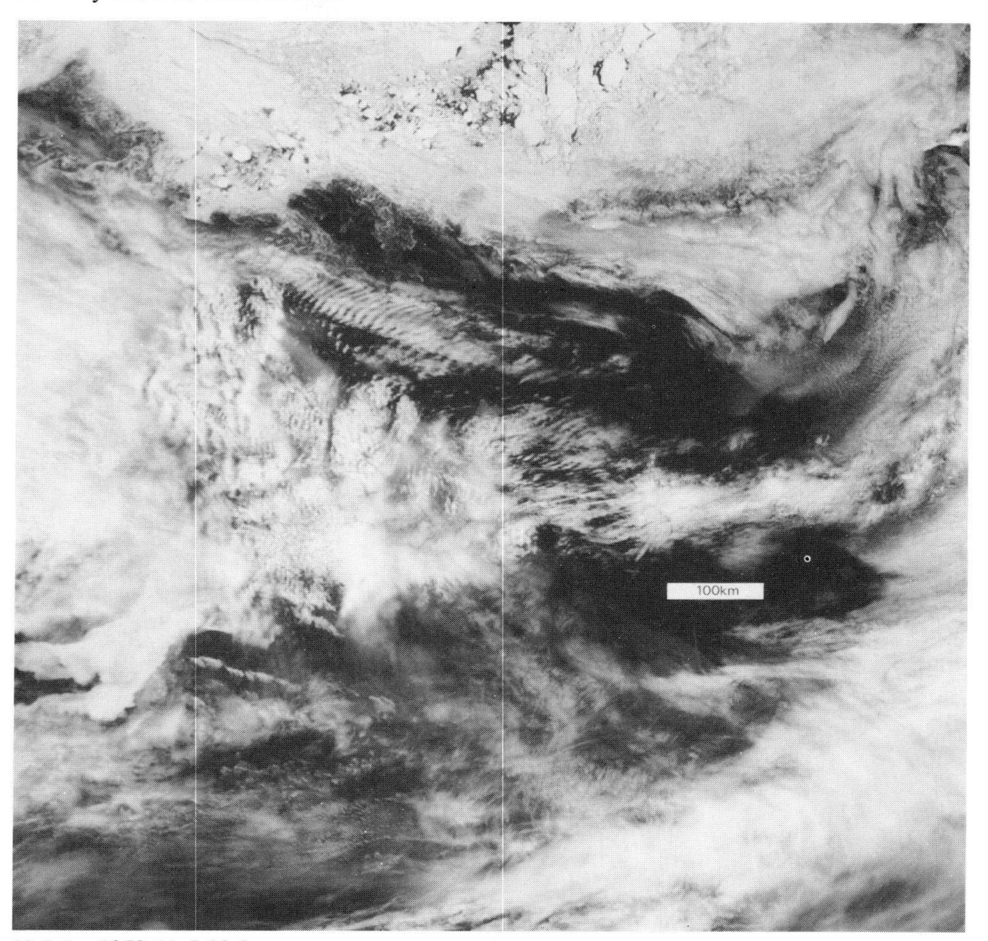

12.2.1 1352,11.5.82,2

12.2.1, 12.2.2, 12.2.3
In Ch 2 we can see two main layers of cloud crossing Iceland. The lower layer contains lee waves with a wavelength of about 8.5 km. In the upper layer the wavelength is about 21 km and some orographic cirrus is carried off downwind. Some major aspects of Iceland's topography show through the clouds, but a single wavelength dominates each cloud layer. More commonly there is only one wavelength. The topography is complicated and is best located in this case by placing the fragments of coastline which are visible between clouds.

The Greenland ice pack is seen to the north in Ch 2.

12.2.2 1352,11.5.82,4

12.2.3 1352,11.5.82,3

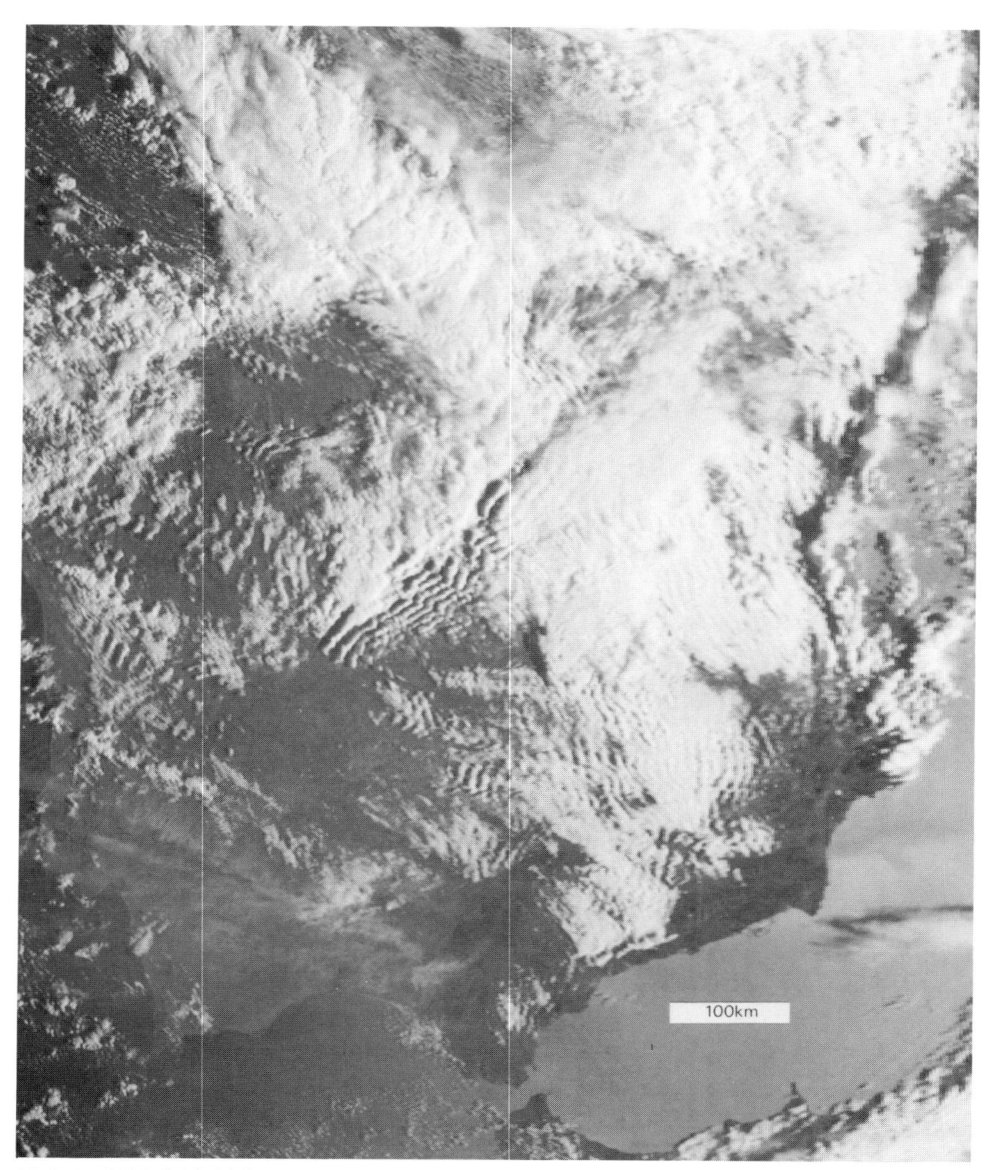

12.2.4 0828,5.10.84,2

12.2.4, 12.2.5, 12.2.6, 12.2.7
Early in the day the effect of the Iberian mountains is to generate lee waves lying across the wind almost everywhere. There is presumably an inversion at the top of the low cloud layer. Over the sea there is the same air mass, but the convection produces beady cumulus, rising to produce showers and cloud-free areas where the downdrafts have spread out over the sea.

The presence of the inversion which facilitates the waves produces a situation favourable

(*continued next page*)

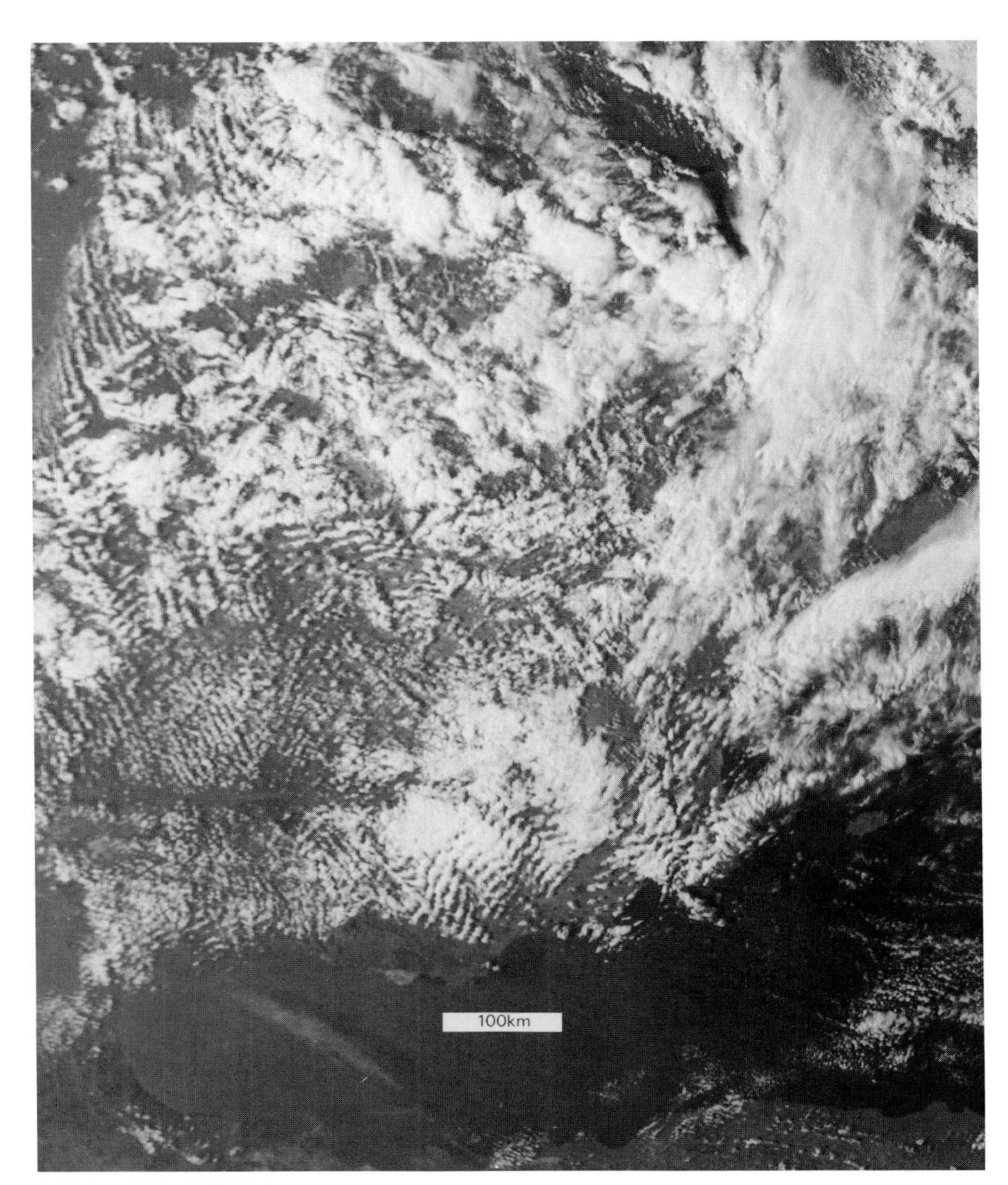

12.2.5 1512,5.10.84,2

12.2.4, 12.2.5, 12.2.6, 12.2.7 (*continued*)
for the formation of streets of small cumulus growing up to the inversion as the sunshine warms the ground. The streets lie perpendicular to the lee waves and they are sometimes present together. Streets are preferred where the air first moves inland across the coast, and it is noteworthy that there is a strip of cloudless air on the ocean side of the coast, implying subsidence there (see **12.4.5**). Over the lower ground the streets are preferred where the air

(*continued next page*)

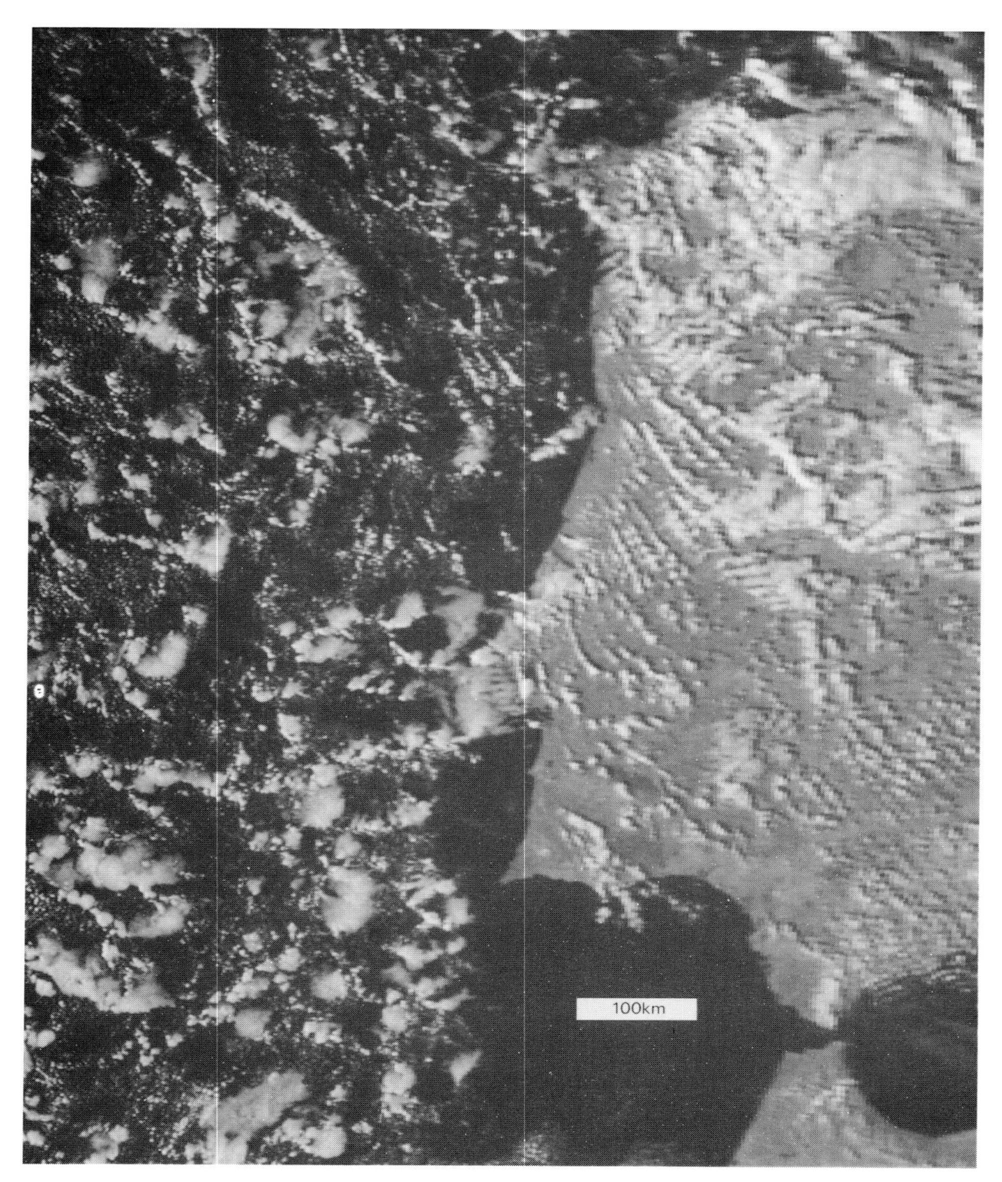

12.2.6 1654,5.10.84,2

12.2.4, 12.2.5, 12.2.6, 12.2.7 (*continued*)
has evaporated the layer of cloud; but that layer and waves are dominant over the areas of higher ground.

The following orbit gives us a good picture of the same air mass over the sea and land 100 min later. But three hours after that (**12.2.7**) the streets have died out, and waves predominate and extend almost across the Alboran Channel (east of Gibraltar) where the shorter

(*continued next page*)

12.2.7 1948,5.10.84,4

12.2.4, 122.5, 12.2.6, 12.2.7 (*continued*)
wavelength probably indicates a weaker wind. In Central Spain (Sierra de Gredos, west of Madrid (M)) there is a good example of the first lee wave being parallel to the mountain range but arranged as a series of three steps because the waves tend to lie perpendicular to the northwest wind (as behind the Guadarramas north of Madrid). As the heating cools in the evening the ocean cloud invades the coastal cloud-free strip.

12.2.8 1514,12.2.87,2

12.2.8, 12.2.9

This is a similar situation in which the air mass characteristics can change as the air crosses the coast. In the north the small showers (notice how dark the large cumulus are in Ch 3) are converted into streets with no showers, but probably with a much more rapid time scale for the generation of individual clouds, as the air crosses the French coast from Biscay. It does not easily cross the Pyrenees, but travels into Spain further west where lee waves with ladders are produced (typical wavelength 16 km). Each ladder consists of an area of lee waves in which the

(*continued next page*)

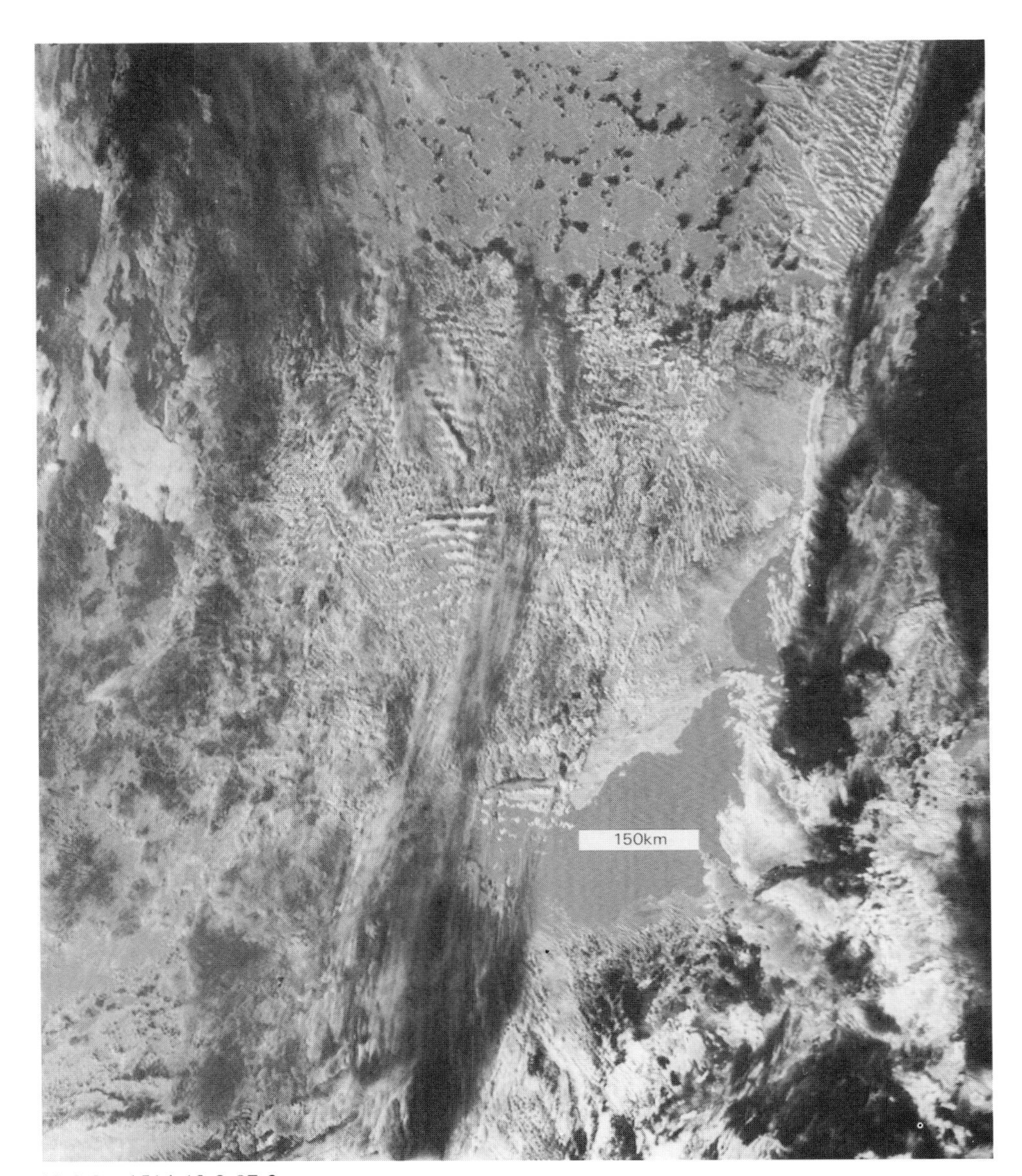

12.2.9 1514,12.2.87,3

12.2.8, 12.2.9 (*continued*)
amplitude, phase, and wavelength are preserved, but are not necessarily the same as in the neighbouring ladder travelling over different topography. Ladders cross the Alboran Channel but become streets of very small cumulus in Morocco.

Orographic cirrus covers the central strip across Spain from north to south, and far into Morocco.

Sometimes there is a strip of unwaved cloud between the sides of two abutting ladders, and some waves indicate ship-wave patterns in the northwest.

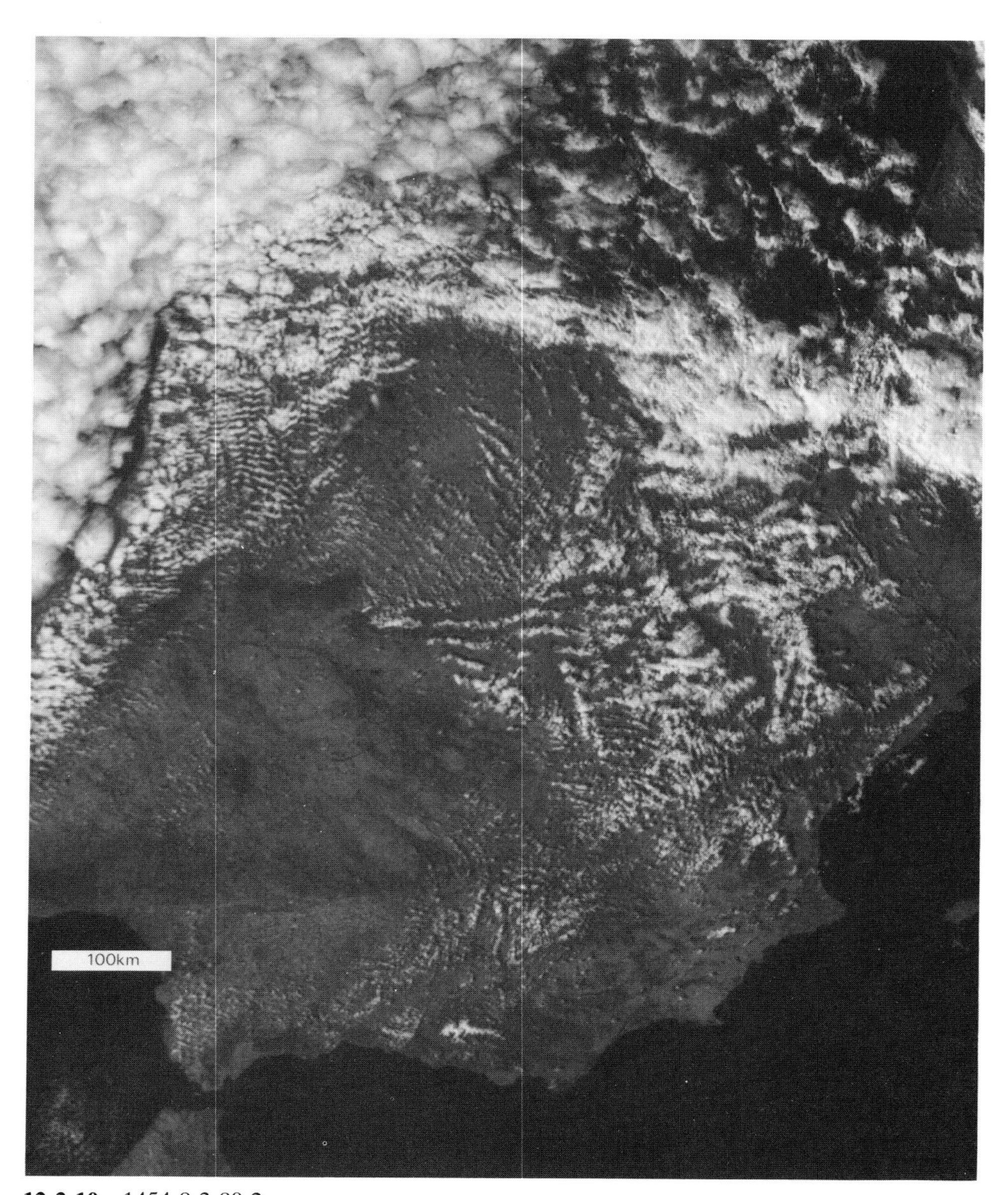

12.2.10 1454,8.3.80,2

12.2.10, 12.2.11

The Atlantic air mass has a uniform cloud layer with a cellular pattern and the convection over land breaks up the cells (which, in Ch 4, have a uniform top temperature) into progressively smaller ones, at the same time setting up wave ladders. Further east they have become streets, but are not perpendicular to the waves although their orientation is varied so that they become

(*continued next page*)

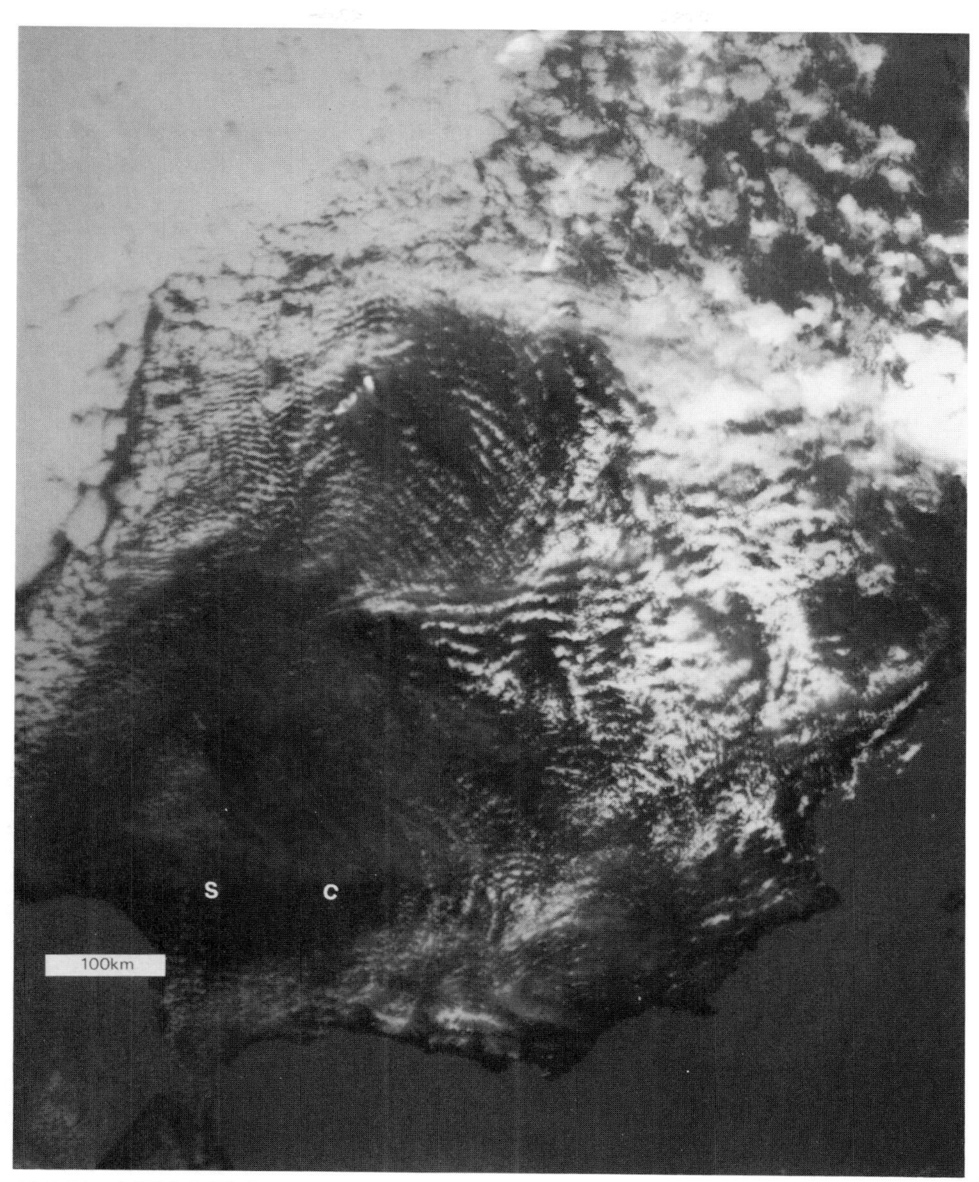

12.2.11 1454,8.3.80,4

12.2.10, 12.2.11 (*continued*)
waves in the lee of the Guadarramas. The formations further east and south do not suggest uniformity.

There is a transformation across the river Guadalquivir, which flows through Cordoba (C) and Seville (S), as if the flow off the plateau north of the river produced a significant warming, until further cooling is produced due to the ascent of the remaining mountains to be crossed before the sea is reached. The Ch 2 picture hints at a change of vegetation across the river.

12.3 INVERSIONS

Cold air drains into the bottom of valleys at night. Some valleys are quite narrow and we can see in Ch 4 that the bottom is much colder than the side in sunshine. We also we know from glider pilots that there are anabatic winds up the sun-warmed slopes, especially when the air is stably stratified ([2]; 9.8 (5)).

Much larger valleys, such as that of northern Italy, become full of very stable air which drains down to the sea; the Mediterranean itself lies in a valley surrounded by mountains, and is frequently filled with very stable air.

The top of a cloud layer generates an inversion by radiative cooling into space and scattering back upwards much of the incoming sunshine. This stable layer causes the convection below to take the form of streets, but when it flows over mountains the stable layer is the level at which the lee wave amplitude is a maximum. This means that, over the sea or a wide valley, there may be lee waves without any clouds to reveal their presence.

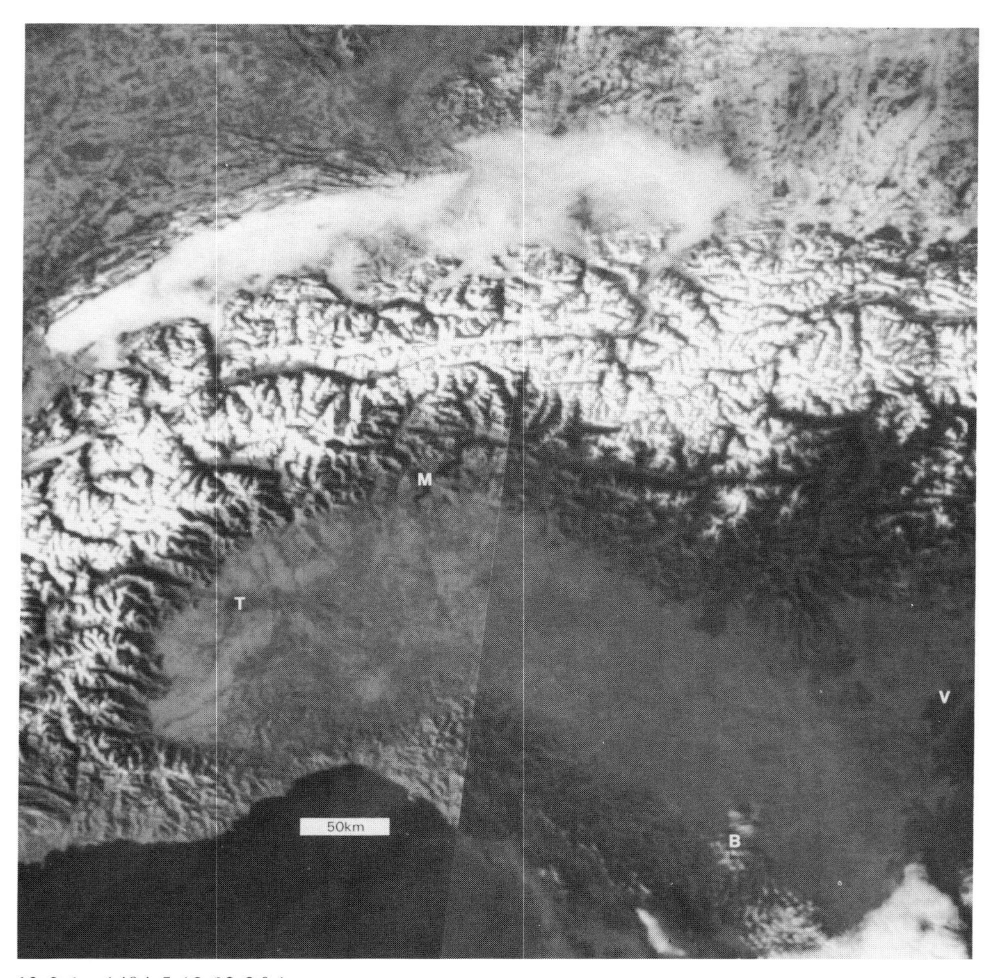

12.3.1 1404,5.12.83,2&1

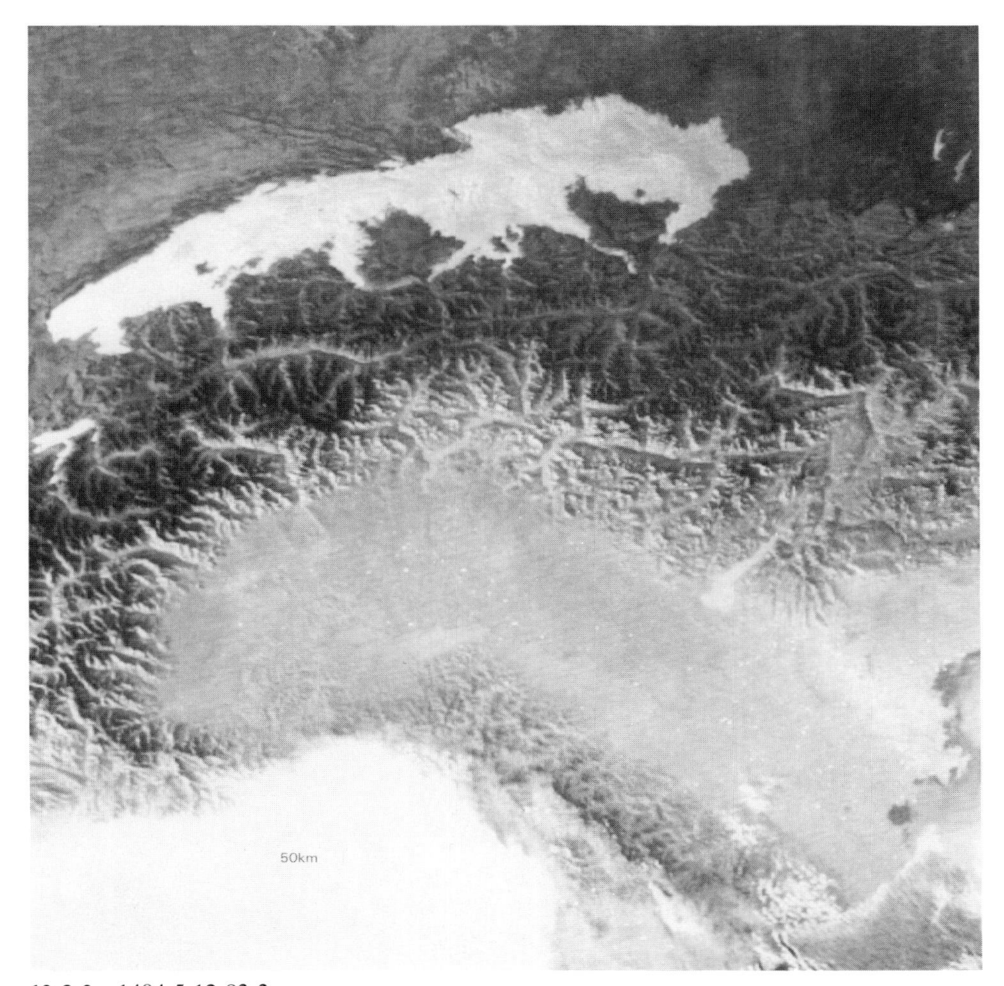

12.3.2 1404,5.12.83,3

12.3.1, 12.3.2, 12.3.3
In Ch 1 (right) and Ch 2 (left) the mountain snow-line and the shadows in the Alpine valleys are obvious. The north Italian plain is hazy, and in Ch 2 where the land is brighter, half of Milan occupies a dark patch, while in Ch 1 the haze glow smothers the illumination of many land features.

The exact boundary of the stratus on the north side of the Alps is rather uncertain in Chs 1 and 2, whereas it is very clear in Ch 3, which also shows a multitude of hot spots in the plain and in which the drier cities glow more brightly than the surrounding land. There is a prominent feature (autostrade) stretching southeastwards from the foot of the Alps about half way from the southern end of Lake Maggiore (M) to Turin (T). Bologna (B) and Venice (V) are also indicated.

The air up to the snow-line is presumably capped by an inversion which prevents the upward spread of the valley haze; this is undoubtedly fed by cold katabatic flow down the snow

(*continued next page*)

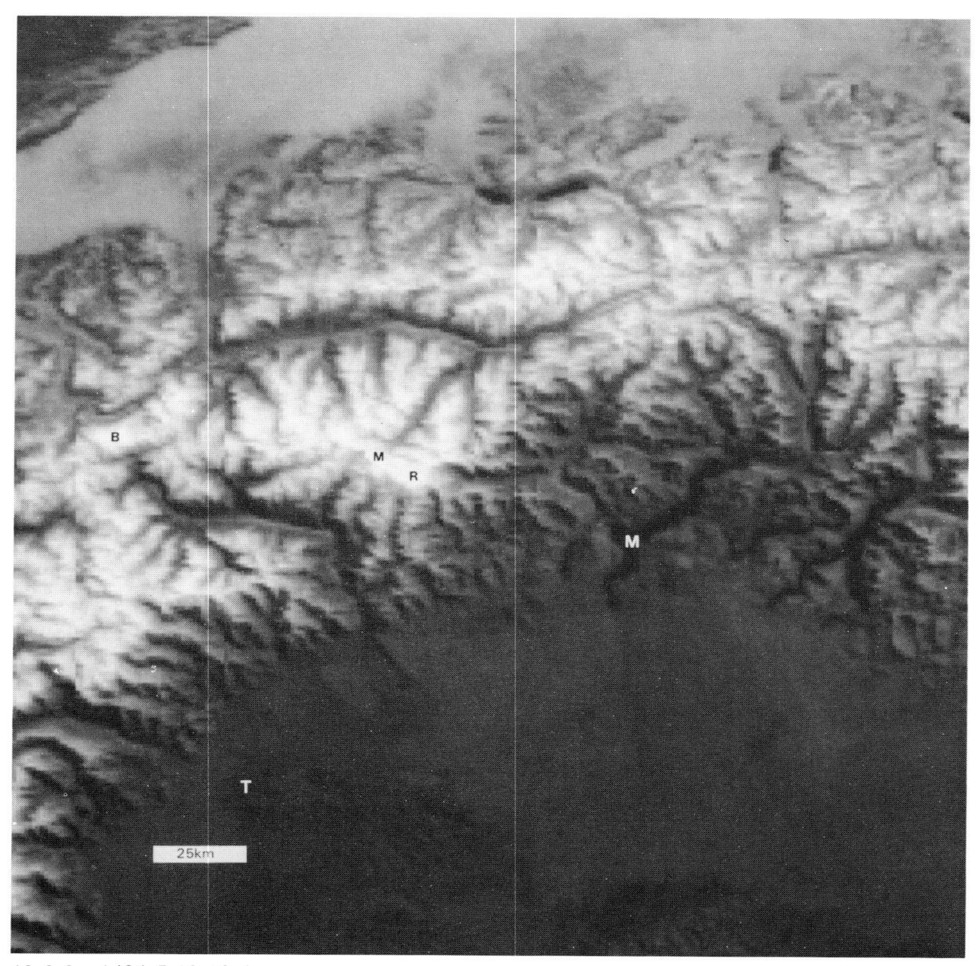

12.3.3 1404,5.12.83,4

12.3.1, 12.3.2, 12.3.3 (*continued*)
slopes and is a common feature where there is a persistent and fairly uniform snow line. Below this line the land is drier and appears to glow in Ch 3, and the big lakes are warm (no glint in Ch 3). Sun-facing mountain sides are warm and so, remembering that this is December and that the proportion of the Ch 3 radiation received by the radiometer is an emission according to temperature rather than a scattering of the midwinter sunshine, the general brightness of the lowlands is not surprising. But it is not really glow of haze.

In Ch 4 where the warm lowlands are very dark, the Swiss valleys may be examined in the 2× enlarged Ch 4 picture, where the cool valley bottoms contrast with the warm hillsides. Estimates of the depth and magnitude of the valley inversions can be made from the emissions of single pixels (e.g. at Chamonix on the northwest side of Mont Blanc (B 4807 m). Matterhorn (M 4478 m) and Monte Rosa (R 4634 m) are also marked.

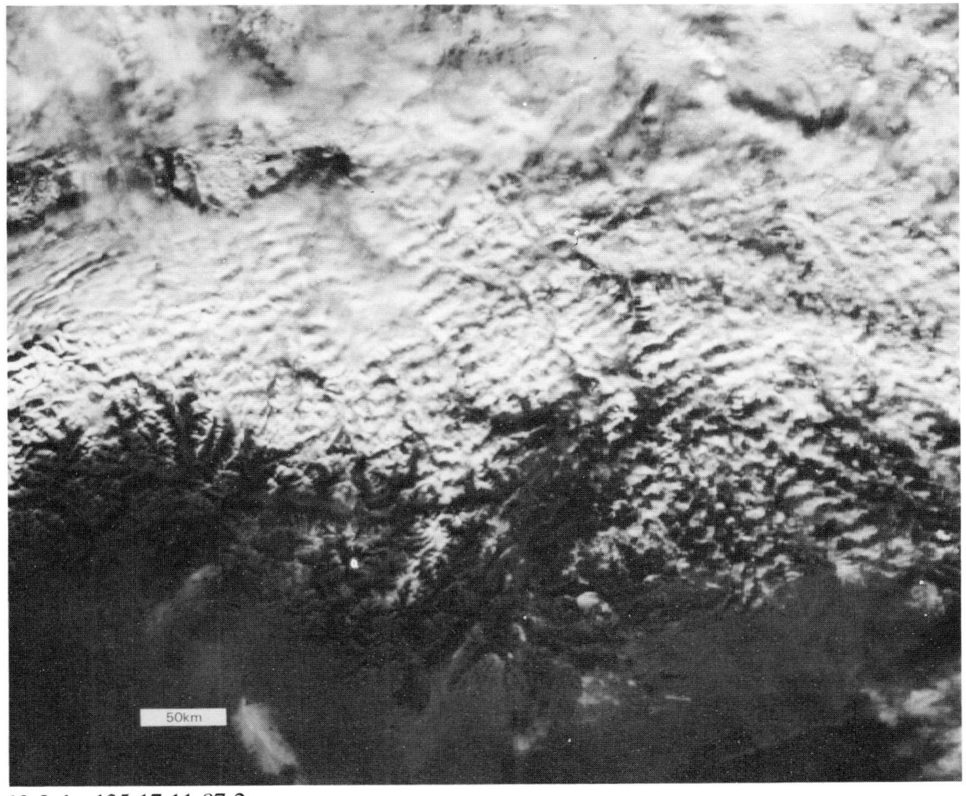

12.3.4 135,17.11.87,2

12.3.4

Air descending the lee side of the Alps, or any other large mountain mass, such as the Pyrenees, Iceland, Greenland, and many others, is warmed so that cloud evaporates. But, because of the multitude of irregularities in the topography, lee waves nearly always exist. The waves remain after the clouds have evaporated, and the associated dryness and variability of the wind from place to place are a cause of continual personal irritation which has long been associated with föhn winds (the wavelength in this case is about 8 km).

In many cases, above the region where the low cloud is evaporating, the upper air rises. This is associated with the deceleration of the airstream as it moves away from the mountain which blocks its flow; a rise in pressure is required for the deceleration, and this is produced by a raising of the streamlines at cirrus level and consequent cooling. It is frequently accompanied by the appearance of orographic cirrus, which often fills the sky where the low cloud has been cleared. It is not difficult to understand that the cirrus is associated with a cooling of the air and a rise of pressure beneath it, but it is not obvious that this must happen simply because the lower levels of air are descending from the mountain top. This is a characteristic of many 'solutions' to problems in fluid mechanics, and all that the mathematics shows is that it describes a possible state of steady flow. It is not explicable casually in terms such as 'if I push something it tends to move in the direction of the push'; but it is necessary to remark that gravity provides a force-field connection between the upper and lower layers. If a different steady-state mathematical solution is put forward there will be observers willing to search for the cloud forms which would be expected to accompany its flow. The case illustrated shows a cool air mass blocked on the north side of the Alps, spilling over the top into northern Italy. Although there are a few patches of cirrus in the wider scene, there is not a strong flow at that level, and no orographic cirrus in this case. This may be taken to mean that the down-flow on the lee side is not decelerated and is producing a föhn strongly illustrating the inconveniences of unpredictable gusts, calms, dryness and other features for which Trieste is renowned (including hand-rails in the streets, so it is reported).

12.3.5 1402,13.6.82,2

12.3.5
Following the above discussion we reproduce a picture of unsteady flow in the same area a few days earlier. In this there is a gap in the higher layers of cloud on the lee side of the Alps and, although some features are clear (e.g. Lake Garda) the low-level clouds have not settled into a pattern and one can imagine two doughnuts of cloud north of Trieste. When these winds blow from the east or northeast they have the name Bora.

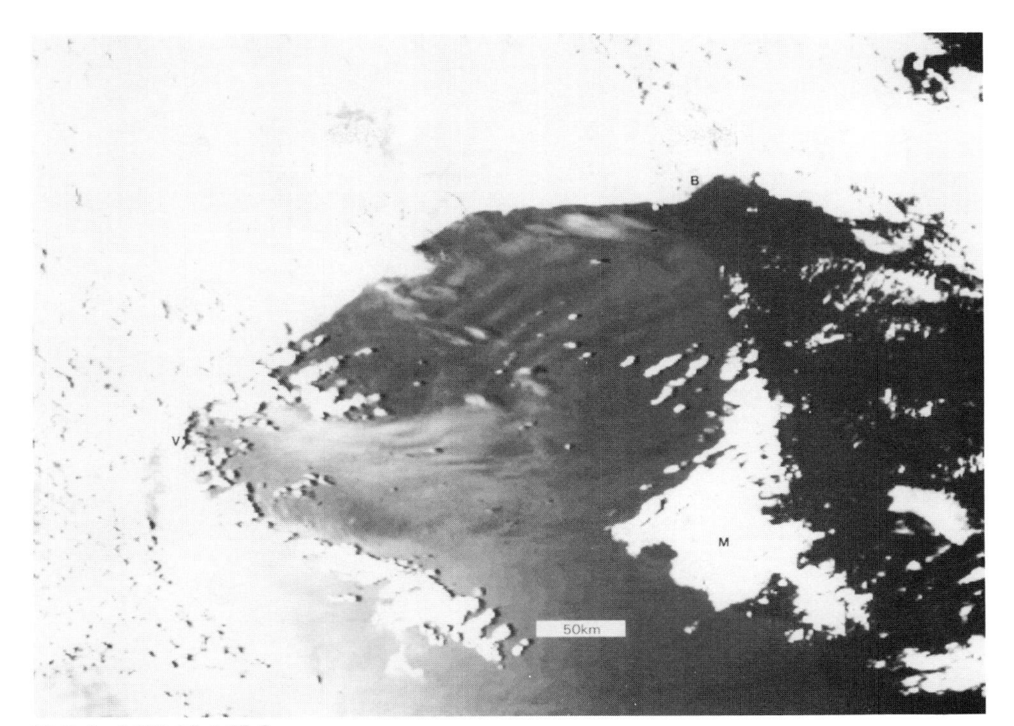

12.3.6 1402,13.6.82,2

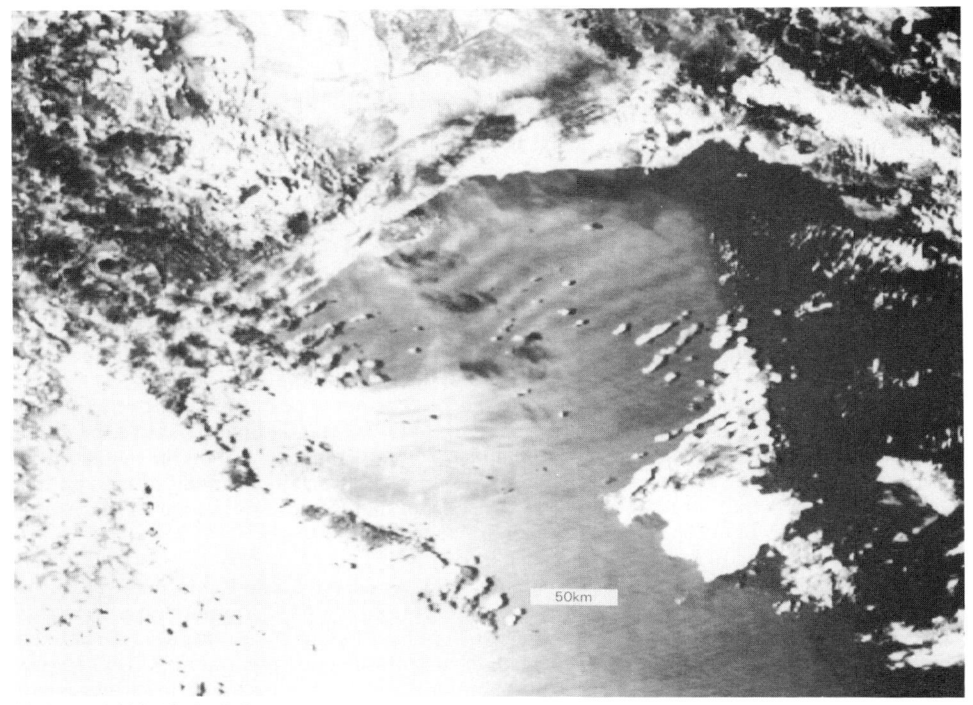

12.3.7 1402,13.6.82,3

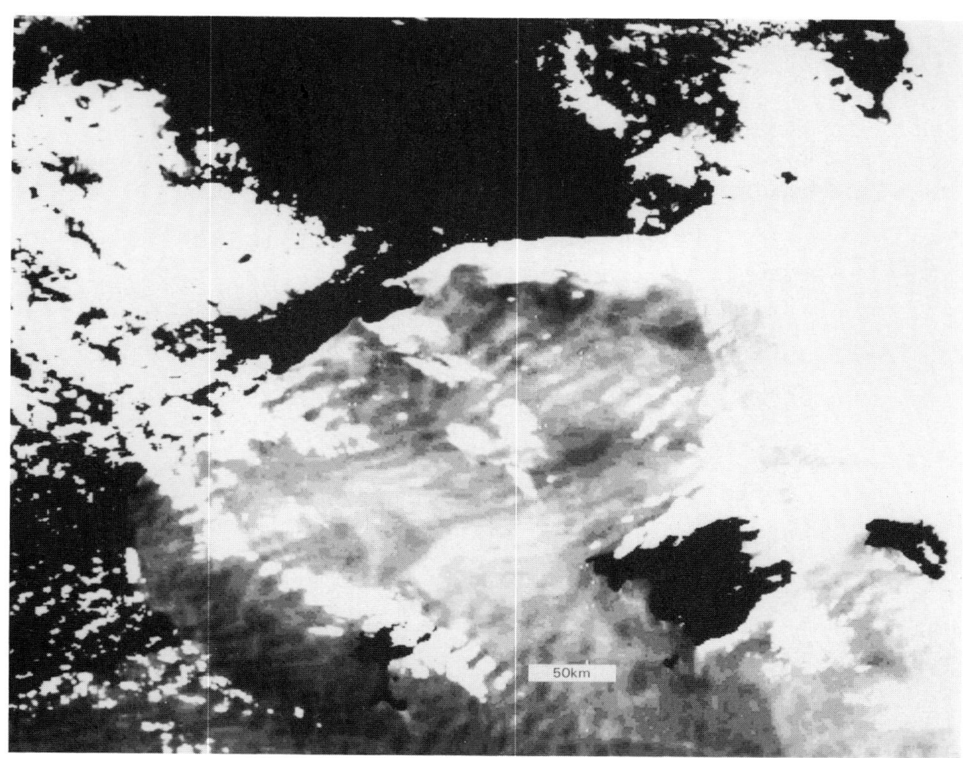

12.3.8 1402,13.6.82,4

12.3.6, 12.3.7, 12.3.8

The Ch 2 picture has been processed so as to show up the variations of glint in the lee of the Spanish coast between Barcelona (B) and Valencia (V), the wind blowing towards Majorca (M). Some wave crests have clouds which show their position (average wavelength about 12 km).

In Ch 3 it is not necessary to narrow the grey range displayed because the glint is much brighter relative to other illuminated objects. It is seen that the glint is brighter under the wave clouds. This would be expected if the glint is reduced by the waves raised in the lee wave troughs where the surface wind is stronger than under the wave crests.

Special interest therefore, attaches to the prominence of the waves over the sea in Ch 4, which does not record any glint. We can see that whatever it is that determines the brightness of the sea makes it brighter under the wave crest clouds. The effect is pervasive, with lee waves surrounding Majorca but not appearing to do so in the other channels.

The clouds appear in only some of the waves, which would be expected if the equilibrium level of the air was below cloud base with a very sharp reduction of humidity above the cloud top. There is therefore a deeper layer of the moist air below the inversion in the wave crests, the layer being colder than the sea and colder in the wave crests. Transmission through this layer modifies the apparent temperature of the sea and thereby reveals the lee wave system. The explanation given in [1], 5.9, is now seen to be incorrect in respect of Ch 4.

This is the opposite to a similar effect seen in very calm conditions in anticyclones (see [1], 4.6, 5.8, 5.19 and 5.20), where a black patch is seen in a larger area of glint because of the extremely calm conditions in the high pressure centre, where the top millimetre or less of water becomes hotter in sunshine when the sea is undisturbed and the warmth is not mixed downwards to a greater depth. The very calm patch therefore appears warmer in Ch 4. There is no raising of the inversion in the centre of the anticyclone. Under lee waves the water is not calm enough to produce this effect to an observable degree.

12.4 SEA BREEZES

Sea breezes are most commonly revealed by the absence of cloud from the coastal strip in which the incoming air is being warmed before cloud appears. This is influenced by the prevailing wind, which may reinforce or oppose the sea breeze. There is a countercurrent at the top of the stable layer in which the sea breeze is being formed, and this is often visible in the cloud over the sea, particularly in the tropics.

12.4.1 1407,12.9.86,1

12.4.1
The wind 'prevailing' over the North Sea is light and from between north and west. Over Denmark we see streets beginning close to the coast, but more slowly in north Germany and Holland. On the east coast of England there is a gap at the coast, the cumulus forming inland, but showing very little effect of drift in its formation pattern. The drift is slight and westwards, as indicated by the cloud lines in some places and the fact that the cloud crosses the west coast before evaporating. So there appears to be a high pressure region extending into the western part of the North Sea from Scotland where there is no indication of wind direction in the clouds.

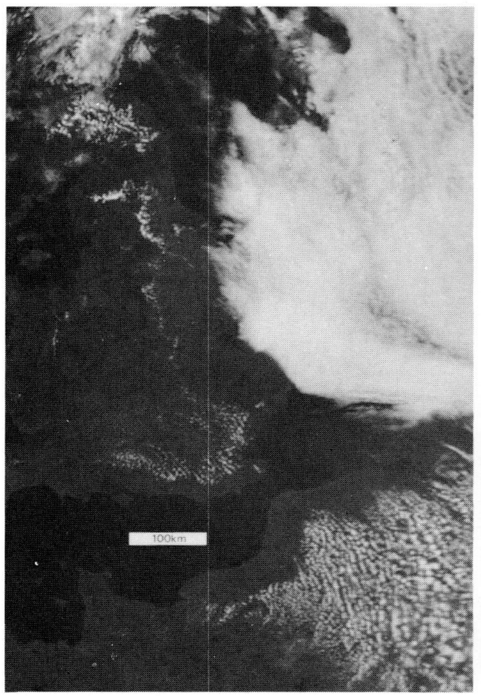

12.4.2 1543,5.7.84,1

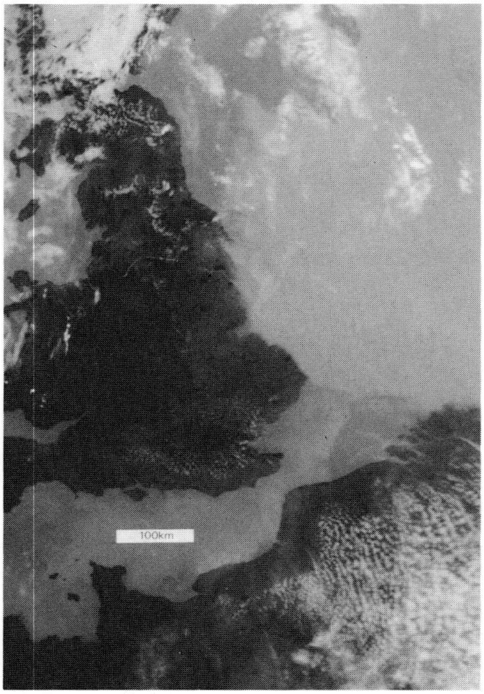

12.4.3 1543,5.7.84,4

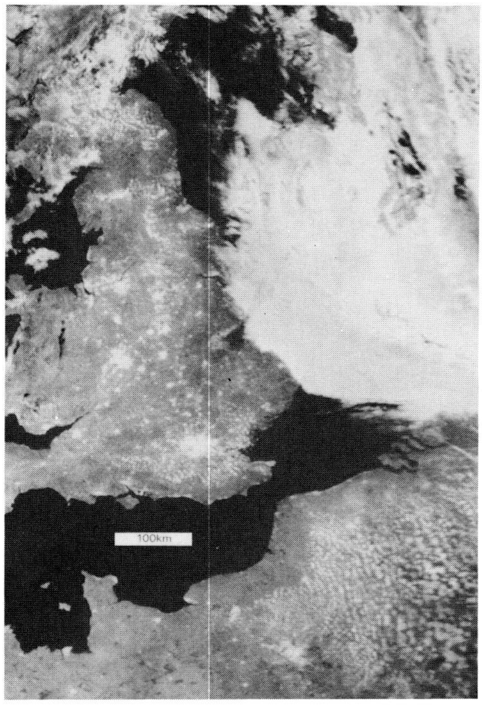

12.4.4 1543,5.7.84,3

12.4.2, 12.4.3, 12.4.4
Sea breezes are easily recognized by the absence of cloud very close to the coast. In this case breezes are flowing into eastern and southern England and northeast France and Belgium at the same time. The wind from the North Sea across the east coast of England has proceeded farthest inland, and the Ch 4 picture confirms that the sea air is cooler than that beyond the sea breeze front. The air southwest of Paris is cool, probably because of a very recent clearance of cloud, while the west of England and France are warm. The large towns are much warmer than the cool sea air. The temperature hot spots are shown even more clearly in Ch 3, because of the greater reflection from dry surfaces, but they are confused with small cumulus.

12.4.5 0735GMT,14.9.66,hand-held Ektachrome (by courtesy of NASA)

12.4.5
The full nature of the sea breeze was revealed by this remarkable picture, taken by astronauts Conrad and Gordon (Gemini XI). It shows the spread inland of the cloudless air at the surface, but more interestingly the spread out to sea of an upper clearance at the coasts of India. This may be a common phenomenon in the sense that it accompanies all sea breezes, but it may be complicated or obliterated by the pre-existing wind structure. It has been observed in this same area several times by satellite, mostly in the so-called quiet periods of March and April.

12.5 AIRFLOW UNDER INVERSIONS

Like the ocean, airflow under inversions is often slow compared with ordinary winds, and is therefore affected by the coriolis force rather dramatically. It fills valleys like a flood and flows through mountain gaps, filling lower regions and shaping up to the mountain contours. At coasts it produces land breezes which are sometimes seen in glint (e.g. **4.3.2**).

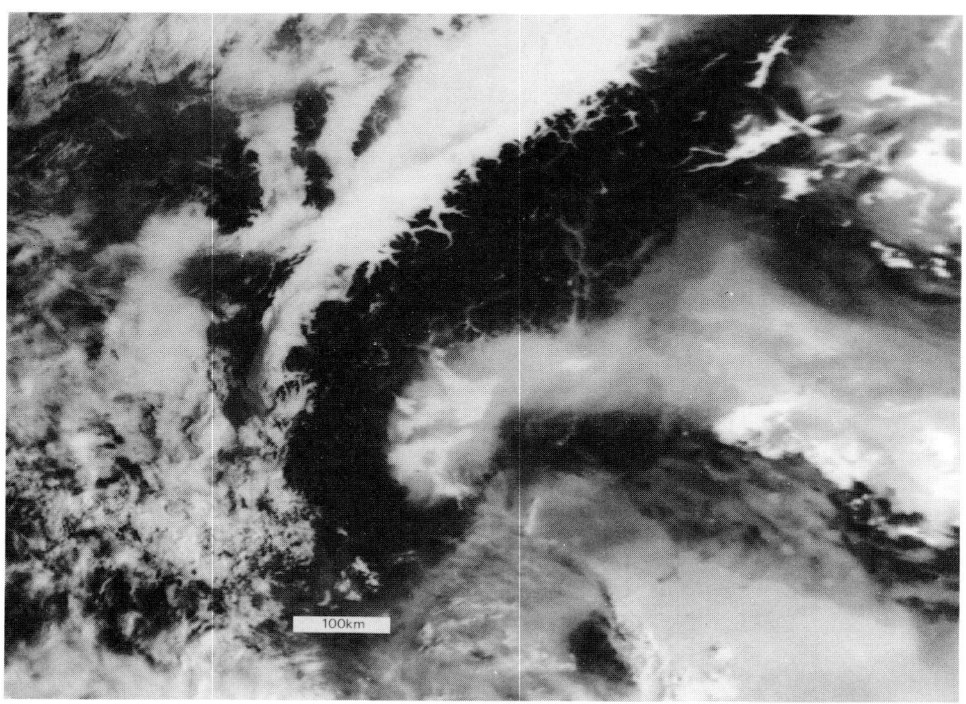

12.5.1 0815,17.10.86,1

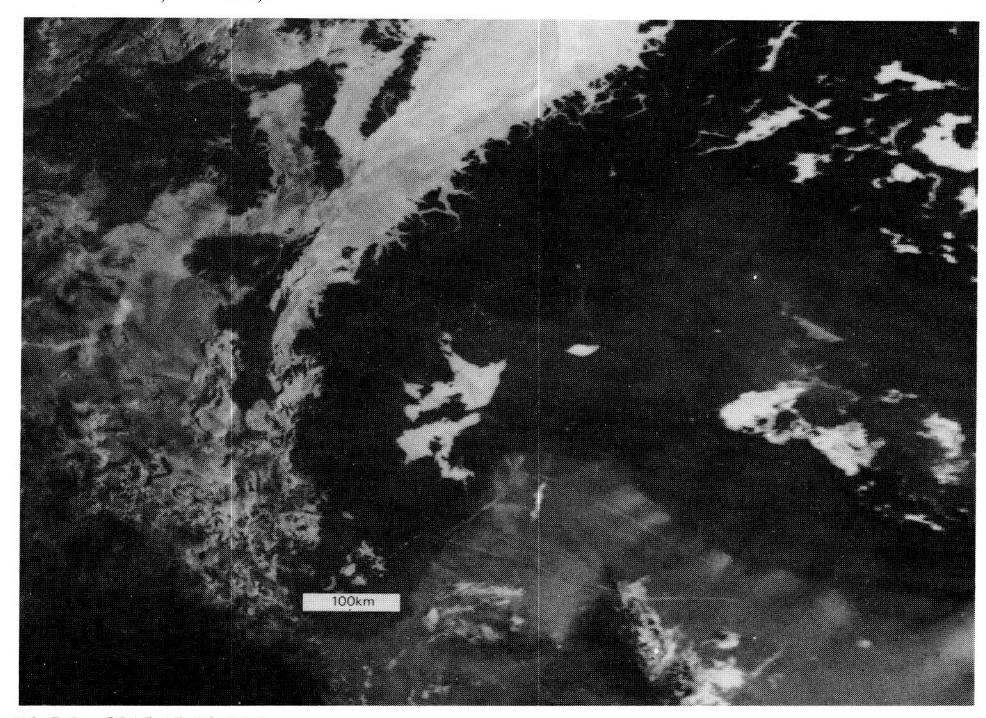

12.5.2 0815,17.10.86,3

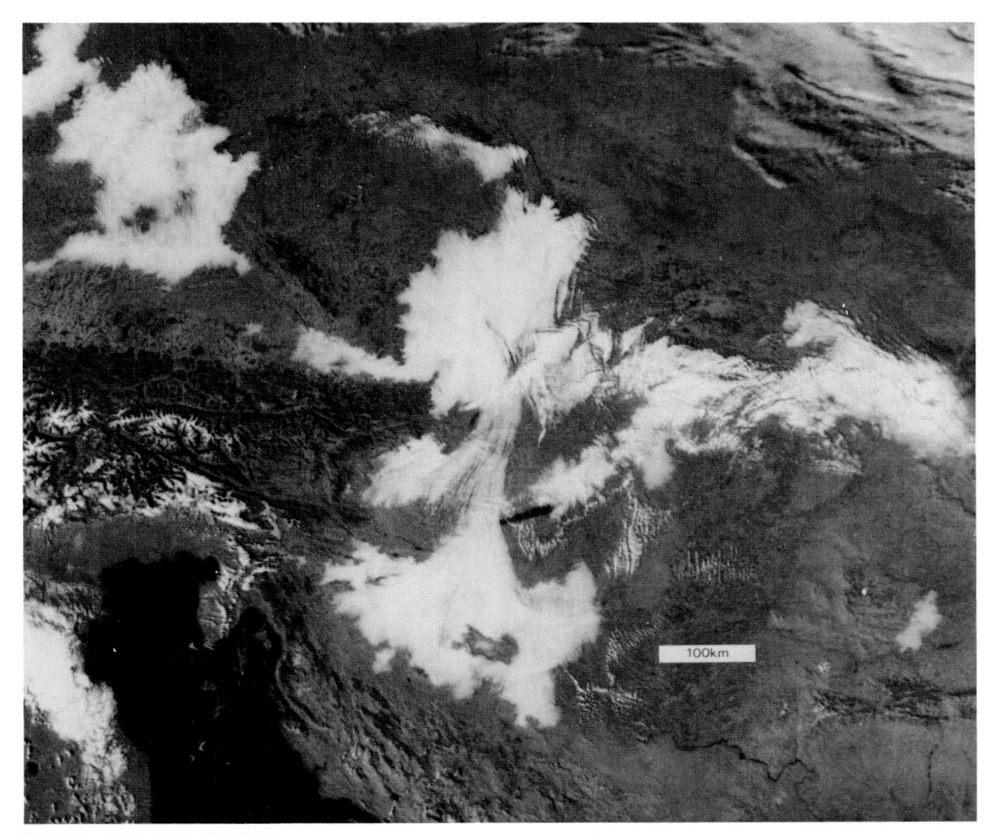

12.5.3 1318,21.10.82,2

12.5.3
The shape of stratus often indicates movement. Thus in this case the air cooled on an autumn night drains through the gap between the Alps and the Carpathians at the Vienna–Bratislava gap, from the higher area of Moravia into the plain of Hungary and Slovenia.

12.5.1, 12.5.2
The edge of the stratus/fog on the north side of the Alps is determined by the way the cold air penetrates into the valleys (see also **12.3.1–3**). In the Po valley and the seas around Italy there is extensive dense haze which is well revealed by Ch 1. Ch 2 (not shown here) adds very little information and shows the west Italian coast opposite Corsica less well because the land is brighter and the haze less bright, relative to Ch 1. We see the haze stretching up many Italian Alpine valleys, particularly above Lake Garda, while cloud (fog) fills the Swiss valleys.

Ch 3 makes a very clear distinction between stratus, which is bright, and haze, which is almost unobserved. Lake Garda appears warm. The stratus edges are sharper than in Ch 1. There is a drift of air from the Italian mainland into the sea from Genoa to Naples, with areas of the strongest surface wind being shown up by brighter glint, which is scarcely seen in Ch 1 (or 2). The contrails, which appear white in Ch 3, and are not visible in Ch 1, appeared white in Ch 4 also, whereas L.Garda appeared black (warm) in Ch 4 (not shown here).

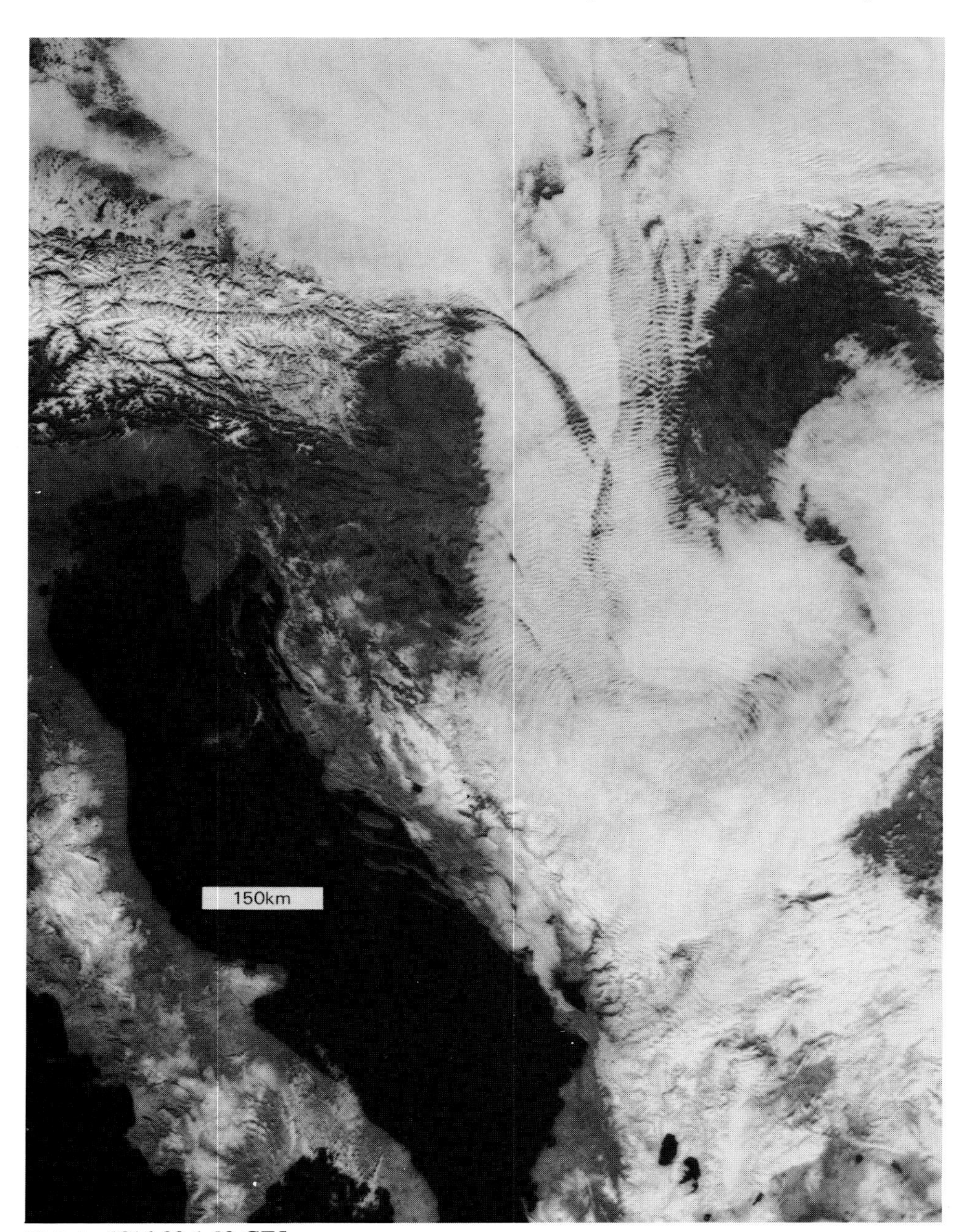

12.5.4 1016,23.1.83,CZ5

12.5.4, 12.5.5

On this occasion the drainage was more forceful and reached the mountains of southern Yugoslavia, generating some reflected waves in addition to the lee waves evident in the strong southward flow in Hungary. In the north the appearance of layers overlapping is not readily explained. Lee waves also occur in the overflow into Montenegro, Albania, and Greece, in the south.

(*continued next page*)

12.5.5 1016,23.1.83,CZ1

12.5.4, 12.5.5 (*continued*)

The CZ 1 picture is saturated, so that detail in the lee waves is completely lost in the Balkans. The drainage of hazy air from the Po valley down the east (Adriatic) coast of Italy is dramatically revealed. This clinging to the Italian coast of air subject to an accelerating force is quite typical and is caused by the Coriolis force. Similar behaviour occurs in the cold polar ocean current down the east coast of Greenland, which turns the corner and continues up the west coast from Cape Farewell. The Southerly Buster is another coast-clinging wind, in New South Wales. See **9.1.20,21**.

12.6 GROWING CUMULUS

Small cumulus are limited in vertical extent by an inversion and become arranged in streets along the wind. But when they grow uninhibited by such barriers they begin to form lines, or fronts, across the wind.

12.6.1 1413,29.6.82,2

12.6.1

At the height of the solar-heating season, convection begins by generating streets along the wind direction, but as the stability of the lower layers is reduced some clouds grow significantly bigger and, as can be seen over Ireland, Britain, and the western part of the European continent, the big clouds begin to form lines across the wind, but even these tend to disappear during a short sea passage. The tendency to form lines along the wind is still present in West Germany.

12.6.2

The tendencies described above are present in this case where the cumulus disappears over the Baltic sea, except downstream of Gotland (G). The north wind is stronger over Lithuania than over Sweden, and the orientation of the lines of large cumulus over Byelo-Russia can be seen as a consequence of the strong circulation of the low which settled over northern Russia four days earlier. There is no development of anvils or glaciation over Poland, and the air is stabilizing as an anticyclonic belt develops from the Azores to Norway.

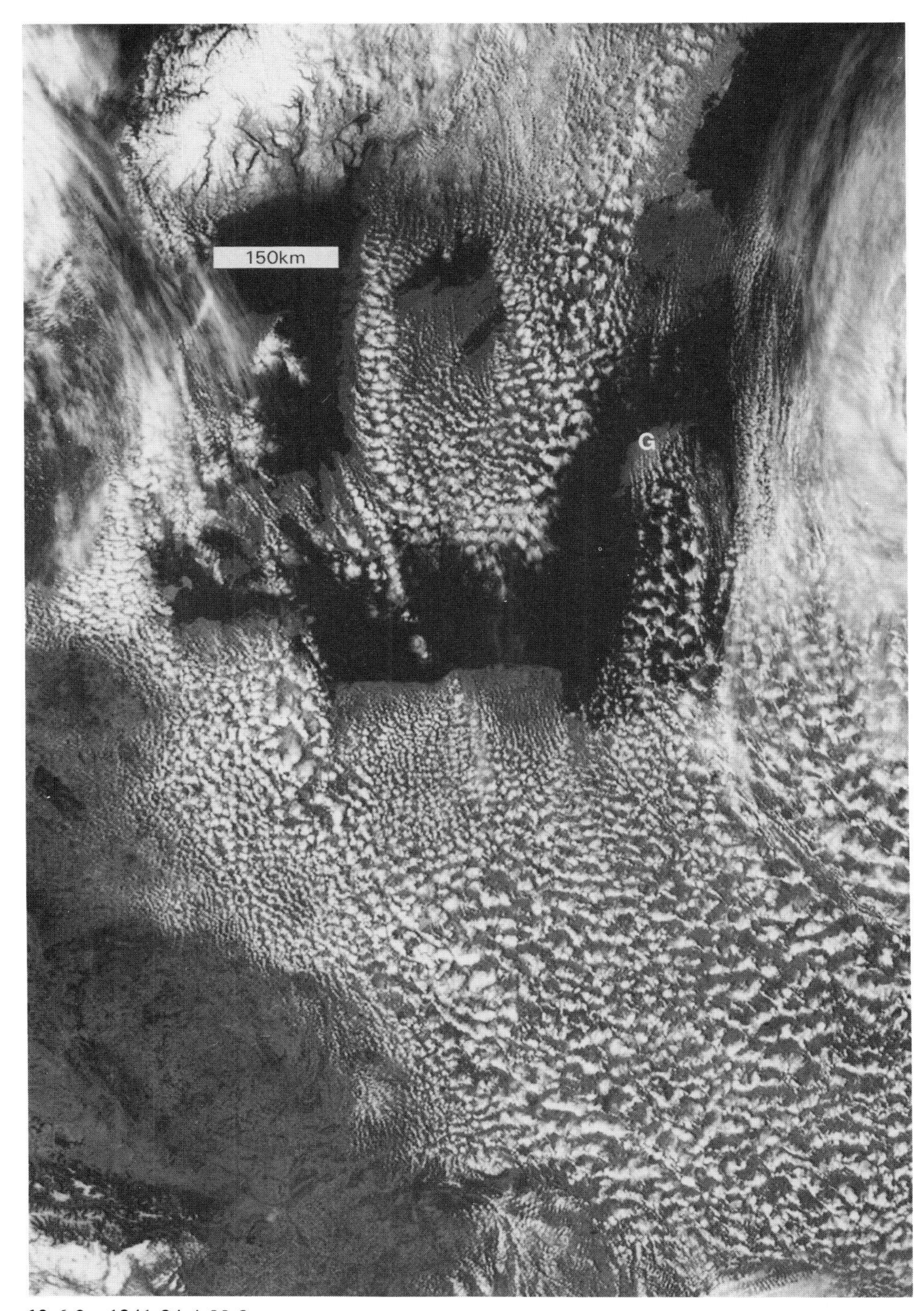

12.6.2 1341,24.4.88,2

12.7 OLD FRONTS

Fronts cause the transformation of air masses as they move them around, and the subsidence or storm activity in an air mass can alter the characteristics of the fronts at their boundary. Satellites are the best method of following these developments and for transforming them into an appreciation of the weather which will accompany them, the weather being the final product.

12.7.1 1548/1406/1227,6.4.87,4

12.7.1
When cyclonic activity is greater at the ends of a front, the part in-between becomes narrowed by stretching, but is not destroyed by subsidence because there is no advance of an anticyclone. One low is between Stornoway (S; in the outer Hebrides off northwest Scotland) and the Azores (A); the other is over the Gulf of Finland, between the North Cape (N) of Norway, and Guryev (G) where the river Ural enters the Caspian Sea. A high over the western Mediterranean is suppressing the convective activity due to sunshine in the surrounding lands, and a front lying southwards from Biscay, west of Barcelona (B) is a greater generator of clouds. The col over Denmark is cloudless. Greenland (G, Godthab) is unaffected.

In the sense that fronts have moods, this one is dying passively, with more subsidence at middle levels, leaving the dense strip of high cloud to mark its position but not as a precursor of rainy weather, in this case. There is some residual rope west of Guryev.

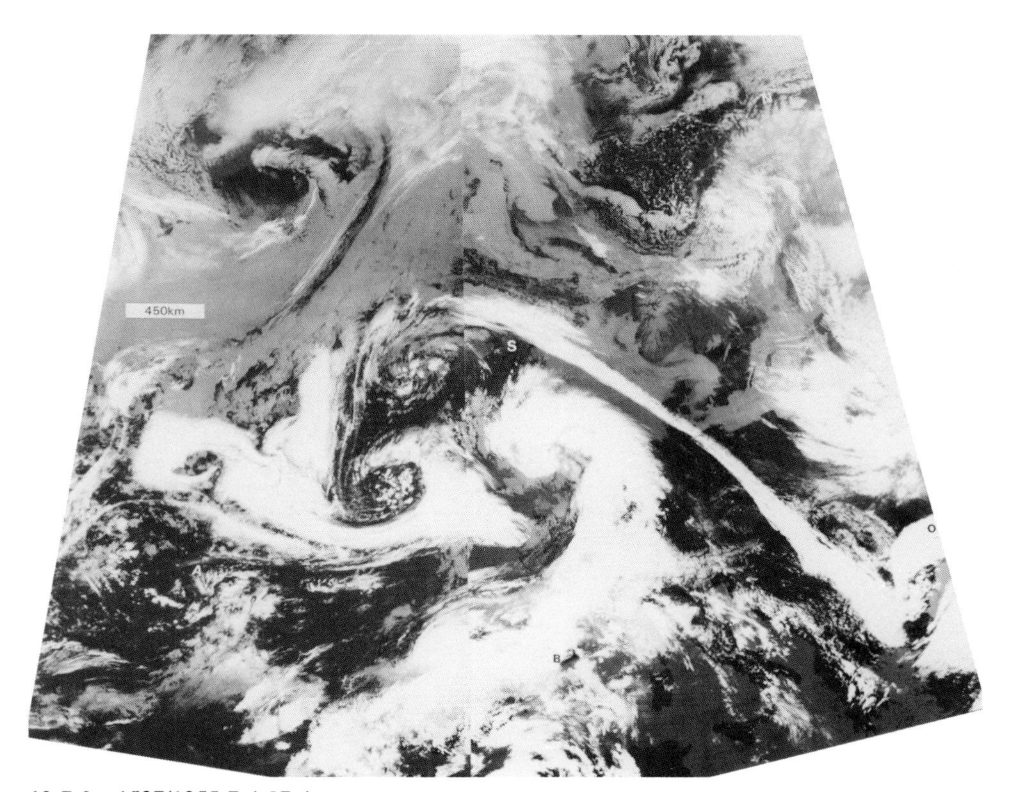

12.7.2 1537/1355,7.4.87,4

12.7.2
The lows south of Stornoway have drawn up the Biscay front into the North Sea and across Spain, while at the other end the front has been drawn down into Bulgaria, and it is stretched very thin in between. One wonders what frontal analysis would have been made of this unusual scene without the assistance of satellites!

13

Newly discovered cloud forms

The nature of satellite observations is such that new discoveries can be made almost daily. Some of these, such as the occurrence of small cyclones in the Arctic, or the scale of lee waves, vortex streets, and other such arrangements of clouds, represent improvements in our knowledge of these occurrences, but in them no essentially new cloud physics or mechanics appear. But looking down from far above, instead of skimming along the top of the troposphere as many passenger aircraft do, makes it possible to see structures which have not been expected anywhere and are unfamiliar in pre-satellite experience.

13.1 SPIKES

These clouds appear at the top of frontal cloud systems as groups or rows of spikes or bullet-shaped individuals. They have lifetimes usually at least as long as the 100 min between passes of the same satellite, and perhaps much longer. Observation by geostationary satellite could determine their duration. The official name, according to the old international Latin classification, should be 'cirrus spissatus spicus' (dense cirrus spikes); but that is not very helpful.

Typical dimensions are shown in the pictures of such clouds. They seem to appear more readily on warm fronts and over the Atlantic more often than over Europe, but that must be taken as a subjective impression. Once seen they appear to become more common, which is true of so much to which we have our attention drawn! The first example here is the picture which introduced them to me, on a day when they were associated with a storm which blew up at Dundee during one of my visits.

They have a vivid three-dimensional appearance, as if with a circular cross-section, and in the Ch 3 picture the sides appear to be viewed tangentially and to be brighter, with the result that sometimes they appear as V-shaped clouds in this and other channels.

In the first published picture of them spikes are called 'transverse bands of cirrus' ([6], Figs 40 and 41, the latter of which is very like the second picture shown here). But in 1966 the picture resolution was not enough to distinguish it from the much more common transverse bands of fibrous cirrus commonly seen from the ground in frontal, and particularly jetstream, cirrus. Nor could it be said that they were not billows, although the clear definition of billows as groups of unstable waves was not widely used at that time. And, even at the time of this writing, no picture of spikes as seen from the ground is known to have been published. Those in refs [1], [11], and [vii] are from the present satellite collection.

No laboratory experiment has produced an example of fluid flow which resembles spikes, nor has any theoretical treatment of fluid mechanics indicated the likelihood that such a pattern would be produced. This may be because the necessary balance between density stratification, initial vorticity, and Coriolis forces has been put together as the starting conditions. The one-sidedness of spike groups, having their sharp ends pointing towards the sharp edge of the wedge of warm air held between the tropopause and a frontal surface, gives a likely starting configuration for the motion which produces spikes.

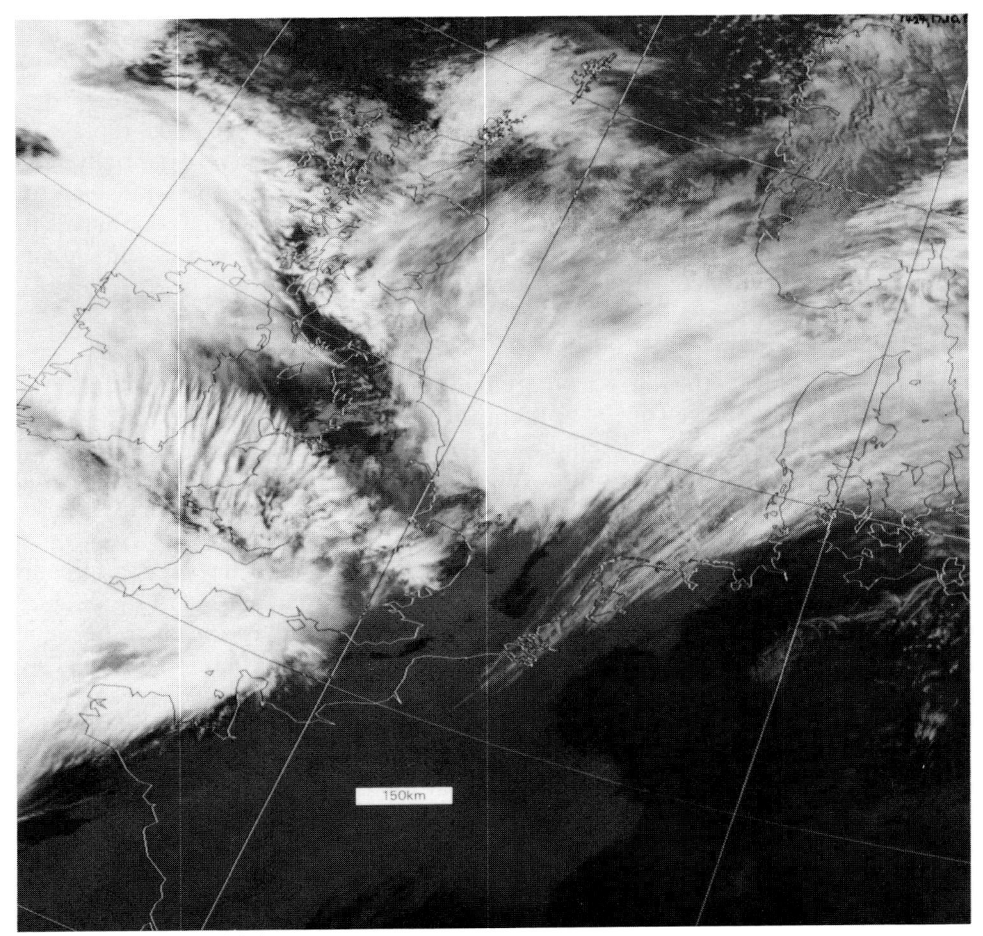

13.1.1 1424,17.10.84,4

13.1.1

An earlier pass at 0840 gave no hint that this group of spikes would appear, and the rest of the scene shows quite ordinary changes of configuration. Strong cyclonic development followed in the next 18 h. Close-ups of the seemingly three-dimensional structure in other channels appear in [1], Chapter 22.

There are some aircraft loop-trails in the southeast corner of the picture over northwest Germany.

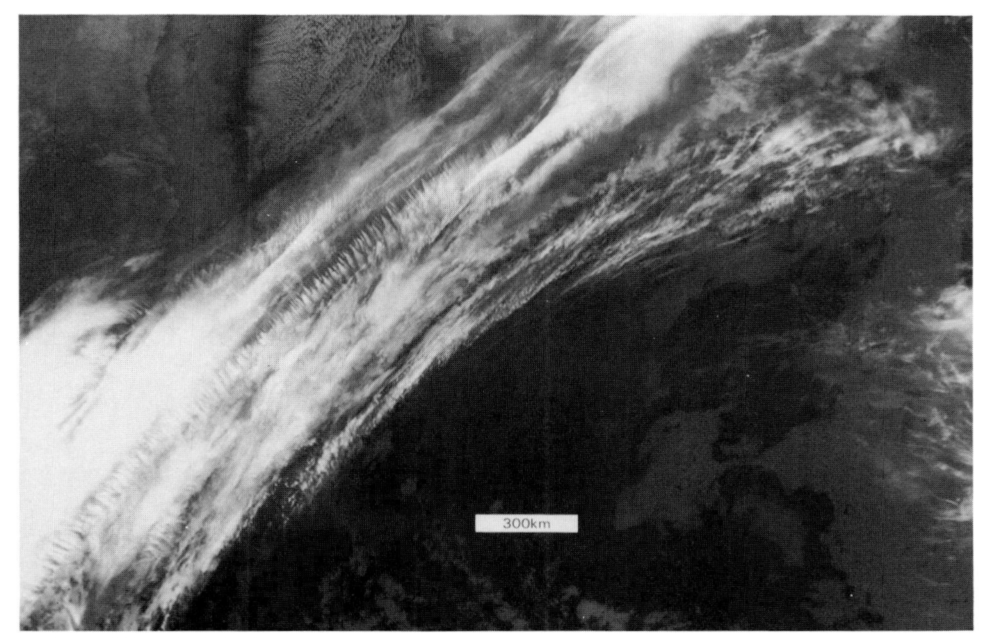

13.1.2 1731,26.2.88,4

13.1.2

A search of the Atlantic quickly reveals other examples of spikes. This second case occurred south of Greenland on the almost stationary part of a wide frontal cloud belt, on the west side of a blocking anticyclone which was centered at 52°N 22°W in mid-Atlantic, and extended from east of Newfoundland to Iceland. The Labrador coast is seen in the northwest.

Although this has been called frontal instability and is sometimes accompanied by the development of strong winds, this is not the case in many instances.

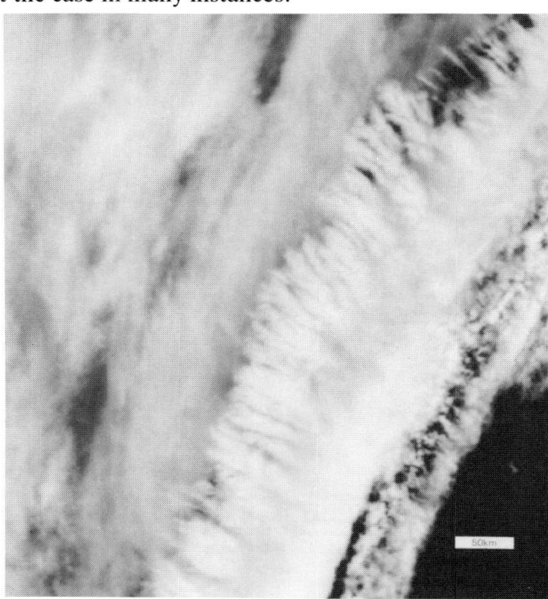

13.1.3 1614,13.4.87,4

13.1.3

There is a feathered variety, of which this is an example from the central North Atlantic, and is seen without spikes.

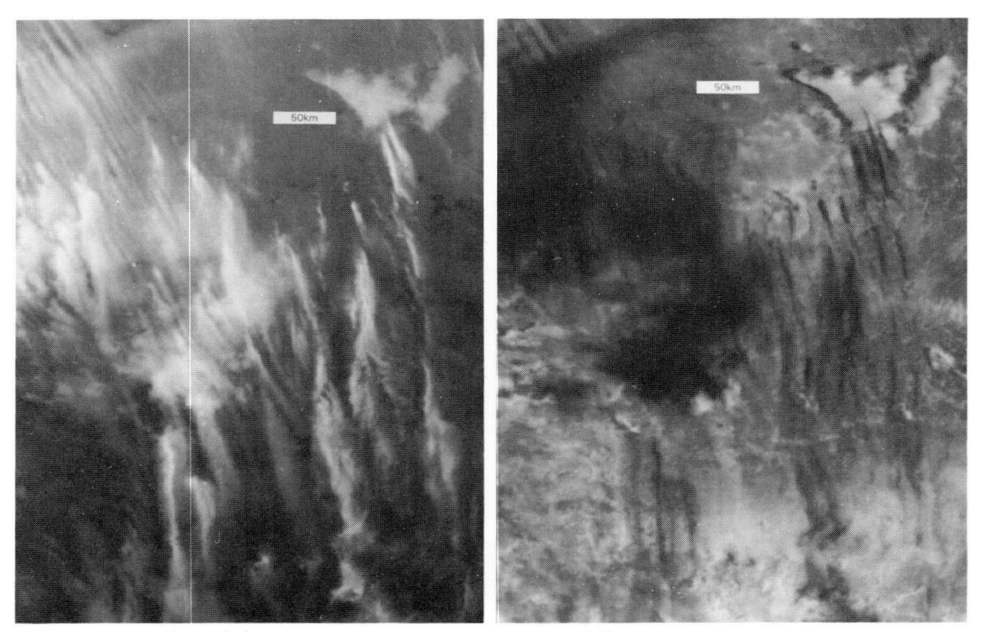

13.1.4 0856,13.4.87,4 **13.1.5** 0856,13.4.87,3

13.1.4, 13.1.5

This case, between Iceland and Ireland, is much more scattered, and in Ch 3 the cloud and its shadows are both black. A considerable amount of detail of the cloud below can be seen through the clouds in this and the next example **13.1.6**.

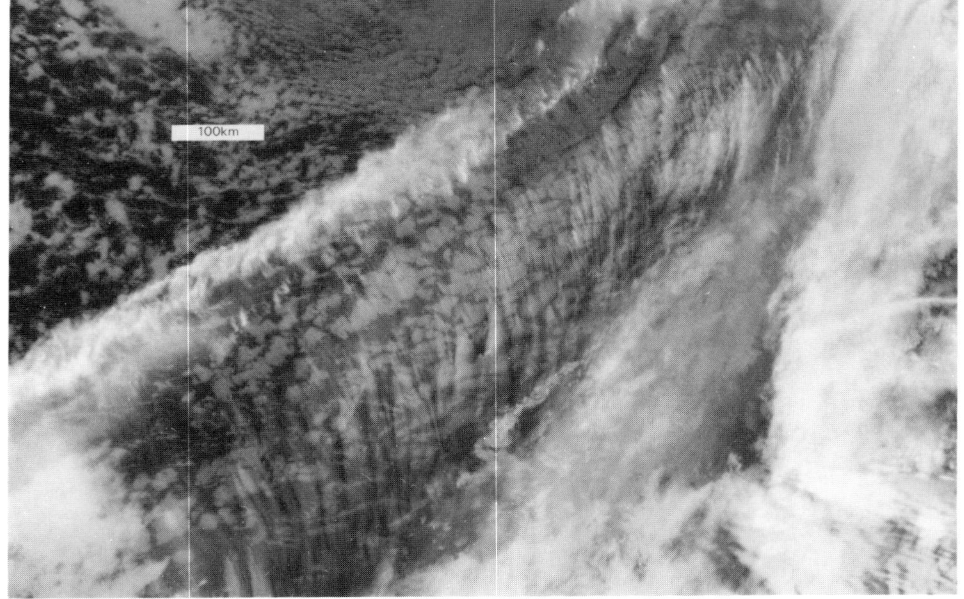

13.1.6 0419,27.3.87,4

13.1.6

The cellular pattern in the lower cloud can clearly be seen through the spike formations in this nighttime Ch 4 picture.

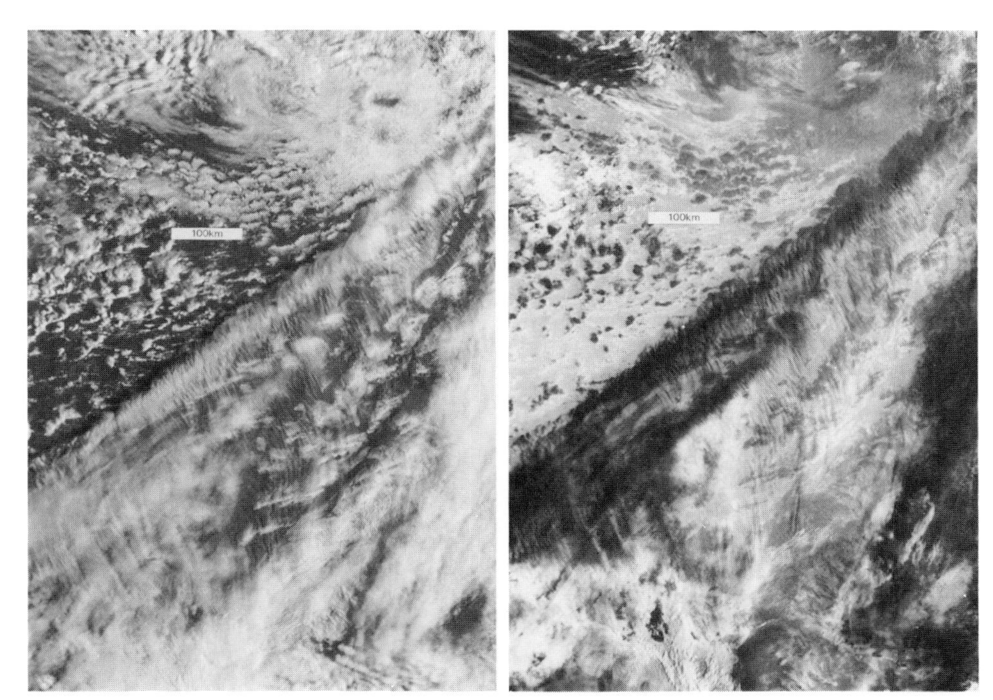

13.1.7 0825,27.3.87,1

13.1.8 0825,27.3.87,3

13.1.7, 13.1.8, 13.1.9

We show the Ch 1 picture here, but it is almost indistinguishable from the Ch 2 picture (not shown) of the same scene, which is at 47°N 5°W. In Ch 3 the illumination is quite differently distributed; the sea is bright because it is warm, not because of glint, and there are no shadows on it. Only the small low clouds scatter Ch 3 sunshine. The high cloud seems less transparent in Ch 4, and this may be so, but it should be noted that the emissions from the low cloud are much more subdued relative to those from the high cloud than in Ch 3.

13.1.9 0825,27.3.87,4

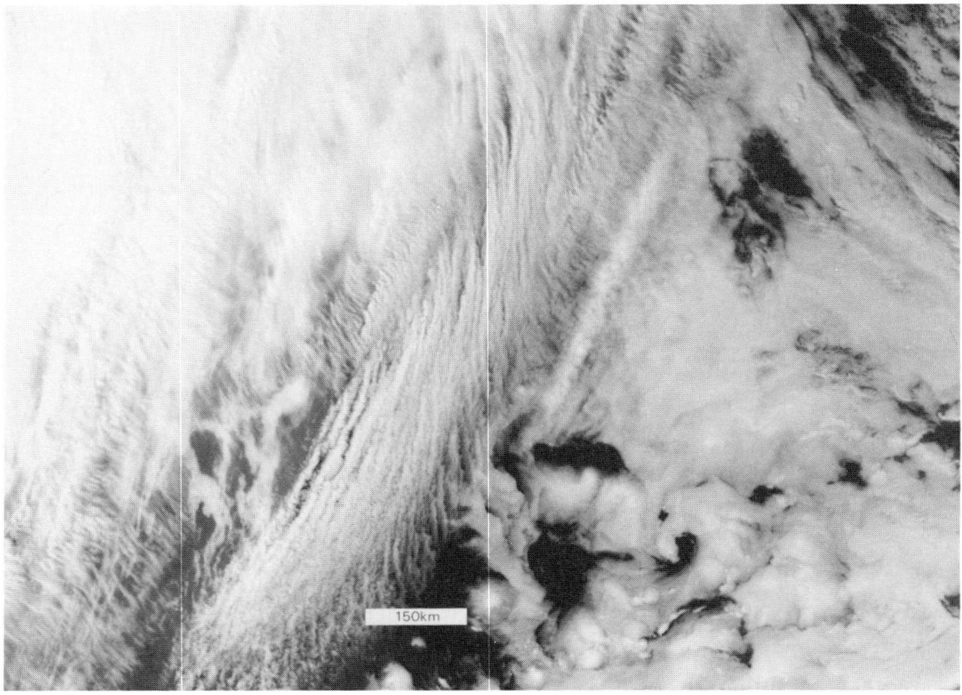

13.1.10 1631,22.2.88,2

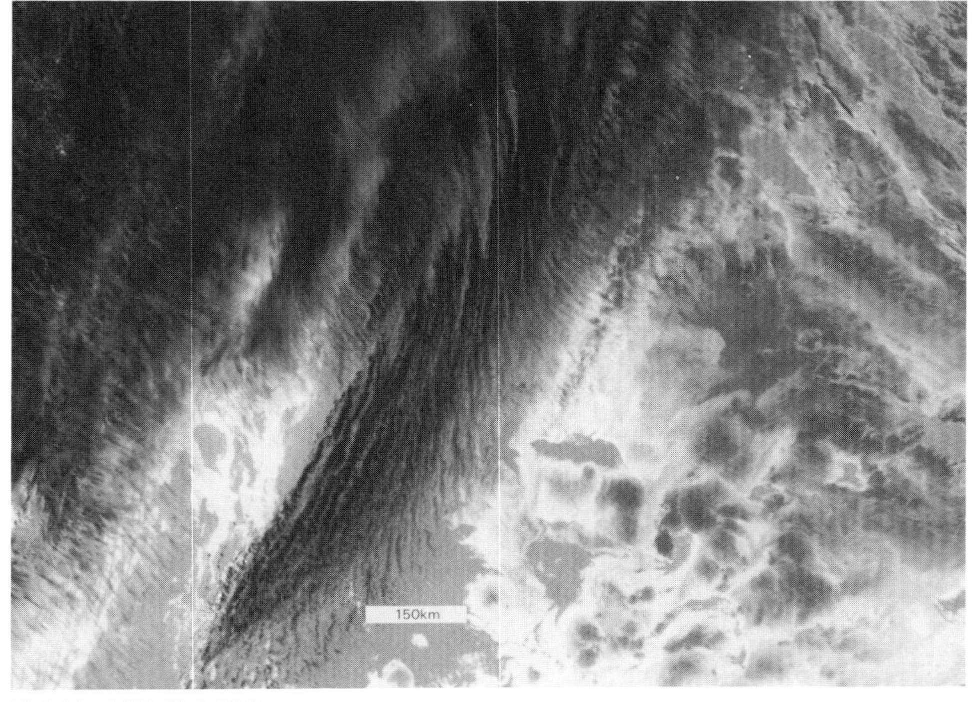

13.1.11 1631,22.2.88,3

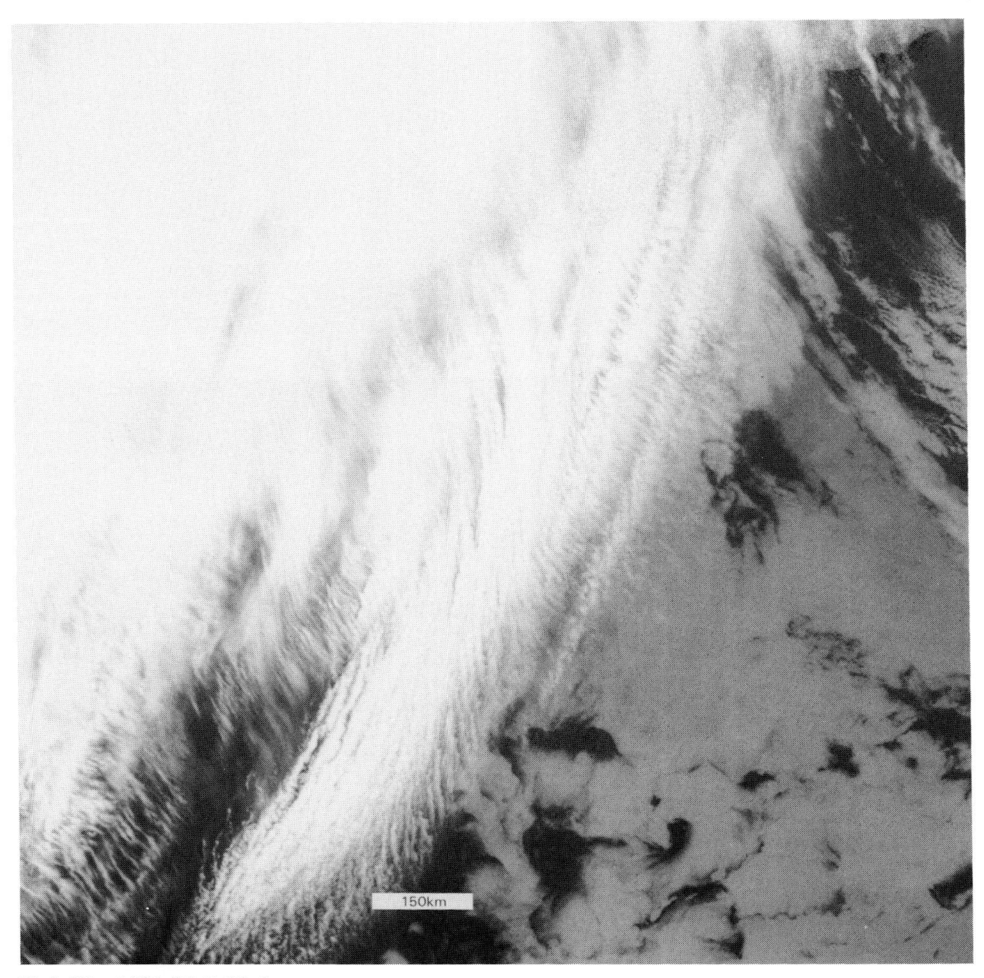

13.1.12 1631,22.2.88,4

13.1.10, 13.1.11, 13.1.12
The spikes are seen best in Ch 4 when the low cloud cells are uniformly dull. (In Chs 2 and 3 the cells show opposite shading (see **5.6.1/2/3**.)

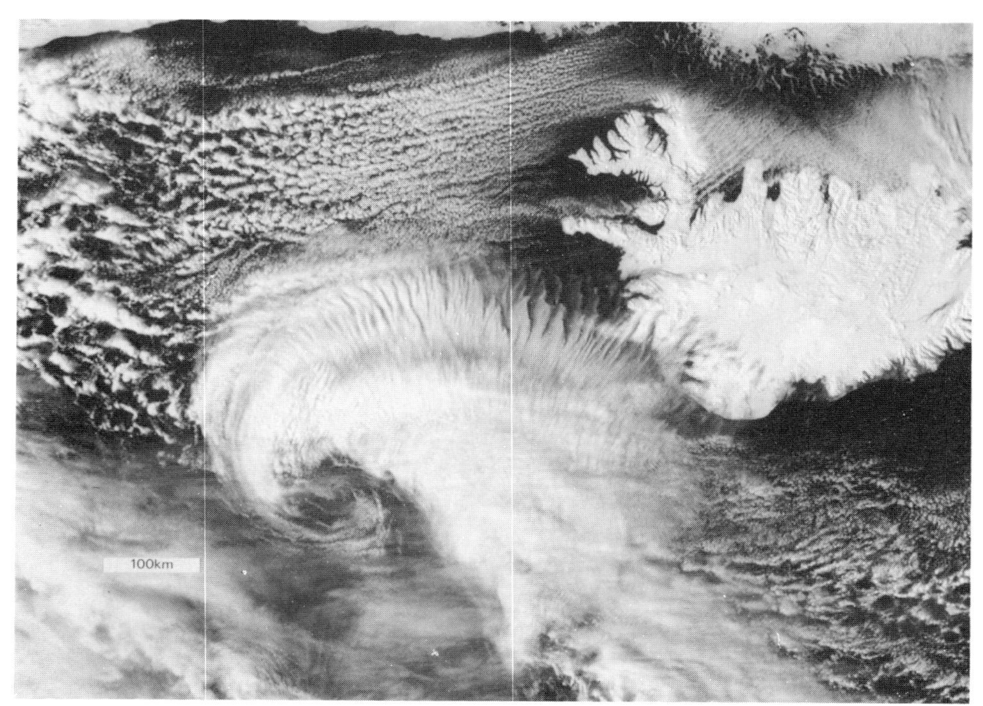

13.1.13 1602,11.4.88,2

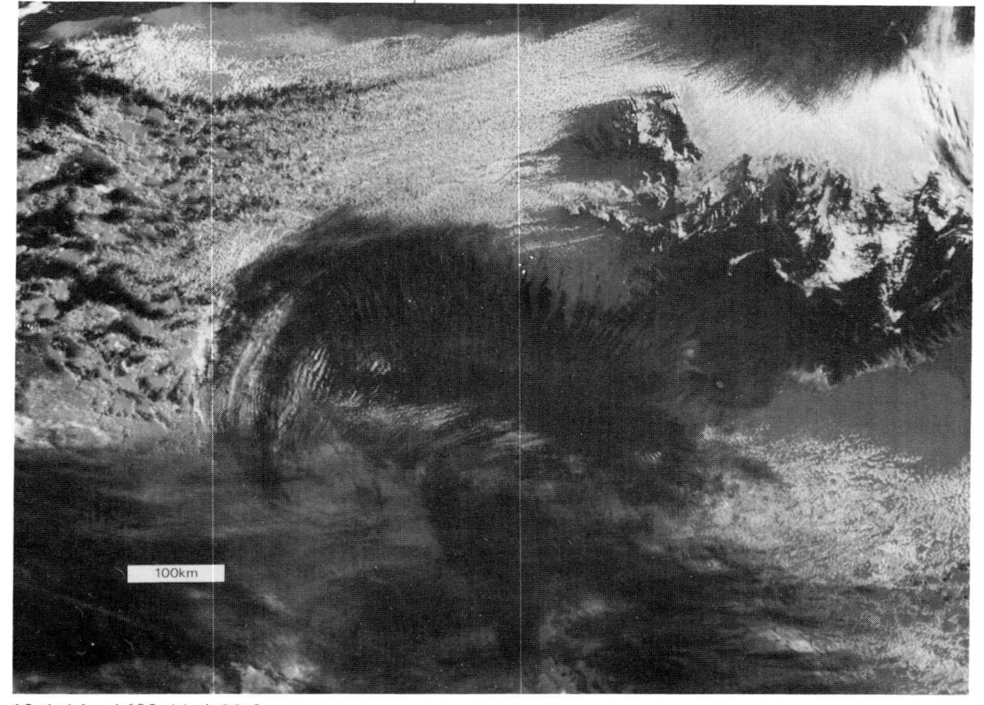

13.1.14 1602,11.4.88,3

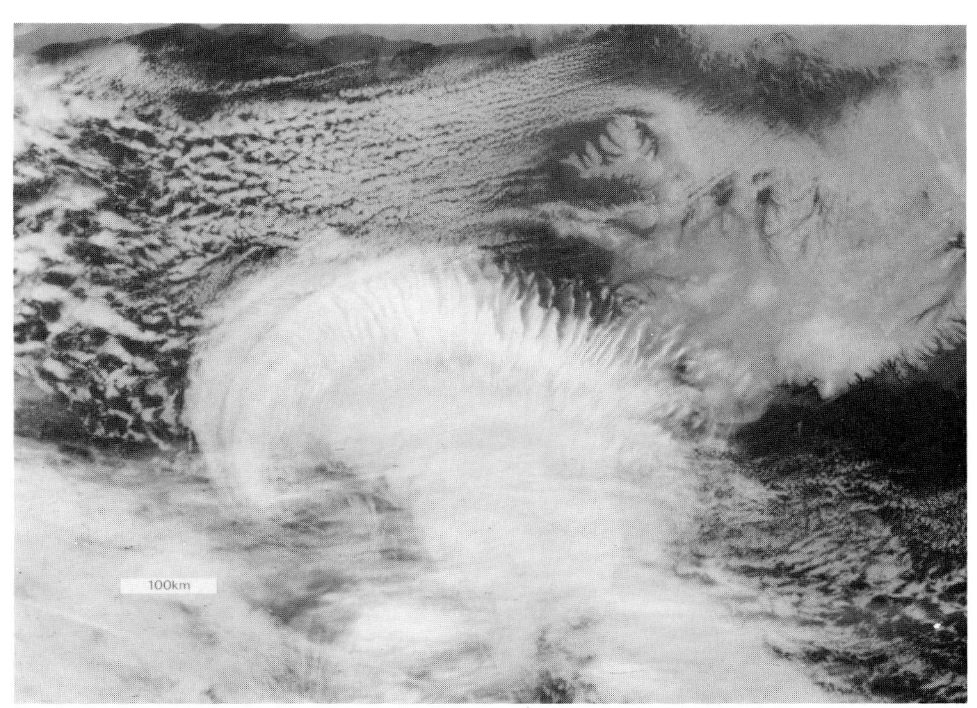

13.1.15 1602,11.4.88,4

13.1.13, 13.1.14, 13.1.15

This set of pictures is of special interest because all three channels confirm the deep three-dimensional nature of the cloud, which is transparent, particularly in Ch 2, over the peninsula west of Reykjavik. Over the adjacent sea the top edges of the clouds which appear to be in vertical sheets are more brightly illuminated in Ch 2, colder in Ch 4, and more absorbent in Ch 3, and in all cases less distinguishable from the sea below in their lower parts.

There are billows in the outflow system in the west, and a small patch is seen in the east most easily in Ch 3. In this channel the street structure further north is scarcely detectable through lack of contrast with the sea.

The chart analysts called this a trough line, but promoted it to an occluded front the next day!

13.2 ROPE AT THE SURFACE

This term, taken from the Japanese literature on the subject, is not to be confused with nuages en 'cordes' (plural) meaning cloud streets; nor with the 'rope' stage of a tornado funnel ([5], 14.3), which is the usage of T. Fujita. In [6] Fig. 50, the description used is 'Cold front with thin cloud line along leading edge'. Schietecat [9] gives a good example in Fig. 194, saying of un front froid 'Sa position en surface est determinee par une ligne de nuages cumuliformes bien visibles en VIS.' What has been discovered by satellite is the long endurance of this thin line and the fact that such lines may become stationary and still persist over a length of many hundreds (even thousands) of kilometres.

Ropes at the surface seem to be quite common when air masses entering the Mediterranean through different gaps in the surrounding mountains abut against each other. They are also common when a cold air mass invades the northwest Pacific Ocean from the great land mass of northeast Asia.

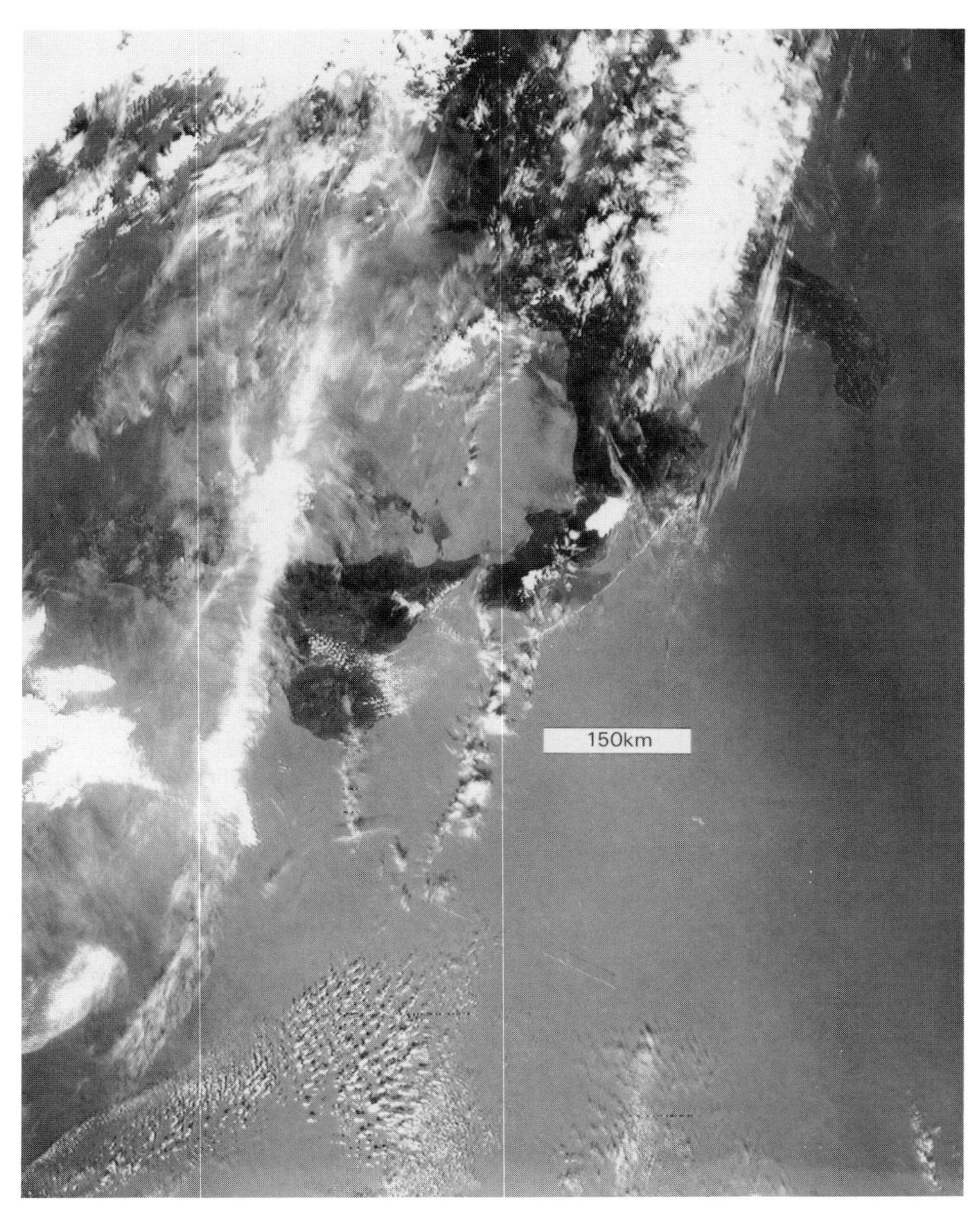

13.2.1 1023,27.9.81,CZ1

13.2.1

In this scene the blue channel, CZ1, depicts the calm Mediterranean with a thread of cumulus marking the boundary of an air mass which has crossed the toe of Italy. We see no shadow of this cloud on the sea because it is very low down; but the sea is glowing more than usual with this satellite and this may indicate that the camera was not fully tilted to remove glint. Cumulus and fragments of trail and other cirrus are well separated from their shadows. The land is very dark in this channel and the cloud very bright, so that western Sicily is almost obscured.

Similar loops of rope are often seen over the sea between Marseilles and Genoa, usually indicating the penetration of a small amount of an air mass from central France (e.g. **9.1.21**).

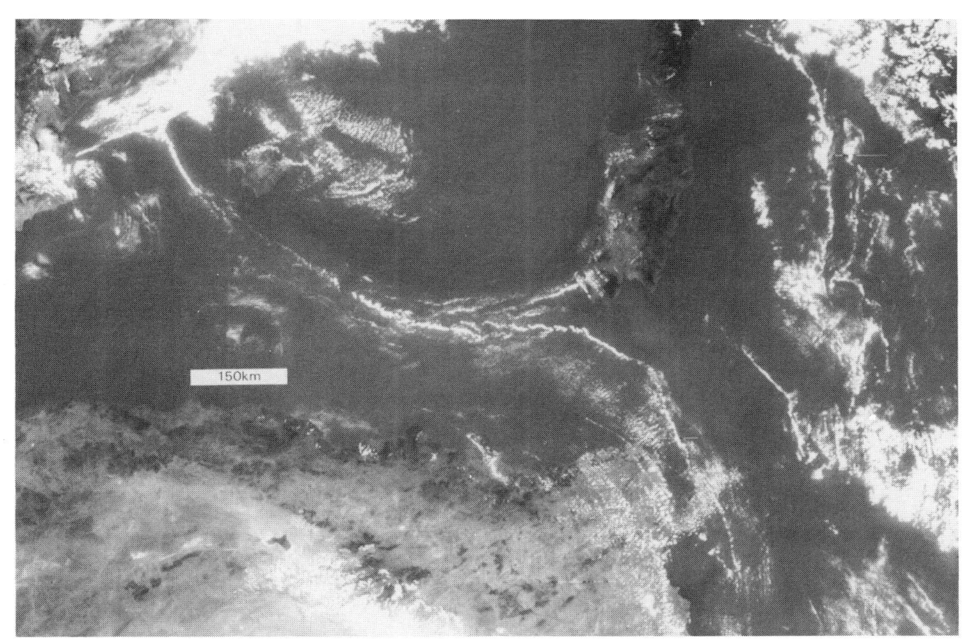

13.2.2 1101,3.9.83,CZ1

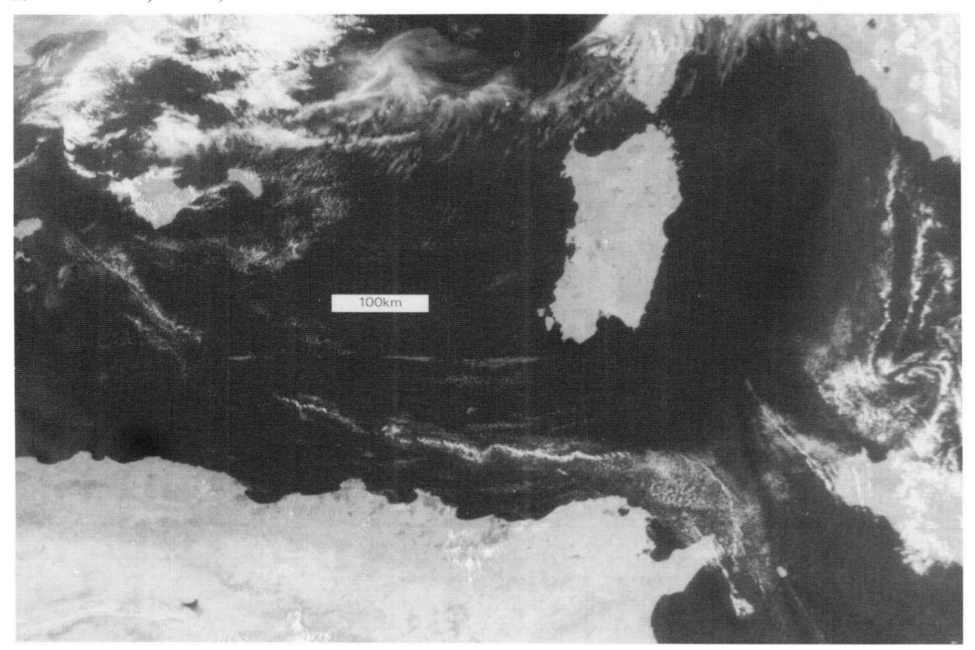

13.2.3 1434,3.9.83,1

13.2.2, 13.2.3

In the Western Mediterranean very long rope clouds are often seen associated with the boundary of a hazy area, but not always at the edge of the haze! In this pair of pictures we have used the shortest wavelength available to detect haze, and they show us typical movements of rope. It appears that the Cape Bon peninsula (Tunisia) is blocking the low-level flow.

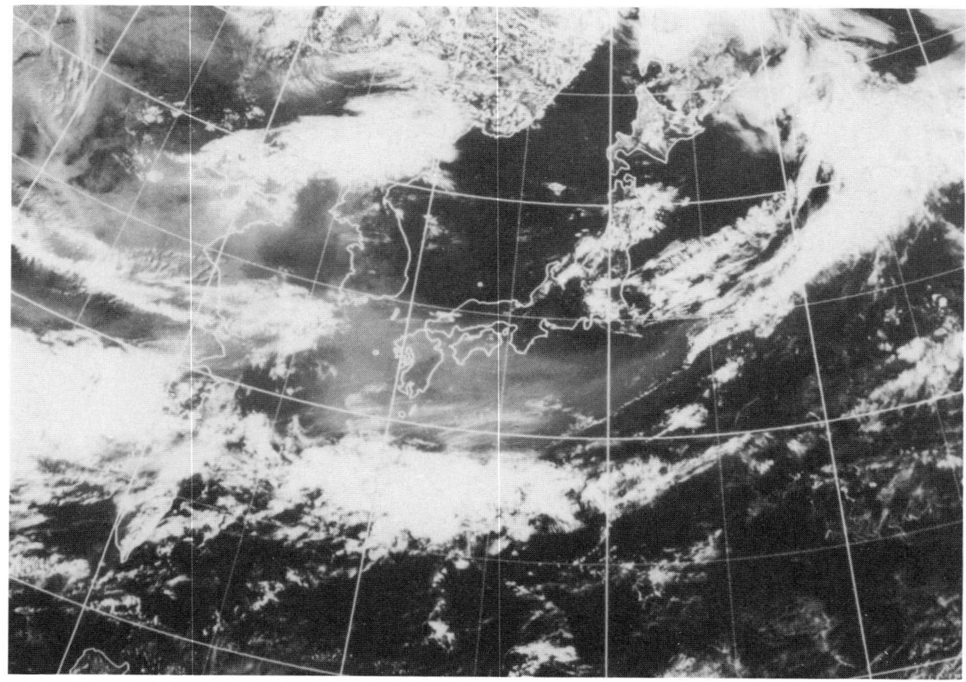

13.2.4 0553,15.4.79,VIS (GMS1)

13.2.4

In this case the rope is at the boundary of dust-laden air from the Gobi Desert, which has been carried across southern Japan (for more descriptive detail see [1], 13.21a).

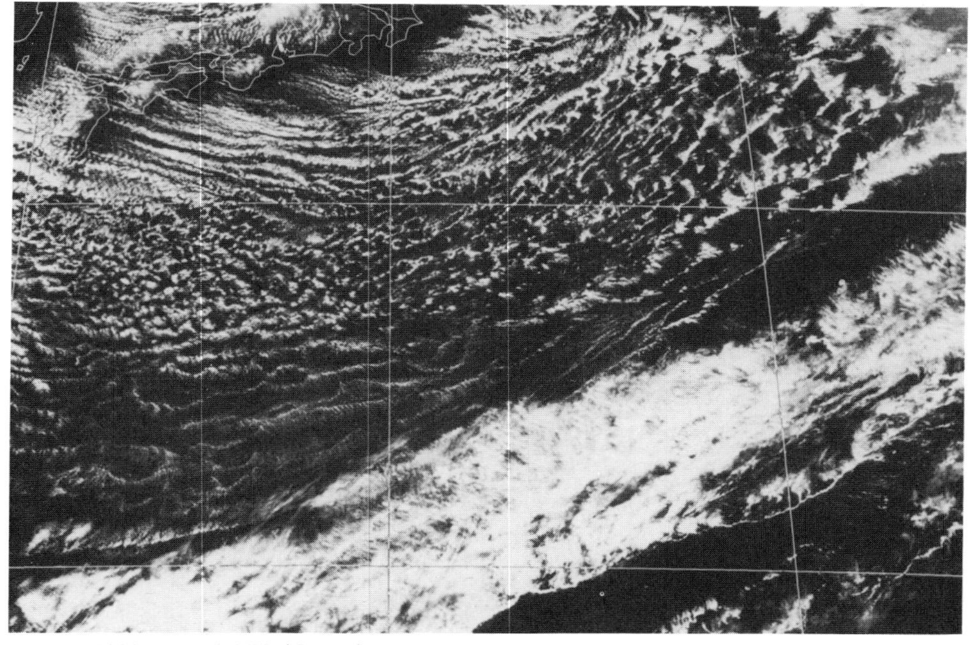

13.2.5 0300,19.2.83,VIS (GMS1)

13.2.6 0833,17.9.81,2

13.2.6

Ropes are often seen at the southeastern boundary of a cold air mass, in a col, northwest of Cape Finisterre. The rope is all that is left of the frontal cloud of a cold front that has almost ceased to advance, (41°N 17°30′W) and it is continuous for over 1000 km. Such rope may persist for a day or two.

13.2.5 (oposite)

This picture shows part of the cold outflow from northeast Asia into the northwest Pacific across Japan, with typical rope clouds at the front of each fresh burst of cold subsiding air. The middle one is at the front of the main cold air mass, and often such rope clouds stretch without a break for over 2000 km. (The full lines are at 10° intervals of latitude and longitude. Japanese coast top left.)

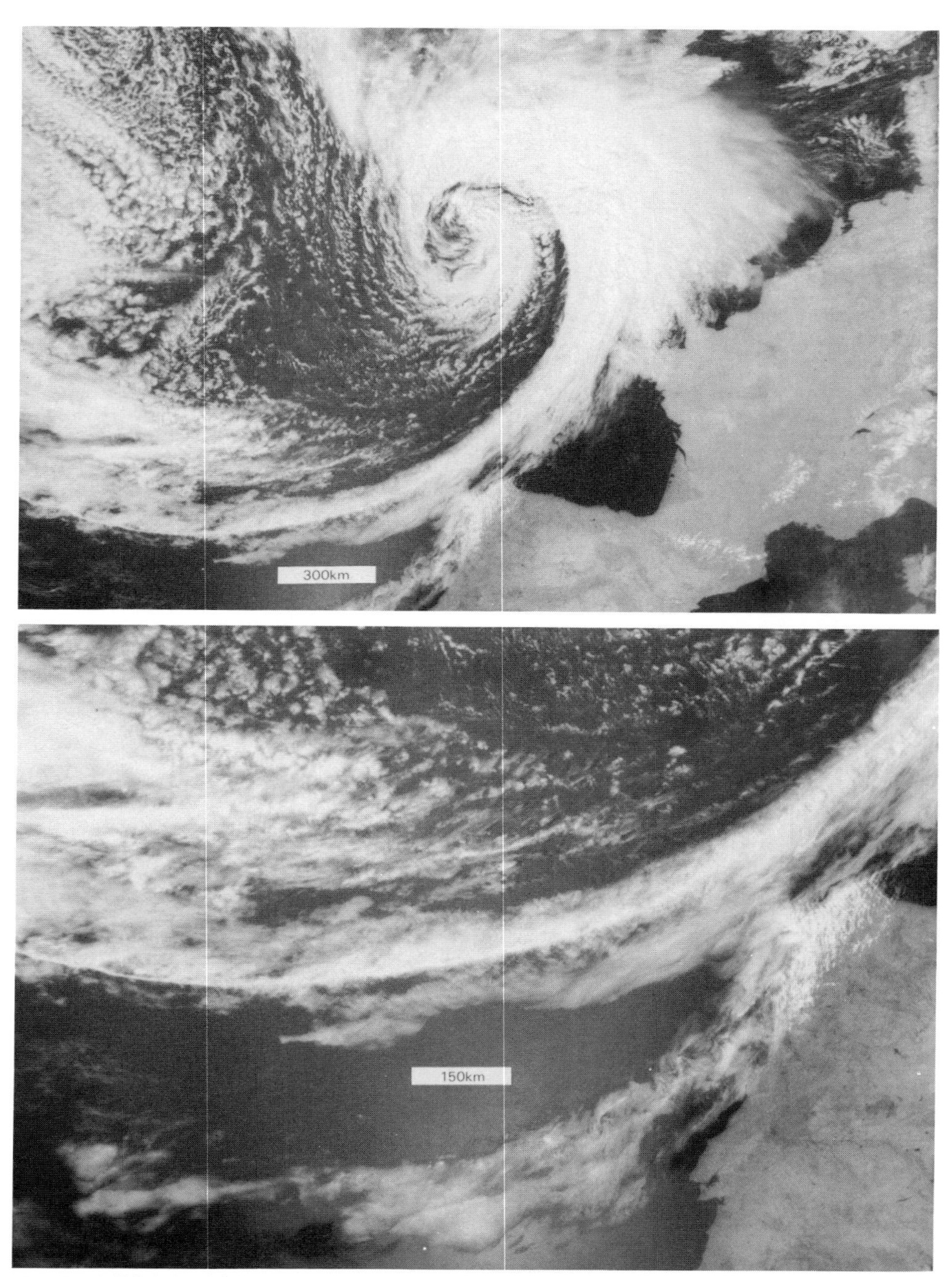

13.2.7 1455,1.8.86,2

13.2.7

This case is of a more actively advancing cold front with the rope best developed on the slower moving part which is shown in enlargement.

In Chs 3 and 4 (not shown here) the rope is identifiable where it stands isolated from the rest of the low cloud, which has a uniform shade.

13.3 ROPE IN THE SKY

Cold air spreading out at the surface as it subsides as a gravity current has a rope-like cloud defining its boundary; similar behaviour of warm air spreading out at an inversion as an anvil cloud is to be expected. In that case the air is squeezing in between a lighter layer above and a cooler, heavier layer below. The essential difference is that the spreading warm air is pushing the upper layer upwards as well as pushing the lower layer down. At the head of a gravity current there is a bulge in the advancing air of which the upper half is in motion similar to that of a cold air mass advancing along the rigid ground ([12] and [4, Chapter 13]).

Consequently there may often be a rope cloud in this upper half of the advancing edge of a cumulonimbus anvil. The same argument leads one to suppose that the same may happen at the front edge of the wedge of warm air at a warm front. If the warm air ceases to advance it is likely that the bulge, or nose, at the front edge will collapse and the rope cloud will disappear, and this may have value in analysis. In any case the advancing wedge must be very thin if the rope cloud is to become detatched from the main anvil cloud. Thus the most likely situation for the appearance of a rope

13.3.1 0300,5.4.82,IR (GMS1) (See also [1; 13.23ii])

13.3.1

The rim of the outflow, at the top of this southern hemisphere typhoon off the coast of Queensland (seen in the southwest corner), is marked by a tropopause rope cloud.

cloud is that a rapidly expanding anvil, or warm front advancing cirrus, should cease to be supplied with warm air, leaving the thick nose at the front to advance away from the main cloud and be visible as a rope for a short time before it becomes flattened by gravity between the upper and the nether air mass.

The detatched pieces of cloud sometimes seen on the edges of cumulonimbus anvils have been noticed for a long time. The vertical section of the edge when not detatched often has the appearance of an upturned hook (see [5], 2.2.3), but the same has not been remarked upon in the case of stratocumulus anvils formed by the spreading out of cumulus, probably because water droplets evaporate more readily and are therefore not as good tracers of the airflow as ice crystals. Equally, however, when a cumulus anvil spreads out in a layer already occupied by a thin layer of stratocumulus anvil, the downward motion in the lower half of the advancing nose is revealed by the evaporation of the existing cloud at the junction with the advancing anvil (e.g. [5], 2.1.3, where the gap is 'explained' as being due to glaciation induced by the advancing anvil causing fallout as in a fallstreak hole). The persistence of small gaps between two pancakes or regions of stratocumulus could be due to one or other of them advancing.

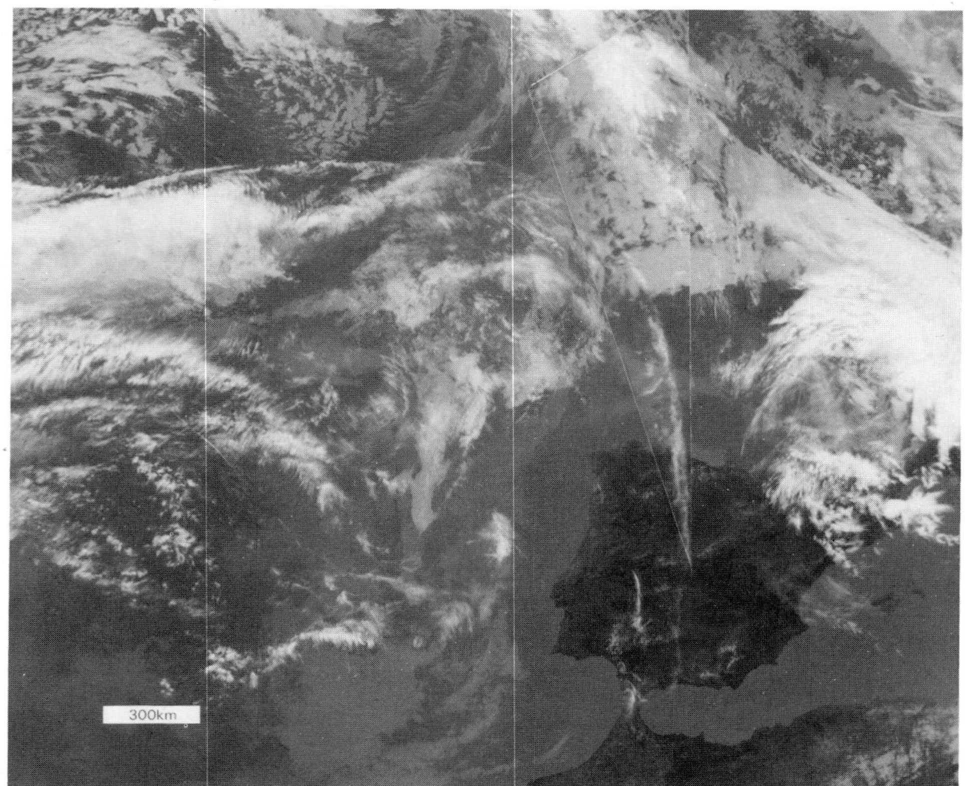

13.3.2 1607,23.3.88,4

13.3.2
A tropopause rope of a front from Malaga to 50°N 10°W to 45°N 35°W. This rope was quickly evaporating and the central section, northwards from Madrid, was taken from the previous pass, at 1426.

13.4 WAKES, WAVES, AND CELLS

The first reaction to the discovery of vortex streets in the lee of the islands of Guadeloupe and Madeira was to remark on their analogy with the well-known laboratory experiments, as if to say 'we know all about that phenomenon'. But further investigation, particularly of the wake of Jan Mayen ([1] Chapter 9 and [i]), soon showed that the same island sometimes produced lee waves, sometimes a vortex street, sometimes a mixture of the two (e.g. [4; 3.1a]), and sometimes much more complicated lee wave patterns. Sometimes it produced a thickened cloud line, and occasionally there appeared a line clear of cloud ([1], 9.18).

It is not always possible to distinguish between stable waves and billows. Stationary waves are nearly always lee waves and the obstacle setting them up can usually be identified. Billows are unstable waves, but a group of them might look very like lee waves, except that the wavelength is usually shorter. But when the wind is weak the wavelength ranges of billows and lee waves overlap, and one instantaneous picture is not enough to determine the cause of a group of waves. Occasionally a mountain system produces curly waves, so called because they have an inflection in the middle. These have been seen caused by Lanzarote and Fuerteventura in the Canary Islands and by Jan Mayen in a wind from the southwest, which meant that the wave patterns of the two peaks interfered ([1, 9.31 and 9.36]). In the Canary Islands case the configuration was almost identical on two successive days and so it might have been a long-enduring stationary pattern. But confidence in such implications is disturbed by two much earlier satellite observations of groups of waves over the Eastern Atlantic scarcely near enough to land to have been caused by it ([9, p. 105] and [13, p. 26]). In the second case the appearance was of 10 thin streamers with a spacing of about 5.5 km at 44°N 12°W coming from a point in northwest Spain; in the first case the waves look as if they might have been set off by the Canary Islands as lee waves with a similar wavelength, but the islands are about 20 wavelengths away to the northeast and the waves have the same orientation as in the other case. In both, the wind direction is doubtful but the strength was light. There is no space here to illustrate all the possibilities.

13.5 ODDITIES

Occasionally a very bizarre cloud is seen. A shortlived 'double doughnut' was observed to the west of Portugal and is illustrated here by the series of pictures made because the event was provocative (the other channels not shown are, of course, available in the archive).

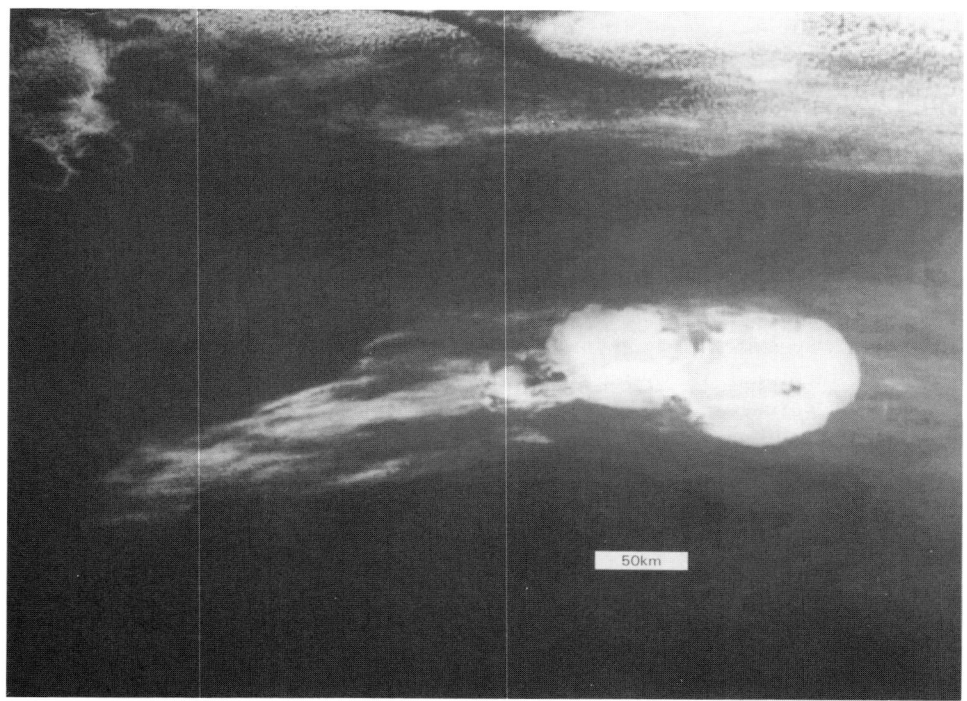

13.5.1 0920,28.11.81,2

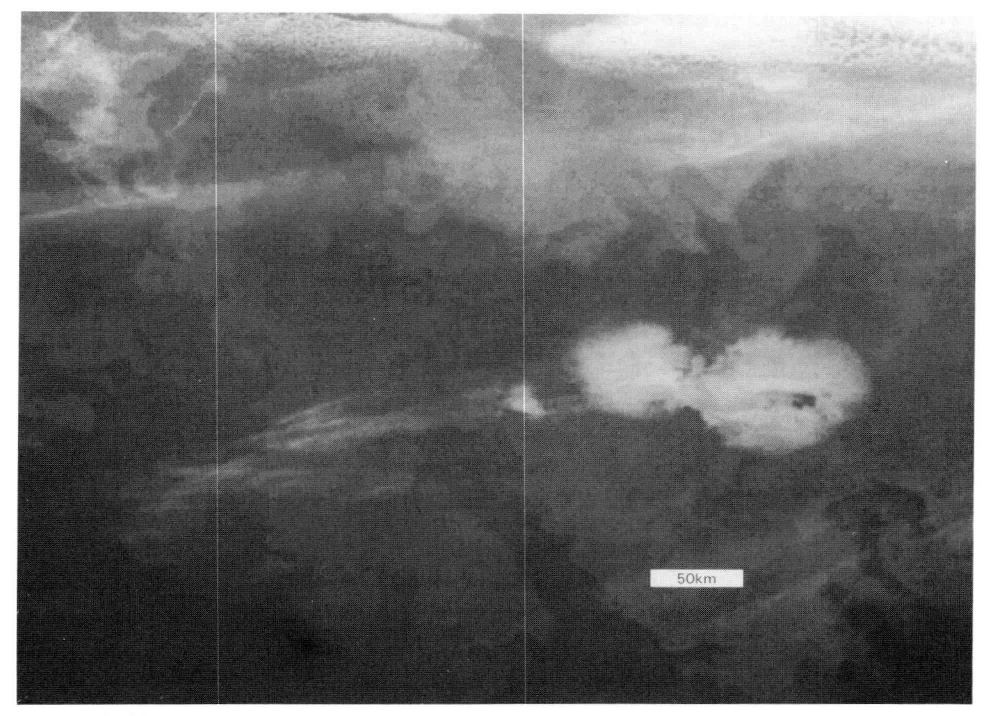

13.5.2 0920,28.11.81,4

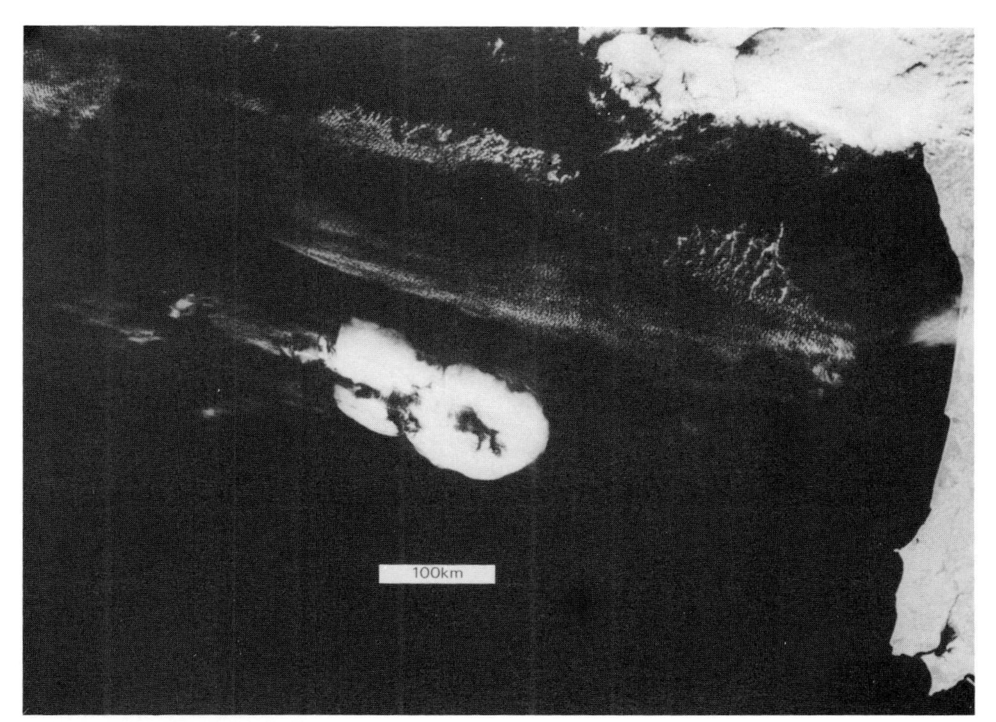

13.5.3 1214,28.11.81,CZ5

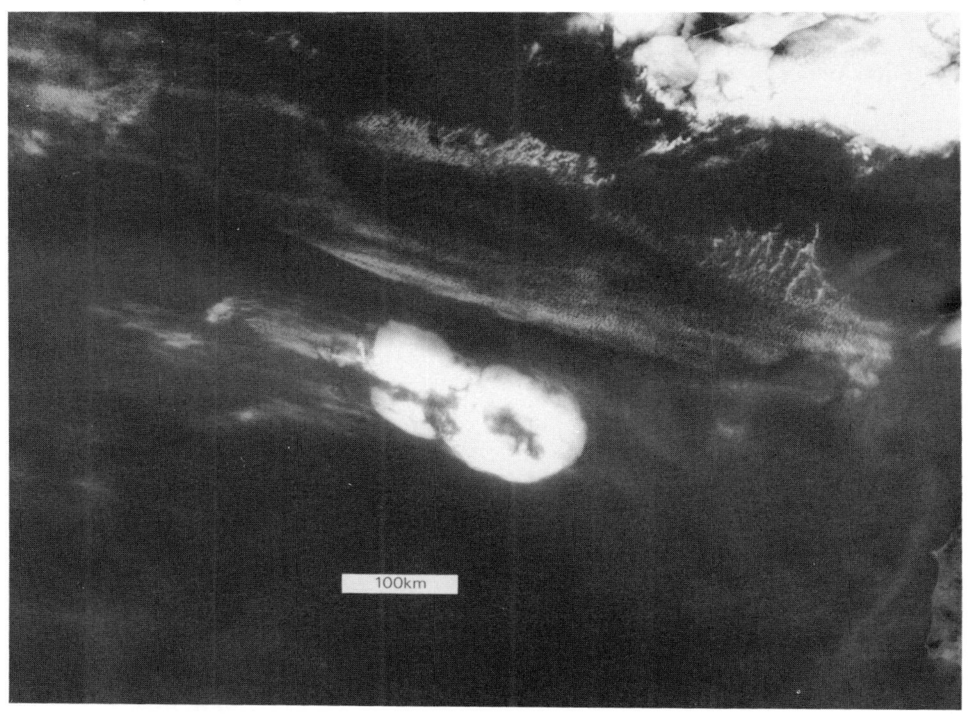

13.5.4 1213,28.11.81,CZ1

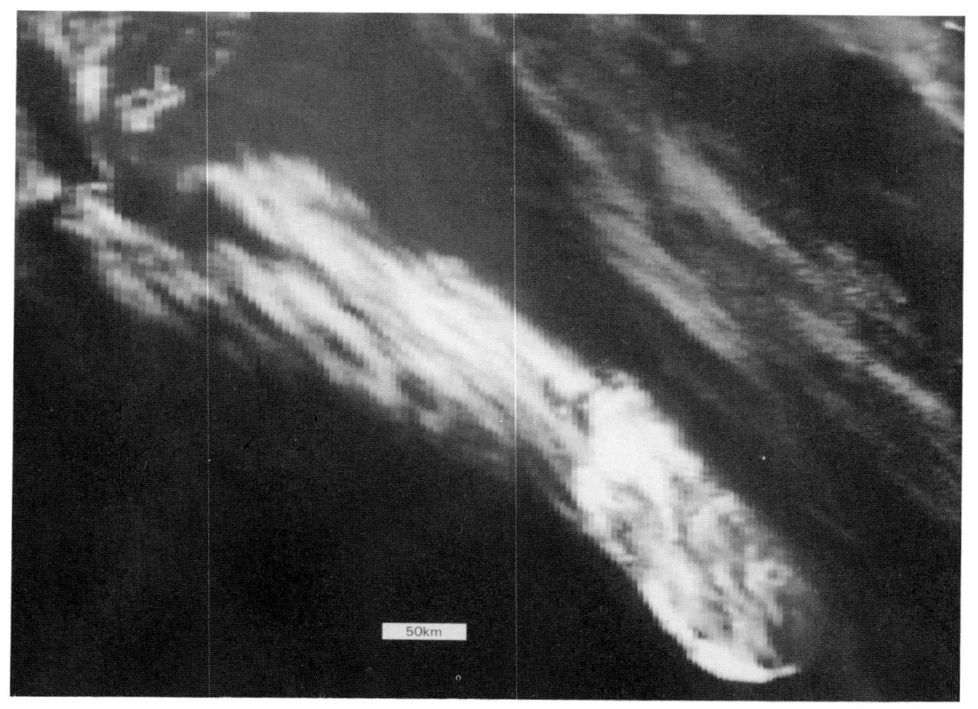

13.5.5 1408,28.11.81,2

13.5.1, 13.5.2, 13.5.3, 13.5.4, 13.5.5
The transient lone doughnut cloud seen in this picture series at 41°N 12°W and heading for Lisbon had a short existence of less than a day, and there was no obvious cause for it. It *appears* to have been travelling through the air because a wake of debris seems to be left in its track. The Ch 4 picture indicates that it was composed of low cloud. **13.5.5** is on the extreme western edge of the pass.

13.5.6, 13.5.7, 13.5.8
There is bright glint on the lee (west) side of Corsica and Sardinia, with some low cloud in the lee of Corsica. But what is the bright centre of the ear-shaped area east and south of Sardinia? Not glint; it is not bright in Ch 3, and the coast is not sharp which indicates that it is smothered by sea fog. The bright area is deeper cloud with a cooler top. The brighter glint patches in Ch 2 coincide well with the warmest coastal sea areas in Ch 4. The curly wave from southwest Sardinia is well recorded in Ch 3, in spite of general poor quality in that channel.

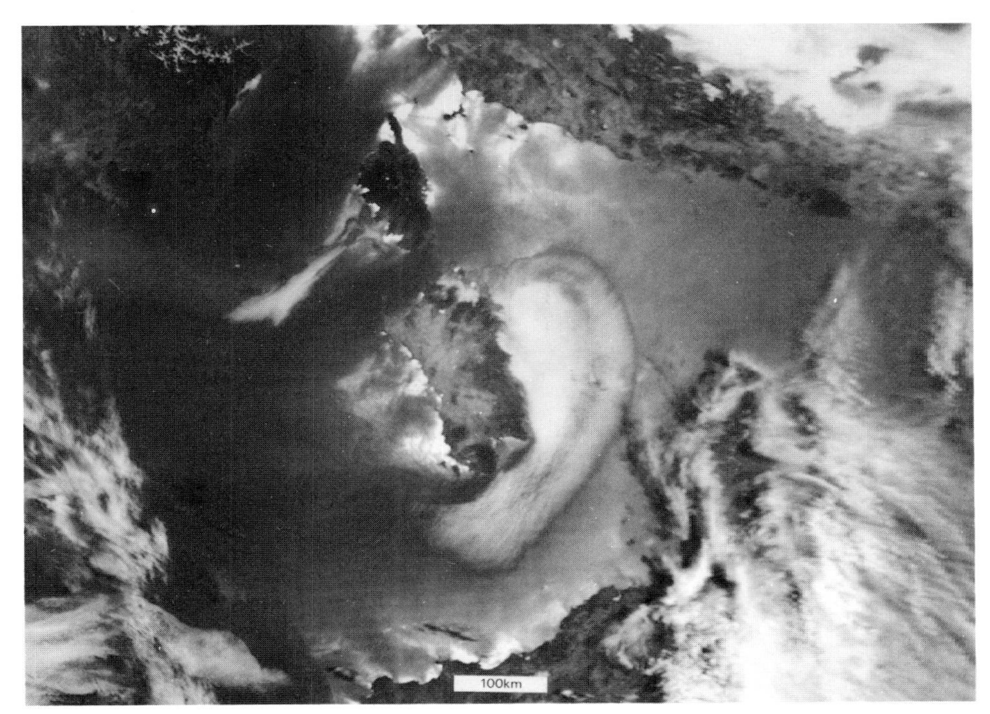

13.5.6 0806,6.5.88,2

13.5.7 0806,6.5.88,3

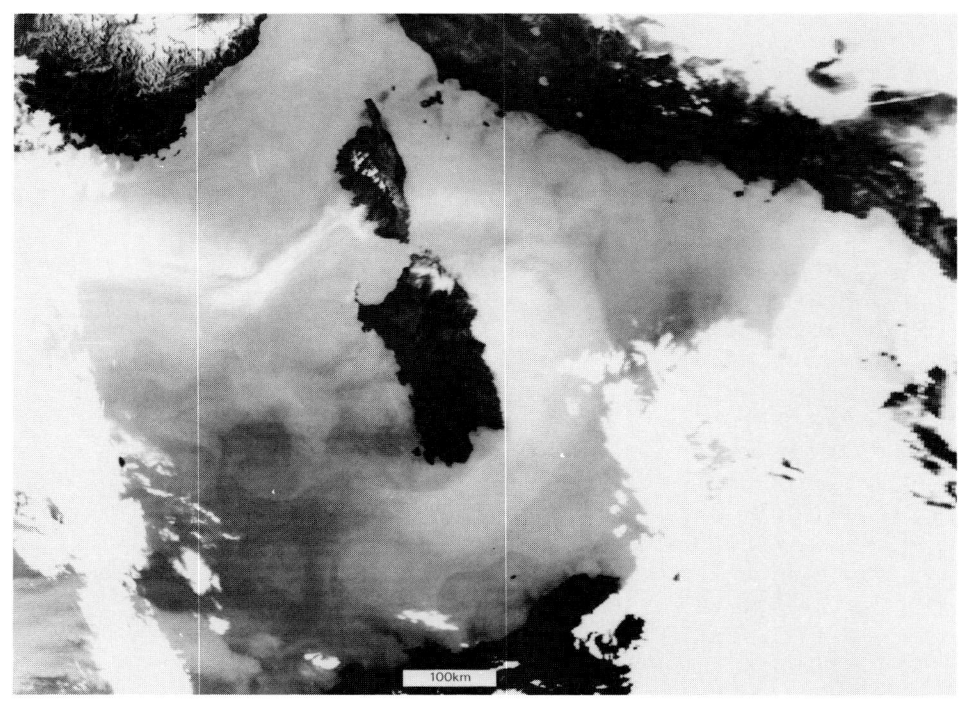

13.5.8 0806,6.5.88,4

13.6 OTHER POSSIBILITIES

It is not to be inferred that the fairly trivial cloud forms presented in this chapter, and for which the explanation may be incomplete, contain any suggestion as to probable future discoveries. Some clouds seen for the first time are attributed incorrectly to human influence. Thus political suspicion caused a plume cloud from Novaya Zemlya, which was actually a not unusual piece of orographic cirrus, to be explained as a gigantic nuclear or military experiment, or as a volcanic explosion. In such a complicated subject as the fluid mechanics of the atmosphere there is no substitute for wide-eyed and alert observation as a source of new knowledge. That is what this book is intended to stimulate.

References

[1] *Cloud investigation by satellite* (1986) R. S. Scorer, Ellis Horwood, Chichester.
[2] *Atmospheric radiation* (1964) R. M. Goody, Clarendon Press, Oxford.
[3] *The nature of light and colour in the open air* (1954) M. Minnaert, English translation, Dover.
[4] *Spacious skies* (1989) Richard Scorer and Arjen Verkaik, David and Charles, Newton Abbot.
[5] *Clouds of the world* (1972) Richard Scorer, David and Charles, Newton Abbot.
[6] *The use of satelite pictures in weather analysis and forecasting* (1966) Technical Note No. 75, World Meteorological Organisation.
[7] *Weather satellite picture interpretation* Vols 1 & 2 (1982) T. A. Marshall, DNOM memo No. 1/82, Naval Weather Handbook Unit.
[8] *The Earth's atmosphere viewed from space* (1979) R. R. Fotheringham, University of Dundee.
[9] *Les photos meteorologiques* (1984) G. D. Schietecat, Institut Royal Meteorologique de Belgique.
[10] *Environmental aerodynamics* (1978) R. S. Scorer, Ellis Horwood, Chichester.
[11] *Satellite and radar image interpretation* (1987) EUMETSAT reprints of workshop at Meteorological Office College, Reading.
[12] *Gravity currents* (1988) J. E. Simpson, Ellis Horwood, Chichester.
[13] *The best of Nimbus* (1971) Allied Res Assocs — Contract NAS 5-10343 NASA, Greenbelt, Maryland.
[14] Radiation hormesis (1989) J. H. Fremlin *Atom* **390** April 1989, p. 4.

RESEARCH PAPERS BY R. S. SCORER

[i] The wake of Jan Mayen. *Bull. Inst. Math. and Applic.* **19** 135, July/Aug 1983.
[ii] Etna: the eruption of Christmas 1985 as seen by meteorological satellite. *Weather* **41** (12) December 1986.
[iii] Hot spots and plumes: observation by meteorological satellite. *Atmospheric Environment* **21** (6) 1427–1435, 1987.
[iv] Ship trails. *Atmospheric Environment* **21** (6) 1417–1425, 1987.
[v] Cloud reflectance variations in channel 3 *Int. J. Remote Sensing* **10** (4, 5) April/May 1989.
[vi] The use of visible wavelengths in the study of particulate air pollution using regular meteorological satellite observations *Atmospheric Environment* **23** (4) 817–829, 1989.
[vii] Lessons from observations of chaotic flow in the atmosphere. *Journal de Méchanique théoretique et appliquée* **7** (Supplement 2), 145–165, 1988.
[viii] Sunny Greenland (climatological features as seen by satellite) *Quart. J. Roy. Meteorol. Soc.* **114** (479) 3–39, 1988.

Index

The references are to section numbers (e.g. 4.2) or to picture numbers (e.g. 12.6.1) and where several are together, a comma represents "and" (e.g. 8.3.1,2). Separate references are separated by semi-colons (e.g. 5.4; 12.1.1).